The Essential Oils

BY

ERNEST GUENTHER, Ph.D.

Vice President and Technical Director
Fritzsche Brothers, Inc., New York, N. Y.

VOLUME FIVE

INDIVIDUAL ESSENTIAL OILS
OF THE PLANT FAMILIES

ROSACEAE, MYRISTICACEAE, ZINGIBERACEAE, PIPERACEAE, ANACARDIACEAE, SANTALA-CEAE AND *MYOPORACEAE, ZYGOPHYLLA-CEAE, LEGUMINOSAE, HAMAMELIDACEAE, DIPTEROCARPACEAE, ANONACEAE, OLE-ACEAE, AMARYLLIDACEAE, RUBIACEAE, MAGNOLIACEAE, CAPRIFOLIACEAE, VI-OLACEAE, RESEDACEAE, SAXIFRAGA-CEAE, CARYOPHYLLACEAE, PRIMULA-CEAE, TILIACEAE,* AND *COMPOSITAE*

D. VAN NOSTRAND COMPANY, INC.
PRINCETON, NEW JERSEY

TORONTO   LONDON

NEW YORK

D. VAN NOSTRAND COMPANY, INC.
120 Alexander St., Princeton, New Jersey (*Principal office*)
24 West 40 Street, New York 18, New York

D. VAN NOSTRAND COMPANY, LTD.
358, Kensington High Street, London, W.14, England

D. VAN NOSTRAND COMPANY (Canada), LTD.
25 Hollinger Road, Toronto 16, Canada

PRINTED IN THE UNITED STATES OF AMERICA

Dedicated
to Mr. Frederick H. Leonhardt,
President of
Fritzsche Brothers, Inc.,
whose vision and generosity made this
work possible

PREFACE

The present work, like the two previous volumes of this series, describes the production, physicochemical properties, chemical composition, and uses of *individual* essential oils. For the reason noted in the Prefaces to Volumes III and IV, these oils are grouped within the botanical families to which the corresponding plants belong, but the families themselves are not arranged according to any taxonomic system.

Some of the oils treated in this volume are of great importance to the flavor chemist: nutmeg, cardamom, ginger, pepper, cubeb, star anise—to mention only a few. Others will be of interest to perfume, cosmetic, and soap perfumers: among these are rose, sandalwood, cananga, ylang ylang, jasmine, tuberose, and violet.

Originally it was hoped that the fifth volume would complete the treatment of individual essential oils. In the course of preparation, however, the book grew to such size that division into two parts became necessary. The present volume, therefore, will be followed by a sixth—the last of the series.

Ernest Guenther

New York, N. Y.
January, 1952

NOTE

All temperatures given in this work are expressed in degrees Centigrade unless otherwise specified in the text.

ACKNOWLEDGMENT

In the preparation of this series, it has been the author's practice, where methods of production, physicochemical properties of oils, and questions of quality are discussed, to draw intensively upon his own experience in the essential oil industry. However, as a result of far-reaching political and economic events, such profound changes have occurred recently in this field, that no individual could possibly hope to keep personally abreast of developments without the cooperation of leading producers and experts throughout the world. Fortunately, the author has been able to rely upon personal and business friends, in many countries, as uninterrupted sources of pertinent and up-to-the-minute data. He wishes, therefore, to take this opportunity to express his sincere thanks to these amiable co-workers, and particularly to the following:

Mr. Ramon Bordas, Destilaciones Bordas Chinchurreta, S. A., Sevilla, Spain.

Mr. João Dierberger, Dierberger Agro-Comercial Ltda., São Paulo, Brazil.

Mr. Jean Duclos, Les Fils de Maurice Duclos, Paris, France.

Dr. Teikichi Hiraizumi, Takasago Perfumery Co., Ltd., Tokyo, Japan.

Mr. F. J. Ippisch, Asociacion de Productores de Aceites Esenciales (formerly Oficina Controladora de Aceites Esenciales), Guatemala, C. A.

Mr. G. O. Krauch, F. Krauch & Cia., S. A., Asunción, Paraguay.

Dr. Francesco La Face, Stazione Sperimentale, Reggio Calabria, Italy.

Dr. Karl H. Landes, Karl H. Landes, Inc., New York, N. Y.

Dr. Yves-René Naves, Research Laboratories of Givaudan & Cie., Geneva, Switzerland.

Dr. J. A. Nijholt, Laboratorium Voor Scheikundig Onderzoek, Buitenzorg, Java.

Mr. S. G. Sastry, Government of Mysore Sandal Oil Factory (Retired), Bangalore, India.

Mr. Carlos Schaeuffler, Schaeuffler Hermanos, Retalhuleu, Guatemala, C. A.

Mr. Edwin H. Sennhauser, Volkart Brothers, Inc., New York, N. Y.

Data contained in the monographs on most of the natural flower oils produced in Southern France and Morocco (jasmine, rose, cassie, mimosa, violet, etc.), and on several other essential oils represent the result of years of closest collaboration with the author's esteemed friend, Mr. Pierre Chauvet, of Seillans (Var), France.

With Mr. Robert Garnier, of Paris, the author had the privilege of studying rose oil production over a period of several seasons at the very source—the "Valley of the Roses," in Bulgaria.

Mr. Charles Muller, of Ambanja, Madagascar, initiated the author into the complexities of ylang ylang distillation in Nossi-Bé some years ago, and has continued to be an invaluable informant on the subject up to the latest moment.

Mr. A. R. Penfold and Mr. F. R. Morrison, both of the Museum of Applied Arts and Sciences, Sydney, Australia, wrote the section on Australian Sandalwood Oils. The author is greatly indebted to these leading experts in Australian essential oils for this interesting contribution.

Dr. Theodor Philipp Haas, of the Philadelphia College of Pharmacy and Science, has generously lent his extensive botanical knowledge wherever any question of nomenclature has arisen.

To the Staffs of the Libraries of the Chemists' Club and of the Bronx Botanical Garden, both in New York, the author is grateful for their unfailing courtesy, patience, and efficiency.

As usual, the author has leaned heavily upon his associates at Fritzsche Brothers, Inc., New York—particularly Mr. Edward E. Langenau, Director of Analytical Laboratories, who furnished most of the data on the physico-chemical properties of oils examined in the laboratories of Fritzsche Brothers, Inc. Sincere thanks are also due to Mr. W. P. Leidy and Mr. A. H. Hansen, of the Library of the same Company, for their help with every phase of the literature on the subject. And once more, the author must single out for special praise his faithful secretarial staff—Mrs. Ann Blake Hencken, Miss Catherine McGuire, Mrs. Agnes Clancy Melody, and Mrs. Elizabeth Adelmann, who have for so long shared the burden of a complicated manuscript!

ERNEST GUENTHER

CONTENTS

ILLUSTRATIONS

xv

CHAPTER I

ESSENTIAL OILS OF THE PLANT FAMILY *ROSACEAE*

OIL OF ROSE

Essence de Rose   *Aceite Esencial Rosa*   *Rosenöl*   *Oleum Rosarum*

History.—Of all the scents used in perfumery that of the rose is one of the oldest and best known. Since time immemorial poets have been inspired by the delightful aroma of this flower, and historians have given many an account of the methods by which its alluring fragrance can be captured. Earliest mention of distillation of roses occurs perhaps in the "Ayur-Vedas" of Charaka and Susruta,[1] a series of books on the "good life," written in ancient Sanskrit. In the "Iliad," Homer [2] describes how Aphrodite anoints the body of the slain Hector with rose oil (by which Homer probably means a rose-scented ointment). When the classic writers— Dioscorides,[3] for example—speak of "Rose Oil," they undoubtedly refer to fatty oils heavily perfumed with rose. Most probably such pomades were prepared by maceration of rose petals in hot fat. (Small quantities of these *pommades* are still manufactured in the Grasse region of Southern France —cf. Vol. I of the present work, p. 198.) Rose pomades were extremely popular in ancient Rome, being used on many festive occasions.

As regards actual distillation of roses, the first description—aside from the somewhat vague account in the Indian Vedas—is probably that of the Arabian historian Ibn Chaldûn,[4] who reported that in the eighth and ninth centuries rose water was an important article of trade, being exported to India and China. From the tenth to the seventeenth century the rose industry was centered in Persia, particularly in Shiraz, that fabled city of poets and oriental culture. From here the industry gradually spread to India, Arabia, North Africa, and—with the help of the conquering Moors —as far as Spain. Returning crusaders brought the rose from Asia Minor to Southern France.

Legend has it that the Mogul emperor Jehángir [5] in India had the canals of his palace gardens supplied with distilled rose water. One day his con-

---

[1] Sri Madhusudana-Gupta, "The Susruta," Sanskrit College, Calcutta (1835), 2 Volumes.
[2] "Iliad," Chapter 23, V, 186.
[3] "De Materia Medica Libri Quinque," Editio Kühn-Sprengel (1829), Vol. I, 41, 56, 123; II, 399–404.
[4] "Notices et Extraits des Manuscrits de la Bibliothèque Impériale à Paris," Tome 19 (1862), 364. Istakhri, "Das Buch der Länder," Ed. Mordmann, Hamburg (1754), 73.
[5] Gildemeister and Hoffmann, "Die Ätherischen Öle," 3d Ed., Vol. I, 69. Cf. *ibid.*, 17 and 150.

sort noted a thin, oily membrane on the surface of the water in one of the canals and ordered it collected; this highly aromatic substance she named "Atr-i-Jehángiri." Thus, possibly, began the distillation of roses for their essential oil. However, in view of the very large quantities of rose water produced at that period in Persia and neighboring countries, it appears more likely that the separation of the semisolid rose oil on the surface of distilled rose water had been observed before this, and that rose oil had actually been used for the perfuming of aromatic oils and beauty preparations for some time. Thus, an early reference to distilled rose oil is probably contained in Harib's [6] chronicle of the year 961, written in Cordova, Spain. Italian and German literature [7] of the fifteenth and sixteenth centuries describes several methods for the actual distillation of rose water and rose oil. Nevertheless at this period Persia was still the principal producer of rose water, which was exported on Dutch and Portuguese ships via ports in the Gulf of Persia or via Aden in Arabia.

At the beginning of the seventeenth century the conquering Turks introduced the rose industry into the Balkan countries, particularly to Bulgaria, where plantings were started near the newly founded town of Kazanlik, on the southern slopes of the Balkan Mountains. However, it was only in the nineteenth century that the Bulgarian rose oil industry began to prosper and acquire a leading, indeed an almost monopolistic, position, supplying the world with precious rose oil. This condition prevailed until the end of World War II brought about the most profound political and economic changes; Bulgaria's rose industry has now been nationalized, and production greatly reduced. What may take place behind the "Iron Curtain" in regard to rose oil is not clear at the time of this writing. According to some sources, a new rose industry is being created on the Crimean Peninsula, and in the Caucasus, but no detailed report has become available to the Western World.

Aside from Bulgaria, Anatolia in Turkey has been producing limited quantities of rose oil, but compared with the Bulgarian, the quality of the Turkish oil has in general been quite inferior.

The rose industry of Southern France (Grasse region) began early in the nineteenth century. Since the type of rose grown near Grasse is not particularly suited to the production of volatile (essential) oil of rose by distillation, the flowers are extracted with volatile solvent, yielding the so-called concretes and absolutes of rose (see below).

Attempts to cultivate roses in North Africa for extraction purposes have been made on several occasions, but only partly with favorable results. Lately sizable rose plantations, employing modern agricultural equipment,

[6] R. P. A. Dozy, "Le Calendrier de Cordue de l'année 961," Leyden (1873).

[7] Cf. Gildemeister and Hoffmann, "Die Ätherischen Öle," 3d Ed., Vol. I, 153.

have been established in Morocco; these will be described in a separate section, below.

For ornamental purposes roses are grown on a very large scale, and in many parts of the world, Texas, Oregon, and California among them. However, such roses are bred and cultivated for appearance rather than for exploitation of their essential oil. On distillation or extraction they give a low yield of oil, and an oil of such different quality [8] that the perfume industry cannot use it to replace the Bulgarian or French products.

Botany.—Of the more than 5,000 varieties of roses known to the horticulturist only a relatively few exhibit a marked fragrance; this fragrance, moreover, differs considerably according to plant variety. Thus, the perfume of some types recalls hyacinth or violet flowers, or musk; that of others may be reminiscent of certain fruits—raspberry, for example. Most of the fragrant roses are hybrids.[9] The aroma referred to by perfumers as rose odor in the more narrow sense is found almost exclusively in the roses belonging to the so-called *Centifolia* group (fam. *Rosaceae*); of these only three are exploited commercially for the isolation of their perfume:

1. *Rosa damascena* Mill. forma *trigintipetala* Dieck,[10] the so-called "Pink Damask Rose." Originally probably unknown in the wild state, and most likely developed horticulturally as a hybrid of *Rosa gallica* L. and *Rosa canina* L., it has now escaped cultivation and can be found growing wild in the Caucasus, in Syria, Morocco, and Andalusia. *Rosa damascena* is very fragrant and contains a relatively large amount of volatile oil, which can be isolated by steam distillation or by extraction with volatile solvents. By far the most important of all perfumery roses, it is planted extensively in Bulgaria, and to a smaller extent in Anatolia (Turkey).

2. *Rosa alba* L. or *Rosa damascena* Mill. var. *alba*, the so-called "White Cottage Rose," contains much less volatile oil than *Rosa damascena*, the oil being also of lower quality. Being considerably hardier and more resistant to unfavorable climatic factors than the pink damask rose, the white cottage rose is grown in Bulgaria as a hedge around the pink roses, and in the higher and colder altitudes where *Rosa damascena* does not flourish. Distillers in Bulgaria object to the white rose as being of inferior quality, but farmers insist on growing it for protection and for marking the boundaries of their fields of pink roses.

3. *Rosa centifolia* L., the so-called "Light Pink Cabbage Rose," is grown extensively in the Grasse region of Southern France, and lately also in North

[8] Sorrels and Ratsek, "A Study of Rose Oil Production in Texas," *Agr. Mech. Coll. Texas, Bull.* No. 84 (1944).
[9] *Perfumery Essential Oil Record, Special Issue* **16** (1925), 283–328.
[10] Tschirch "Handbuch der Pharmacognosie," Vol. II, Part 2, Table XXV. Cf. Gildemeister and Hoffmann, "Die Ätherischen Öle," 3d Ed., Vol. II, 802.

Africa (particularly Morocco), where it is known as "Rose de Mai." It contains a good amount of flower oil which, however, cannot economically be isolated by steam distillation. Limited quantities are used for the production of the fragrant rose water so popular in Mediterranean and Latin countries. The bulk of *Rosa centifolia* L. serves for extraction with volatile solvents, which yields concretes and absolutes of rose, one of the most important products of the natural flower oil industry in the Grasse region. Concerning the botanical origin of *Rosa centifolia* L., several theories have been advanced in literature.[11] The plant appears to be closely related to *Rosa damascena* Mill.

I. THE BULGARIAN ROSE OIL INDUSTRY

Producing Regions.—As has been pointed out, the cultivation of roses for industrial purposes was introduced to Bulgaria by the conquering Turks several hundred years ago. Centers of production lie to the north of Plovdiv, in the valleys of the Toundja and Struma, both tributaries of the Maritsa River. To the north, these two valleys are bordered by the Central Balkan Mountains, to the south by the Sredna-Gora. Of the two sections, the Struma Valley, with Karlovo as center, is now the more important one; in 1938 about 7,700 acres were under rose here. The Toundja Valley in the east, with Kazanlik as center, is the older of the two "Valleys of the Roses;" in 1938 it counted 3,600 acres of rose fields. These two valleys are not the only rose producing areas in Bulgaria; rose plantings can also be found near some villages located on the southern side of the Sredna-Gora. Bulgaria's rose industry reached its peak in 1916 with a total of 22,000 acres. After World War I it went through a critical period, with a substantial reduction of the acreage. In 1938 rose plantations in Bulgaria totaled about 15,000 acres, planted near 142 villages (Karaivanoff [12]). During the years of world-wide depression many rose growers switched their crops to tobacco and peppermint, which promised better returns. World War II brought a further decline in Bulgaria's rose industry; in 1945 less than 5,700 acres were under rose cultivation.[13]

The roses flourish at the foot of the hills or on the southern slopes of the Balkan Mountains, which protect the plantings from cold north and northeast winds. The altitude here ranges from 900 to 1,500 (or even 2,500) ft. The section lies between the grain fields in the valleys and the mountain forests of the Central Balkan range. The soil consists of fertile gravel,

[11] Cf. Gildemeister and Hoffmann, "Die Ätherischen Öle," 3d Ed., Vol. II, 803.
[12] *Drug Cosmetic Ind.* **42** (1938), 580.
[13] *Am. Perfumer* **48** (June 1946), 49.

limestone, sand and loam, mostly free of acids, well drained and easily permeable to water.

Planting, Cultivating and Harvesting.—As was mentioned in the section on "Botany," two types of rose are cultivated in Bulgaria, viz., *Rosa damascena*, the pink or light red rose, and *Rosa alba*, the white rose. Of these, the former is by far the more important, as it yields a much higher percentage of oil, and an oil of superior quality. It grows on shrubs 4.5 ft. high. The bushes of the white rose are higher and stronger—hence their use as hedgerows for protection of the pink roses. Plantings of white rose are also found in the colder areas of the higher altitudes, where the pink rose does not survive.

Prior to planting, the ground must be well plowed; then parallel ditches, 3.5 ft. deep and 1.5 to 1.75 ft. wide, and 7 to 8 ft. apart, are dug. The earth shoveled up from the ditches is piled in equal amounts on both sides of the ditch. This work should be undertaken in the fall, or not less than two months prior to planting, so that the earth can be exposed to air and moisture sufficiently long for the organic matter to decompose. About the end of February or the beginning of March, a part of the earth piled up along the ditch is shoveled back so that the ditch has now a depth of only 1.5 to 2 ft. (The deeper the bushes are planted, the better they will withstand periods of drought.) The cuttings necessary for planting are procured from an old rose field, usually purchased for this purpose. Cuttings 1 to 2 ft. long are taken from the base of healthy old rose bushes, the upper, green parts being removed. The woody cuttings are placed into the ditches horizontally, in four uninterrupted rows, 2.5 to 3 in. apart. Then a layer of earth (from the piles along the ditches), 2 to 3 in. thick, is placed on top of the cuttings; this is followed by 2 in. of seasoned stable manure. In May the first shoots appear above ground. The soil is then slightly hoed, weeds are removed, and some more earth is shoveled into the ditches and placed around the young plants. During the first year this procedure (hoeing, weeding, and placing of earth around the young plants) has to be repeated at least eight times, preferably after each rainfall. In the second and third year the field must be hoed and plowed five times each year. During the winter the plants are hilled up for protection against frost. After ten years the plantation may show signs of decay; the bushes are then cut down to the ground for rejuvenation. If properly taken care of every ten years, a field may remain productive from thirty to forty years. However, such care requires a great deal of work: weeds must be removed, old branches cut out, and plenty of manure applied.

The first harvest can be gathered in the third year; it amounts to only about 1,500 kg. of flowers per hectare. From then on the quantity increases, reaching as much as 5,000 kg. per hectare at the height of productivity. On the average, a well-kept rose field yields about 4,000 kg. of flowers per hectare per year. Andreeff [14] reported yields of flowers ranging from 1,200 kg. to 2,000 kg. per hectare, varying with the year and the cultural care given to the field.

Blooming begins in the second half of May and continues for three or four weeks. Obviously, date and duration of the crop depend a great deal upon the weather prevailing in a particular year. Mild and humid weather prolongs the flowering period, increasing at the same time the yield of oil and producing an oil of good quality. During very hot and dry weather the harvest may last only two weeks, and the yield of oil may be lowered due to loss by evaporation. Moreover, the quality of the oil is also affected because the large quantities of roses arriving at the distilleries cannot be worked up immediately and may be left lying on the floor for hours. During this interval they are likely to start fermenting or at least to turn so stale that their aroma suffers gravely.

In general, harvesting begins as soon as the flowers begin to open and continues until all have been picked. Flowers are collected by hand, being nipped off just below the calyx; the work is done by peasants with the help of their families. To the visitor the harvest affords a pretty and picturesque sight, characteristic of the "Valley of the Roses." Gathering of the flowers starts at daybreak and ceases, if possible, at eight or nine in the morning, while the dew is still fresh on the flowers. When collected early in the morning, the roses give a much higher yield of oil than when gathered in the afternoon. At the height of the season or in spells of hot weather, each bush produces so many flowers at the same time that the picking may have to be continued into the afternoon in order to gather all the flowers maturing on that day. Yield and quality of the oil will then be poor.

The harvesters drop the flowers into bags that hold from 25 to 30 kg., and then place them into sacks, which are hauled to the distilleries as soon as possible. Transport is via trucks, ox- or horse-drawn carriages, or the backs of donkeys.

Changes in Bulgaria's Rose Industry.—Before discussing methods of distillation and extraction, it may be of interest to review briefly the changes which Bulgaria's rose industry has undergone during the last fifty years.

Until 1902 all the roses were grown on small holdings, by a great number of individual peasants who also distilled their crops in old-fashioned, and primitive portable stills. According to statistics of the Chamber of Commerce in Plovdiv, there existed at that time 2,798 distilleries, with 13,128 small stills.[15] The small lots of oil thus produced were purchased by a number of well-known exporters, who supplied the world market with bulked

[14] *Ind. parfum.* **1** (1946), 168, footnote.
[15] Karaivanoff, *Drug Cosmetic Ind.* **42** (1938), 580.

lots. Some of these exporters enjoyed international reputations. From 1902 on, a number of exporters started erecting distilleries, equipped with large, modern stills, in which they processed flower material supplied by the small peasant-growers. (Large-scale distillers never succeeded in producing their own flower supply for the simple reason that the peasant, with the help of his family, can grow and harvest the flowers at much lower cost than a large landowner, who is always faced with labor difficulties and wage problems.) This enabled these exporters to offer brands of rose oil of their own production, of unquestionable purity, and of a quality superior to that of the peasant oils. Some of the individual brands became well known throughout Europe and America. In 1904 Charles Garnier, of Paris, constructed a plant for extraction with volatile solvents in Kara Sarli; and after 1919 six other large-scale producers equipped their distilleries also with extraction batteries, thus producing not only distilled rose oil, but also concretes of rose (see below). As a result of this development, Bulgaria's rose industry underwent an intense modernization between 1902 and 1922, new installations gradually replacing the old-fashioned field stills. On the other hand, the peasant-distillers, too, had to organize themselves against the competition of the large-scale producers. For this purpose cooperative societies were founded, to process the flowers in their own stills. Some of these cooperatives installed modern equipment, others worked with the old-fashioned small stills. From 1929 on, the number of cooperatives increased gradually, helping thereby the development of the industry. In 1930, more than 75 per cent of Bulgaria's rose oil was produced by large-scale, individual operators, the balance by cooperatives and small peasant-distillers. The flowers have always been grown by the peasants. At the outbreak of World War II there existed in Bulgaria a number of quite modern plants, equipped with 649 open fire stills (see below), 40 steam stills, and 32 apparatus for extraction with volatile solvents.[16]

About 1929 Bulgaria's rose industry and trade underwent some serious disturbances, caused chiefly by the endeavor of a few large-scale producers and exporters to corner the flower supply. As a result of their speculative actions prices soared to such a height that consumers abroad lost interest in natural rose oil, replacing it in their formulas, at least partly, with low-priced synthetic preparations. As the disorder increased in the most alarming manner, with grave consequences for the producers, the Bulgarian Government entered the picture, and after a careful study of the whole rose industry by competent experts, promulgated a number of decrees and regulations (1932 to 1936), which placed the industry under strict government control. These laws aimed at:

[16] *Ibid.*

1. Establishing a reasonable price for the crop every year, the price being held as constant as possible in order to prevent speculation.

2. Insuring and certifying the purity of the oil. For this purpose the government created the "Banque Agricole et Coöpérative de Bulgarie," whose inspectors supervised actual production in every factory and sealed the oil containers. The entire output of rose oil was placed in a specially built warehouse of the bank. Any handling of the oil and the preparation of foreign orders were carried out in the laboratories of the bank, and in the presence of the parties concerned.[17] Every lot of oil was analyzed.

These were the conditions under which Bulgaria's rose oil industry operated at the outbreak of World War II. When, at the end of the war, Bulgaria became a communist country, all rose oil factories were nationalized and the industry was declared a government monopoly. By a law of May, 1945, a state enterprise, "Bulgarska-rosa," was established for the purpose of producing and marketing rose oil and other by-products. Today this state concern is the sole producer of rose oil in Bulgaria, and owns the up-to-date distilleries in the Valley of the Roses. Moreover, it is the only exporter guaranteeing the absolute purity of all rose oil it produces for the world market. The oil is submitted to stringent chemical tests in the modern laboratories which form one of the important features of the plant. After analysis, the oil is stored in containers called "koncoums" (vases). All containers are sealed with the special seal of the "Bulgarska-rosa" and numbered. Each container is accompanied by a certificate of purity (Behar [18]).

A. Distilled Rose Oil
("Otto of Rose" or "Attar of Rose")

Method of Distillation.

(a) *In Old-Fashioned Field Stills.*—As was mentioned above, until about fifty years ago practically all rose oil was produced in small, portable, direct-fire stills operated by peasant-growers and set up in convenient spots adjacent to the fields, or in villages, and near a supply of running water.

These simple and portable apparatus—some are still used, particularly in remote mountain villages—hold from 120 to 150 liters. The retort is wider at the base than at the top. Two handles on the side walls permit lifting of the retort and emptying of the contents after distillation. The top of the still consists of a removable spherical head, the still head, from which a gooseneck or connecting pipe leads through a barrel filled with lukewarm

[17] *Ibid.*
[18] *Perfumery Essential Oil Record* **41** (1950), 248.

water, for cooling of the condensate. (In appearance these stills closely resemble the old-fashioned field stills occasionally employed in the lavender regions of Southern France—see the picture in Vol. I of the present work, p. 112. The shape of the retort, and particularly that of the still head—"Tête de Maure" in French—clearly betrays the Arabian origin of these apparatus.) On the other side of the barrel the condenser tube ends above a glass bottle in which the condensate is collected. The stills are made of copper, tinned on the inside. Placed, usually in a row, on a rough brick hearth, the stills are heated by an open fire beneath each unit, the fire being kindled with wood.

Depending upon their condition (fresh or stale), from 15 to 20 kg. of roses are charged into the retort, and four to five times their weight of water is added. From 20 to 30 liters of water are then distilled over; however, no oil separates from this so-called "First Water" (note, however, that with industrial distillation in modern rose stills such separation does take place—see below). The "First Waters" from several stills are bulked, and 100 liters are placed in one of the empty stills, then very slowly redistilled (cohobated) until 12 liters of "Second Water" have come over. (The residual 88 liters of water remaining in the retort are used for a new flower batch.) The 12 liters of distillate are collected first in a large bottle holding 8 liters, and then in a smaller bottle called "Surier." The latter holds only a little more than 4 liters and is long and slender in shape, without any angular walls to which oil droplets, separating from the water, might adhere. To bring about complete separation of the oil, the "Second Water" in both vessels is kept standing for several hours, at the end of which time the oil can be decanted from the top. This is effected by means of a conical spoon resembling an old-fashioned candle extinguisher. The point of the cone has a small hole through which any water drips off; the rose oil, which is of semisolid consistency, stays in the cone and is transferred into the oil container.

Once distillation is under way, the "Second Waters" (the separated oil having been removed) are combined with the "First Waters" from another still, and redistilled (cohobated). The distillate of the mixed waters yields the above-mentioned "Peasant" rose oil.

As regards the residual water remaining with the exhausted roses in the retort (after the first operation), it is sometimes poured, with the flowers, over a screen, the flowers thus separated then being thrown away. Some distillers return this residual water to the same retort for distillation with a new batch of roses. This practice results in a slightly higher yield of oil— but at the expense of quality. Inorganic substances probably accumulate in the water on repeated use, and raise its boiling point, thus "pushing" distillation of high boiling compounds which do not readily come over in regular water distillation.   On the other hand, products of decomposition also accumulate in the residual water, and on distillation impart an undesirable odor to the distillate.   For this reason most of the small producers strongly object to the use of the residual water, although it may increase the yield of oil.   If it were employed again and again—which for obvious reasons is never done—the distillation of roses in the above-described manner would represent a closed circle, and nothing would be lost, except a certain amount of water by evaporation.

The yield of oil from the old-fashioned "peasant" stills amounts to 1 kg. per 2,500 to 3,200 kg. of roses—a much higher yield than that obtained in modern stills.   The explanation of this paradox will be given later.

(*b*) *In Modern Direct Fire Stills.*—For a better understanding of the difficulties connected with rose distillation two points should be emphasized:

1. The roses cannot be distilled by injecting live steam into the flower charge, because under these circumstances the petals agglutinate to form a compact mass, through which the steam cannot penetrate.   To prevent such agglutination, the flowers must float freely in boiling water.   In other words, the roses must be processed by water distillation (cf. Vol. I of the present work, pp. 112, 120, and 142).   Even under these circumstances the petals have a tendency to form a huge lump which slowly rotates in the boiling water.   It has been suggested that the stills be equipped with powerful mechanical stirrers, but such stirrers are expensive and may give mechanical trouble.

2. Rose oil is relatively soluble in water, and only a part separates as oil on distillation of the flowers.   The distillation waters must be redistilled (cohobated) in order to obtain the water-soluble portion in oil form.   As carried out in Bulgaria, distillation of roses is a clever combination of distillation of the flowers and cohobation of the distillation waters—these latter being used again and again for new flower batches.   Only by adhering to this procedure can a normal yield of oil be obtained.

The following description of the methods actually employed is based upon the author's observations in the course of two producing seasons in Bulgaria: [19]

Most of the industrial stills employed today are directly fired (with an open fire beneath each retort), and of 1,000 to 2,000 liters capacity.   Re-

[19] Guenther and Garnier, *Am. Perfumer* **25** (June, September, October, and November 1930).

quiring no steam generator with accessory steam and water pipes, they offer the advantages of low initial investment and simplicity of operation, important features in the primitive "Valley of the Roses." The stills are made of copper, heavily tinned inside, and of cylindrical shape. Insulation with bricks or some other suitable material permits even distribution and retention of heat inside of the retort. The still top is provided with a manhole through which the flowers can be charged. A grid a few inches above the bottom of the retort prevents contact of the flower material with the directly fired bottom. On the side, and level with the grid, is another manhole (opened and closed by means of levers) through which the exhausted flowers and the residual water can be quickly discharged after completion of the *flower* distillation. The still is also equipped with a pipe and valve, beneath the grid, which permits drawing off the residual waters when the still is used for cohobation.

Generally, the retorts are charged with flowers up to about 10 in. below the still head. The quantity of roses charged depends upon their condition, fresh flowers taking up much more space than those stored in the distillery for a few hours. Retorts of 1,200 to 1,500 liters capacity—the most popular type—are charged with 150 to 300 kg. of roses; these are then covered with 3 to 4 times their weight of water.

For example: a retort of 1,800 liters is charged with 250 to 300 kg. of roses, and 1,200 liters of water are added. The whole mass is stirred thoroughly. Then the manhole in the still top is closed and the fire started. After about 1½ hr., actual distillation starts; this lasts another 1½ hr.—the entire operation thus requiring 3 hr.

At the beginning, great care must be exercised to keep the fire low, so that actual distillation starts slowly. Otherwise the warm expanding air will blow through the condensers with great velocity and carry along the most volatile, highly aromatic constituents of the rose oil, which under such circumstances cannot be recovered. During the operation the inlet for the cooling water in the condensers must be regulated so that the condensate flows at a temperature of 35° to 40° C. At lower temperatures the oil, which contains a good deal of stearoptene, will solidify in the condenser tubes, and no oil will appear in the Florentine flask (oil separator).

Distillation of the *flowers* which, after a slow start, should proceed at a lively pace, is completed when about 140 liters of water per 250 kg. of flower charge have distilled over. The very small quantity of greenish oil separating in the Florentine flask is called "Direct Oil" or "Surovo Maslo" in Bulgarian; this is drawn off with a pipette and poured in a glass bottle. When the oil has been decanted, the 140 liters of distillation water ("First Water") are pumped into a storage tank and bulked with the "First Waters" (freed of *separated* oil) of other operations. As soon as enough "First Waters" have been accumulated, they are redistilled (cohobated), as explained below.

The exhausted flowers in the retort, together with the residual water, are discharged and *thrown away.*

The yield of "Direct Oil" obtained on distillation of the *flowers* is very low, because the greater part of the rose oil is dissolved in the distillation water ("First Water"). To isolate this dissolved oil the "First Water" has to be redistilled (cohobated). For this purpose 1,200 liters are pumped into a retort of 1,800 liters capacity, and 140 liters are very slowly and carefully distilled off. During this operation the condenser temperature should be low. The process requires about two and a half hours: one and a half hours for heating of the water, and one hour for actual distillation (cohobation).

The condensate consists of 140 liters of "Second Water" (see below) and a small quantity of "Water Oil," called "Prevarka" in Bulgaria. The "Water Oil" is drawn off with a pipette and added to the "Direct Oil" obtained from the distillation of the *flowers* (see above). Combined, the two oils constitute the rose oil of commerce. The freshly distilled oil is poured into glass bottles and exposed to sunlight for several days. This causes impurities and colloidal matter floating in the oil to precipitate, and water to separate. The clear oil is then carefully decanted, filtered, and stored in the well-known tinned copper containers.

The ratio of "Direct Oil" to "Water Oil" is not constant and depends upon several conditions, including quality of flower material and method of distillation. In general, the complete oil consists of about 25 per cent of "Direct Oil" and 75 per cent of "Water Oil."

As regards the above-mentioned 140 liters of "Second Water," after the oil has been decanted they are pumped into the storage tank holding the "First Waters" (also freed of *separated* oil) and redistilled with them to remove any oil remaining in solution (see above). The residual water left in the retort after the 140 liters of "Second Water" have been distilled off is kept in the retort and used for the distillation of a new batch of roses.

With the qualification noted some pages back, the entire process of rose oil distillation does, after all, represent something of a closed circle. The distillation waters are used over again and again, only the essential oil and the exhausted flowers, together with their residual waters, being eliminated. The reader will find a flow sheet of the whole process on page 16 of the present volume.

(c) *In Steam Stills.*—The steam stills employed at present for the distillation of roses in Bulgaria closely resemble the large direct-fire stills, except that they are provided with a steam jacket at the bottom for heating with indirect steam, and with a steam coil for the injection of live steam into the boiling still contents. As in the case of direct-fire distillation the flowers must be covered with boiling water and freely move in it. The use of direct steam allows for quicker heating of the contents in the retort, and for much better and easier regulation of the temperature, hence of the rate of distillation. At the same time, the rising steam bubbles effect a certain stirring action on the flower mass. Otherwise the methods of operation are the same. The amount of the flower charge may be somewhat larger in steam stills than in directly-fired retorts of the same capacity.

In this connection it should be mentioned that in 1929 one of the leading rose oil producers in Bulgaria modified his rotary extractors (which usually serve for the extraction of roses with volatile solvents—cf. Vol. I of the present work, p. 208) in such a way that they could also be used for distillation purposes.[20] In these apparatus the flowers were charged into metal baskets rotating around a central horizontal axis, and dipping into the solvent at the bottom of the extractor. Instead of a volatile solvent the bottom of the apparatus, in its modified form, contained boiling water heated by injected live steam. The flowers were thus distilled with boiling water and steam. The great advantage of the system consisted in the fact that by the rotary movement of the baskets through the boiling water the flowers were kept moving and prevented from agglutinating. This permitted the use of much less water than the regular stationary retorts. Distillation could be completed in a shorter time, the rate of distillation was greatly increased, and much smaller quantities of "First" and "Second Water" had to be distilled over to yield the totality of oil. The ratio of "Direct Oil" to "Water Oil" was 1 to 2. The complete oil was of excellent quality, and of strong and lasting odor. It contained considerably less stearoptene than the oils produced in the conventional, stationary stills. Unfortunately, production of this type of oil appears to have been discontinued with the outbreak of World War II.

Some years ago another large-scale producer in Rahmanlaré erected a modern distillery equipped with steam stills in which the pressure could be reduced. Distillation in these stills thus represented a modified water distillation at reduced pressure. This process yields oils of very fine odor, free of any "burnt" off-note. The disadvantages lie in the high cost of the solid and airtight apparatus, and in the fact that hydrodistillation at reduced

[20] Since the publication of Vol. I, a new rotary type of extractor has been introduced. This apparatus, constructed by Dumont and Cie., Grasse, A.M., France, is said to offer several advantages over the older types of rotary extractor. It may be used not only for extraction with volatile solvents, but also for hydrodistillation of aromatic plants.

pressure requires large quantities of steam, resulting in an excessive amount of distillation water which tends to dissolve much oil. Cohobation of the distillation waters is therefore a difficult task.

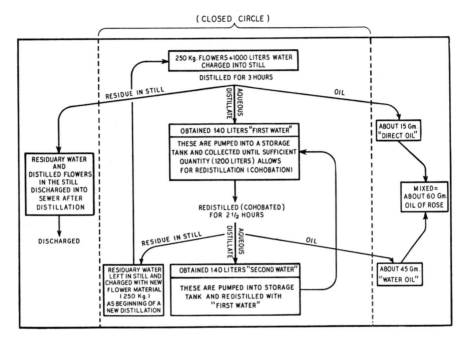

Yield and Quality of Rose Oil.—Yield and quality of the oil depend upon a number of factors, among them the weather, the time of harvest, the condition of the flowers, the type of still used and the method of distillation. The pink *Rosa damascena* yields about twice as much oil as the white *Rosa alba*. The yield is lower at the beginning than at the height of the harvest. In days of mild weather the roses produce much more oil than during spells of intense heat.

On the average, *Rosa damascena* yields 1 kg. of oil per 4,000 kg. of flowers when distilled in industrial stills. Under very favorable conditions only 2,600 kg. of roses may be required per kg. of oil; on the other hand, as much as 8,000 kg. of flowers may sometimes be required to give the same quantity of oil. In directly fired stills the yield is usually slightly higher than in steam stills. In comparative experiments carried out by the author [21] in collaboration with R. Garnier, the yields of oil from flowers of identical origin and quality were:

[21] Guenther and Garnier, *Am. Perfumer* **25** (June, September, October, and November 1930).

(1) 1:3,890 kg. in a directly-fired still

(2) 1:3,972 kg. in a steam still.

In the old-fashioned, primitive, field stills the yield is substantially higher than in large, industrial apparatus. In these small stills only 2,500 to 3,200 —on the average 3,000—kg. of flowers are required to give 1 kg. of oil. This may appear paradoxical, but the explanation lies in the fact that in the simple peasant stills distillation is "pushed" to the utmost, and often the residual water remaining in the retort with the exhausted flowers is used for the next operation (see above).

Obviously, such "pushing" of the distillation has a marked effect upon the quality of the oil. "Farmer Oils" usually exhibit a weaker odor than the oils obtained in modern large equipment; the odor of the former oils is almost honey-like, sweeter, and less pungent than that of the latter. Of the industrial oils those produced in steam stills have a stronger, more powerful aroma than those distilled in directly fired stills.

The development which has taken place in Bulgaria's rose industry during the past fifty years, viz., the gradual transition from small field stills to large industrial equipment, has not remained without profound changes in the physicochemical properties and chemical composition of the oil itself. The problem, which has caused numerous controversies among prominent essential oil chemists, will be discussed later.

To produce rose oil of highest quality the following principles should be adhered to:

1. The flowers should be collected early in the morning and should be distilled as fresh as possible, without delay.

2. In the retort, the flowers should be covered with water, and well distributed by stirring. Very little excess water should be used.

3. Distillation should be started most carefully and slowly; later it should be maintained at a lively pace. Prolonged action of the boiling water upon the oil is harmful.

4. The volumes of water distilling over should be just sufficient to carry over the oil, but not excessive. Too much aqueous distillate "washes out" the oil separating in the Florentine flask, and redissolves some of its constituents.

5. The temperature of the condenser (and hence that of the flow of the distillate) should be watched carefully.

6. In general, distillation must proceed smoothly, without any "Coups de Feu" (boiling and foaming over).

Physicochemical Properties.—The volatile oil obtained in Bulgaria by distillation of *Rosa damascena* Mill. (often mixed with very small quantities of *Rosa alba* L.) may be a light yellow, occasionally slightly greenish, mass, semisolid at room temperature. At 16° to 22° C., lanceolate or lamellate, shiny crystals separate from the liquid; because of their low specific gravity, the crystals rise to the surface and cover it with a thin skin which breaks easily if the oil is gently shaken. On cooling, the oil congeals to a transparent soft mass, which can be reliquefied on warming even at hand temperature.

The odor of the oil is powerful and characteristic of fresh roses, the flavor is strong, sweet, and somewhat honey-like.

In 1929 Gildemeister and Hoffmann [22] reported the following properties for Bulgarian rose oil:

Specific Gravity at 20°/15°....... 0.856 to 0.870
Specific Gravity at 30°/15°....... 0.849 to 0.862
　　　　　　　　　　　　　　　　(The specific gravity decreases with increasing stearoptene content of the oil)
Optical Rotation................ $-1°$ to $-4°$
Refractive Index at 25°.......... 1.452 to 1.466 [23]
Congealing Point............... $+18°$ to $+23.5°$
Acid Number.................. 0.5 to 3
Ester Number................. 7 to 16
Total Alcohol Content, Calculated
　as $C_{10}H_{18}O$ (Geraniol).......... 66 to 75%; in exceptional cases as high as 76%
Citronellol Content (By Formylation—see below)............... 24 to 37%
Stearoptene Content............ 17 to 21%
Solubility..................... Due to its content of sparingly soluble paraffins, rose oil, even in very large quantities of 90% alcohol, gives only turbid mixtures, from which the stearoptene gradually precipitates. The liquid portion of the oil, the so-called "Elaeoptene," is clearly soluble in 70% alcohol

The above properties are those of commercial rose oils produced in Bulgaria prior to 1929. Since then, profound changes in the rose oil industry of that country have taken place (see above): in large part, old-fashioned peasant distillation has been replaced by industrial production, primitive small stills have given way to modern large-scale equipment, and the Bulgarian Government has enforced a strict control over the quality of the oil. Such changes have naturally influenced the physicochemical properties and the chemical composition of the Bulgarian product.

The first deviation from the earlier accepted standards was noted in 1932 by Parry and Seager,[24] who reported that Bulgarian rose oils produced in

[22] "Die Ätherischen Öle," 3d Ed., Vol. II, 821.
[23] Cf. Parry, *Chemist Druggist* **63** (1903), 246. Simmons, *ibid.* **68** (1906), 20.
[24] *Perfumery Essential Oil Record* **24** (1933), 149.

that year had an exceptionally high citronellol [25] content (45 to 61 per cent against 25 to 40, at the most 45, per cent in the previous thirty years). The ratio of citronellol to geraniol in the oil produced in 1932 was 5.2 to 1, against 0.9–1.25 to 1 in the oils examined in 1930 and 1931. Parry and Seager attributed this change either to special climatic conditions prevailing in 1932 or to adulteration with citronellol. The statement of Parry and Seager was soon afterward challenged by Glichitch and Naves,[26] who asserted that since 1924 they had analyzed a great number of Bulgarian rose oils with a rhodinol (*l*-citronellol) content ranging from 34 to 55 per cent. At about the same time Garnier and Sabetay [27] examined four rose oils distilled in Bulgaria under their supervision and noted the properties given in Table 1.1.

TABLE 1.1

Origin and Method of Distillation	Specific Gravity at 15°	Optical Rotation	Refr. Index at 20°	Melting Point	Ester Number	Ester Number after Acetyl.	Total Alc. Content (%)	Rhodinol Content (%)	Stearoptene Content (%)	Ethyl Alcohol Content (%)
Klissoura, Direct Fire Distillation...	0.8650	−3.48°	1.4630	+17.5°	17.5	228	75.6	46	16	1.4
Kara Sarli, Direct Fire Distillation...	0.863	−4°	1.4585	+20°	8.1	218	71.7	53.7	17.8	3.6
Karlovo, Steam Distillation..........	0.875	−2.40°	1.4582	+17°	6.8	239	80	52	7.3	7.7
Kara Sarli, Rotary Distillation.......	0.8704	−4° 14′	1.4563	+17°	4.5	249	84.2	56.8	8.8	7.2

Garnier and Sabetay thus found that the so-called "Rhodinol (*l*-citronellol) Content," viz., the content of compounds which can be esterified by formylation, was higher than had previously been reported and exceeded 45 per cent.

(Note also the relatively high content of ethyl alcohol in some of these oils; for an explanation see below.)

In the same year, C. and R. Garnier [28] more thoroughly investigated the problem of the rhodinol (*l*-citronellol) content in rose oils of various origin and arrived at the conclusion that it ranged from 40 to 63 per cent in oils produced in large modern stills. On the other hand, oils distilled in old-fashioned small stills contained only from 23 to 40 per cent of rhodinol. (The latter compound was determined by hot formylation.) Examining

[25] In the following, the terms *l*-citronellol and *l*-rhodinol will be used interchangeably.
[26] *Parfums France* **11** (1933), 154.
[27] *Compt. rend.* **197** (1933), 1748.
[28] *Bull. soc. chim.* **53** (1933), 513. Cf. R. Garnier, *Perfumery Essential Oil Record* **24** (1933), 370.

various fractions obtained during the distillation of roses in their own factory in Kara Sarli (Bulgaria), C. and R. Garnier [29] observed that the rhodinol content was high in the first fractions and progressively lower in the later fractions. This explains why the earlier rose oils—which were produced chiefly in small peasant stills—contained lesser quantities of rhodinol (*l*-citronellol). In such primitive stills, the quantity of water distilled over is from 1 to 1.5 times greater than the flower charge. In large industrial stills only half the quantity of the water—in relation to the flower charge—is distilled over. Therefore, distillation in small stills gives a better yield of oil than distillation in large, industrial stills. On the other hand, the peasant oils contain less rhodinol (*l*-citronellol) and more inert substances (which weaken the odor) than the industrial oils.

Continuing work on the subject, Garnier and Sabetay [30] in 1934 expressed the opinion that rose oils with less than 40 per cent of rhodinol (*l*-citronellol) should be viewed with suspicion. In some oils of unquestionable purity they observed:

Optical Rotation	$-3° 8'$ to $-3° 44'$
Total Alcohol Content, Calculated as Geraniol	67.8 to 89.9%
Apparent Rhodinol Content	41 to 66.4%
Ethyl Alcohol Content	0.9 to 3.35%

In the same year (1934) Glichitch and Naves [31] examined five genuine Bulgarian rose oils (I distilled at Kara Sarli, in rotary apparatus; II at Karlovo, in steam stills; III at Klissoura, in open fire stills; IV at Kara Sarli, in open fire stills; V at Pavel Banya, in open fire stills), and found these properties:

	I	II	III	IV	V
Specific Gravity at 25°/15°	0.8630	0.8669	0.8579	0.8548	0.8541
Specific Gravity at 15°/15° (Coefficient 0.0008)	0.8710	0.8749	0.8659	0.8628	0.8621
Optical Rotation	$-4° 3'$	$-2° 33'$	$-3° 40'$	$-4° 2'$	$-3° 33'$
Refractive Index at 25°	1.4538	1.4545	1.4603	1.4564	1.4539
Refractive Index at 20° (Coefficient 0.00045)	1.4560	1.4567	1.4625	1.4586	1.4561
Congealing Point	$+15.8°$	$+16.1°$	$+17.9°$	$+19°$	$+17.8°$
Acid Number	2.6	3.5	2.6	2.5	2.6
Ester Number	8.7	11.2	10.6	9.8	8.4
Ester Number after Acetylation	243	248.4	224.3	217.7	222.3
Ester Number after Formylation	196.9	185.1	153.9	178.2	188.8
Ester Content, Calculated as $C_{12}H_{20}O_2$	3.04%	3.9%	3.74%	3.42%	2.93%

[29] *Ibid.*
[30] *Ann. fals.* **27** (1934), 264. Cf. *Perfumery Essential Oil Record* **25** (1934), 347.
[31] *Parfums France* **12** (1934), 7.

	I	II	III	IV	V
Free Alcohol Content, Calculated as $C_{10}H_{18}O$	79.38%	80.04%	70.54%	68.24%	70.49%
Combined Alcohol Content, Calculated as $C_{10}H_{18}O$	2.39%	3.08%	2.91%	2.69%	2.30%
Total Alcohol Content, Calculated as $C_{10}H_{18}O$	81.77%	83.12%	73.45%	70.93%	72.79%
Formylizable Compounds, Calculated as Rhodinol	60.75%	56.75%	46.3%	54.42%	58.0%
Stearoptene Content	8.9%	7.5%	20.2%	20.6%	18.7%

These data led Glichitch and Naves to conclude that Bulgarian rose oils may contain much more than 40 per cent of "Rhodinol" whatever the method of distillation practiced, and in spite of the great variations in content of stearoptenes and total alcohols.

Two years later, Parry [32] published the results of observations on the content of citronellol in rose oils, which he had made in the course of several years:

1. Rose oils remaining liquid even at a low temperature contain more citronellol than oils congealing at a higher temperature (i.e., oils with a higher stearoptene content).

2. Guaranteed pure oils of high citronellol content also have a high content of geraniol, hence a high percentage of total alcohols.

Since the stearoptene content of a rose oil varies with the method of distillation, Parry suggested evaluating the quality of the oil according to the ratio of citronellol to geraniol. Of 28 rose oils examined by Parry and Seager in 1934, only 5 exhibited ratios of citronellol to geraniol higher than 3. In rose oils of high citronellol content, there is a considerable difference in quality between those with a citronellol/geraniol ratio of less than 3, and those with a ratio of about 4 or 5. According to Parry, oils of high citronellol content and a citronellol/geraniol ratio above 3 should be viewed with some suspicion; oils with a ratio of 4 or 5 must be regarded as highly suspect.

In 1938, Karaivanoff,[33] Chief of the laboratories of the "Banque Agricole et Coöpérative de Bulgarie," reported the results of the analyses carried out in this government agency on 243 genuine rose oils distilled in Bulgaria between 1930 and 1937 under the control of government inspectors. The properties of these oils varied within the following limits:

Specific Gravity at 30°.............. 0.848 to 0.8636
Optical Rotation................... −2° 12′ to −4° 24′
Refractive Index at 25°............ 1.4538 to 1.4646

[32] *Perfumery Essential Oil Record* **27** (1936), 278.
[33] *Drug Cosmetic Ind.* **42** (1938), 580.

Congealing Point................... +16° to +22.5°
Acid Number...................... 0.93 to 3.08
Ester Number..................... 7.4 to 16.8
Saponification Number.............. 8.4 to 18.7
Ester Number after Acetylation...... 194 to 240
Rhodinol (*l*-citronellol) Content...... 27.4 to 56.9%
Stearoptene Content............... 16 to 24%

(The data quoted by Karaivanoff for "Alcohol Percentage," "Combined Alcohol," and "Free Alcohol" are not included above, because of the unexplained discrepancies that arise when these values are checked against the Ester Number and Ester Number after Acetylation given by Karaivanoff.)

The conclusion of this laboratory was that the content of rhodinol (*l*-citronellol) in pure rose oils ranged from 35 to 51 per cent, the extreme limits being 26.55 and 56.91 per cent. Regarding the difference in the citronellol content of rose oils examined between 1930 and 1937, as compared with that of rose oils produced in previous years, Karaivanoff [34] entirely agrees with the opinions of Poucher,[35] and of C. and R. Garnier (see above), according to which the ratio of "Rhodinol" to geraniol decreases in the successive fractions of distilled rose oils. Since the distillation of the flowers in large industrial stills is not so complete as in small peasant stills, the oils produced in modern equipment contain more rhodinol than those obtained formerly in the primitive small stills. The conclusions of Karaivanoff were affirmed later by Naves [36] who expressed the view that the increase in the rhodinol (citronellol) content was due chiefly to improved methods of distillation, and to the extension of technical supervision to the Bulgarian cooperatives.

The latest data on the physicochemical properties of Bulgarian rose oil are those reported by Behar [37] from the records of the laboratory of "Bulgarska-rosa," which now produces all of the Bulgarian rose oil:

Specific Gravity at 30°.......... 0.8485 to 0.8605
Optical Rotation................. −2° 18′ to −4° 24′
Refractive Index at 25°.......... 1.4530 to 1.4640
Freezing Point.................. +16.5° to +22.5°
Acid Number.................... 0.92 to 3.75
Ester Number................... 7.2 to 17.2
Saponification Number........... 8.0 to 20.75
Ester Number after Acetylation... 197.87 to 233.33
Free Alcohol Content............ 62.9 to 75.5%

[34] *Ibid.*
[35] "Perfumes, Cosmetics and Soaps," 6th Ed., Vol. II (1942), 229.
[36] *Mfg. Chemist* **19** (1948), 371.
[37] *Perfumery Essential Oil Record* **41** (1950), 263.

Combined Alcohol Content	2.0 to 4.7%
Total Alcohol Content	65.8 to 78.2%
Citronellol Content	30.5 to 58.6%
Stearoptene Content	18.2 to 21.3%

Another subject of controversy, which attracted the attention of several essential oil chemists over a number of years, revolved about the presence of ethyl alcohol in genuine rose oils. About sixty years ago, Poleck,[38] and Eckart [39] stated that rose oil contains small quantities of ethyl alcohol. Shortly afterward, Schimmel & Co.[40] disputed this, claiming that ethyl alcohol is not a normal constituent of rose oil, but rather an added adulterant, or a product of the partial fermentation that may take place during transport and storage of flowers prior to distillation. If the roses are distilled immediately after harvest, the oil, according to Schimmel & Co., contains no ethyl alcohol.

For many years this view was held to be correct, apparently being substantiated by the fact that pure commercial rose oils (which were then produced chiefly in small peasant stills) contained only traces of (if any) ethyl alcohol. However, in 1933 Garnier and Sabetay [41] reported contents of ethyl alcohol ranging from 1.4 per cent—in genuine oils produced in direct-fire stills—to 7.7 per cent in pure oils distilled in modern steam stills (see above). The findings of Garnier and Sabetay were soon afterward challenged by Parry and Seager,[42] who asserted that, with the exception of rotary-distilled oils, none of the rose oils examined by them contained more than 1 per cent of ethyl alcohol. The latter substance, as well as phenyl ethyl alcohol, can easily be removed by washing the oil three times with hot water. However, oils treated in this manner, and tested by Zeisel's method, still show the presence of ethyl alcohol. This, in the opinion of Parry and Seager, indicates that genuine rose oil contains unidentified compounds with ethoxy, methoxy, or similar groupings, which give the Zeisel reaction and are thus erroneously calculated as ethyl alcohol. (Naves [43] has since demonstrated that the compounds in question are eugenol and its methyl ether.) Continuing work on the content of ethyl alcohol in genuine rose oils, Garnier and Palfray [44] demonstrated that neither

[38] *Ber.* **23** (1890), 3554.
[39] *Ber.* **24** (1891), 4205. Cf. Simmons, *Perfumery Essential Oil Record* **16** (1925), 341.
[40] *Ber. Schimmel & Co.*, October (1892), 36.
[41] *Compt. rend.* **197** (1933), 1748. Cf. *Ann. fals.* **27** (1934), 264. *Perfumery Essential Oil Record* **25** (1934), 347.
[42] *Perfumery Essential Oil Record* **25** (1934), 213.
[43] *Helv. Chim. Acta* **32** (1949), 967.
[44] *Perfumery Essential Oil Record* **26** (1935), 259.

the method of Zeisel,[45] nor that of Thorpe,[46] permits definite conclusions as to the presence and percentage of ethyl alcohol in a rose oil. Distilling 5,000 kg. of freshly harvested roses and carefully cohobating the distillation waters, Garnier and Palfray [47] noted the presence of ethyl alcohol in the distillate and concluded that ethyl alcohol is a normal constituent of fresh roses, 100 kg. of flowers containing at least 0.5 liter of pure alcohol. The actual content of ethyl alcohol in the *oil* depends upon the efficiency of the condenser, the temperature at which the condensate flows, and the construction of the oil separator. In the small primitive peasant stills, in which formerly the bulk of rose oil was produced, the ethyl alcohol was either not condensed at all in the short and warm condenser tube, or "washed out" of the oil by the large quantity of water distilled over. Therefore, "Peasant Oils" contained very little, if any, ethyl alcohol. On the other hand, oils produced in modern stills with efficient condensers do contain a small percentage of ethyl alcohol, the quantity depending chiefly upon the construction of the oil separators. Some of these latter are built in such a way that the separated oil is not "washed out" again by the water distilling over; hence oils obtained in this modern type of equipment may well contain a small quantity of ethyl alcohol.

Whether or not ethyl alcohol should be accepted as a desirable (if normal) constituent of commercial rose oil is another question. It may be argued that ethyl alcohol has no odor value whatsoever, that it acts as a diluent, and that hence it should be removed from the oil before shipment to the market. After all, rose oil is a very high-priced product, and the buyer may well insist upon absence of diluents. If all the ethyl alcohol present in 100 kg. of roses (at least 0.5 liter, according to Garnier—see above) were recovered by efficient condensation, the final product would be a solution of 25 g. of oil (the normal yield from 100 kg. of flowers) in 0.5 liter of alcohol. This certainly could not be considered a "Rose Oil." Nevertheless, ethyl alcohol has come to be considered a normal constituent of rose oil, and very small quantities are now usually admitted in commercial lots.

From the above it can be seen that it is most difficult to establish definite limits for the physicochemical properties of genuine Bulgarian rose oils, since they depend upon many factors—among them soil, altitude, weather, condition (freshness) of the roses, type of apparatus used, method of distillation, and, last but not least, quantity of white roses (*Rosa alba* L.) present in the flower material.

[45] A good description of Zeisel's method can be found in W. W. Scott's "Standard Methods of Chemical Analysis," D. Van Nostrand Co., Inc., New York (1939), Vol. II, 2527.
[46] For details of the method, see Garnier and Palfray, *Perfumery Essential Oil Record* **26** (1935), 260.
[47] *Perfumery Essential Oil Record* **26** (1935), 259.

As regards these white roses, they yield an oil with a high stearoptene content. Odor and quality are inferior to those of the oil from pink roses (*Rosa damascena* Mill.). In collaboration with R. Garnier, the author [48] distilled *Rosa alba* L. in a modern steam still at Kara Sarli (Bulgaria), and obtained an oil with these properties:

Specific Gravity at 25°.	0.8516
Optical Rotation at 25°.	−3° 42′
Refractive Index at 25°.	1.4569
Congealing Point.	+22°
Total Alcohol Content, Calculated as Geraniol.	53.35%
Ester Content, Calculated as Geranyl Acetate.	3.57%

Guenther and Garnier [49] also studied the influence which fermentation of the flower material exerts upon the quality and properties of the oil. In a series of experiments a batch of freshly picked roses was distilled in a modern steam still immediately upon harvest, another lot was stored for about 12 hr. and then distilled, and a third batch was moistened, kept for two days in an enclosed tank at a temperature of 36° C., and then distilled. The result of storage and fermentation was a marked increase in the specific gravity, a decrease in the congealing point, and, above all, a surprising increase in the ester content of the oils, i.e., from 1.22 per cent in the case of freshly harvested flowers, to 5.74 per cent in the case of stored flowers, and to 46.30 per cent in the case of the fermented flowers. The remarkably high ester content in the oil from the fermented roses was apparently due partly to the formation of phenylethyl acetate, the odor of which was clearly noticeable in the oil.

Analysis and Adulteration.—Being one of the most highly priced essential oils, rose oil has been subject to adulteration since the time it was introduced on the market. Formerly adulteration was carried out in a somewhat crude way—e.g., by the addition of palmarosa oil (which has a high geraniol content), guaiac wood oil, and spermaceti—the latter to simulate a normal stearoptene content in the adulterated oil. Such additions can be detected by careful analysis. Today, however, rose oil is often sophisticated in such a clever way that the analyst may be faced with considerable difficulties. The adulterator now has at his disposal a number of natural isolates from much lower priced essential oils—compounds, moreover, that are also natural constituents of rose oil. The most important of these are geraniol and rhodinol (*l*-citronellol), the latter derived from geranium oil. If added in moderate quantities, these compounds cannot be detected in rose oil by mere

[48] *Am. Perfumer* **25** (December 1930), 621.
[49] *Ibid.*

routine analysis. In such cases, the chemist will have to rely upon careful olfactory tests—which require much training and experience—and upon a thorough chemical examination of the suspected oil. This may be a time-consuming and quite costly procedure.

The following will briefly describe the most important tests developed for the analysis of rose oil, supplementing information given in Vol. I of this work.

Procedures for the determination of specific gravity, optical rotation and refractive index will be found in Vol. I of the present work, pp. 236, 241 and 244, respectively. Because rose oil may be a semisolid mass at room temperature, the *specific gravity* should be measured at 30° C., compared with water at 15° (d_{15}^{30}). The *refractive index* is determined at 30°, the optical rotation at 25°. Note that the *optical rotation* of pure rose oil is low (about $-2°$ to $-4°$ 30′); therefore, moderate additions of geraniol—which is optically inactive—do not influence it much. Palmarosa oil is either slightly dextro- or laevorotatory. The adulterant most commonly used at present, viz., rhodinol (*l*-citronellol) derived from geranium oil, has only a slight effect upon the rotation of rose oil. Nevertheless, even this slight deviation may indicate addition of geranium rhodinol to rose oil (see below). In the opinion of some experts, the higher the laevorotation of a rose oil—within the limits just noted—the better is its quality.

The *congealing point* can be determined by the procedure described in Vol. I of the present work, p. 329. It gives an indication of the stearoptene content of an oil. As has been pointed out in Vol. I, the congealing point of rose oil has been defined as that temperature at which the first crystals appear when the oil is subjected to slow cooling. (This is quite different from the true congealing point of oils such as anise. Cf. Vol. I, p. 253.) Normal rose oils exhibit a congealing point ranging from $+18°$ to $+22.5°$, the wider limits being $+16°$ to $+23°$ C. Rotary-distilled oils usually have a somewhat lower congealing point than oils produced in direct-fire stills or in steam stills, because the former contain less stearoptene. Oils distilled from white roses (*Rosa alba* L.) are highest in stearoptene, hence exhibit the highest congealing point.

For an assay of the *stearoptene content* and detection of adulterants, see Vol. I of the present work, p. 327. Karaivanoff [50] states that pure Bulgarian rose oils contain from 16 to 24 per cent of stearoptene.

The *ester number* of normal rose oil is low, ranging from about 7 to 17. For determination, see Vol. I, p. 265. Roses that have been slightly fermented prior to distillation yield oils with a much higher ester number than oils from freshly picked flowers (see above).

[50] *Drug Cosmetic Ind.* **42** (1938), 582.

The *total alcohol content* is determined by acetylation (see Vol. I, p. 271) and usually calculated as $C_{10}H_{18}O$, viz., geraniol. According to Karaivanoff, the Ester Number after Acetylation of pure rose oils varies from 194 to 240, which corresponds to a total alcohol content (calculated as geraniol) ranging from 62.4 to 80.5 per cent.[51] The higher the stearoptene content of a genuine rose oil, the lower its total alcohol content. Addition of palmarosa oil—which contains from 75 to 95 per cent of geraniol—usually increases the total alcohol content of a rose oil. The same effect can be obtained by the addition of pure geraniol, or geraniol and citronellol.

The *rhodinol (l-citronellol) content* of genuine rose oils ranges from 27.4 to 56.9 per cent (Karaivanoff [52]); in rotary-distilled oils it may even be higher, for the reasons explained previously.[53] The rhodinol (l-citronellol) content can be determined by formylation (cf. Vol. I of the present work, p. 278). This is a very important assay which should not be neglected whenever a rose oil shows signs of possible adulteration. Two types of citronellol are used today for sophistication of rose oil:

(a) The low-priced citronellol (both *dl-* and *d-*) isolated from sources such as Java citronella oil. Addition of this type of citronellol will markedly decrease the laevorotation of the rose oil.

(b) The higher priced *l*-citronellol isolated from geranium oil. This represents the most annoying form of sophistication now practiced with rose oil. Nevertheless, even such *l*-citronellol can be detected in rose oil, although by a somewhat complex procedure. The method was developed almost twenty years ago by Glichitch and Naves.[54] It is based upon the fact that genuine rose oil contains laevorotatory rhodinol (*l*-citronellol), the rotatory power of which ranges from $-4°$ to $-4° 30'$. It is not possible to obtain an *l*-citronellol of similar high laevorotation economically from any other source. The rotatory power of rhodinol (*l*-citronellol) derived from geranium oil never exceeds $-2° 30'$, and the product from other sources (see above) is either racemic or dextrorotatory.[55] Glichitch and Naves, therefore, recommended isolating the rhodinol (*l*-citronellol) from the suspected rose oil and determining its optical rotation and the dispersion of the latter, after the purity

---

[51] Author's correction of Karaivanoff's figures for "Alcohol Percentage" (59.6 to 82%). Garnier and Sabetay [*Ann. fals.* **27** (1934), 264. Cf. *Perfumery Essential Oil Record* **25** (1934), 347] reported for genuine rose oil a total alcohol content, calculated as geraniol, ranging from 67.8 to 89.9%.

[52] *Drug Cosmetic Ind.* **42** (1938), 582.

[53] According to Garnier and Sabetay [*Ann. fals.* **27** (1934), 264. Cf. *Perfumery Essential Oil Record* **25** (1934), 347] oils with an apparent rhodinol content of less than 40% should be viewed with suspicion. These authors reported an apparent rhodinol content, in pure oils, ranging from 41 to 66.4%.

[54] *Parfums France* **11** (1933), 156. Cf. Naves, *Mfg. Chemist* **19** (1948), 372.

[55] *Perfumery Essential Oil Record* **37** (1946), 120.

of the isolated *l*-citronellol has been checked by measurement of other properties, among them specific gravity, refractive index, and dispersion of the latter. The rhodinol can be isolated from rose oil by means of the technique developed by Barbier and Bouveault [56] in 1896. In the procedure, the rose oil is heated to 140°–160° in the presence of an excess of benzoyl chloride. All the alcoholic constituents are thereby destroyed, with the exception of the rhodinol (*l*-citronellol) and a part of the phenylethyl alcohol, these two compounds being converted into their benzoic esters. The low boiling products are then eliminated by distillation, and the residue is saponified. The alcohols are subsequently rectified by entrainment with steam, the phenyl ethyl alcohol is destroyed by heating at 140° over potassium hydroxide, according to the procedure of D. Sonntag, [57] and the rhodinol (*l*-citronellol) isolated by fractional distillation. Addition of rhodinol (*l*-citronellol) from any source other than rose oil depresses its rotatory power.

The quantity of *ethyl alcohol* present in a rose oil can be determined by the tests described in Vol. I of the present work, p. 339. As has been pointed out, pure rose oils should contain only a small percentage of ethyl alcohol.

As regards several other specific adulterants of rose oil:

Small quantities of *phenylethyl alcohol* are occasionally added. According to Garnier and Sabetay, [58] this compound has a tendency to increase the congealing point and the apparent rhodinol content of the oil (see above).

Oil of guaiac wood is an old adulterant of rose oil; it has a soft odor reminiscent of tea roses. Moderate additions may result in an increase of the specific gravity, a slight increase in the laevorotation of the rose oil, and a decrease in the total alcohol content. The presence of guaiac wood oil can be detected by isolation of its chief constituent, guaiol m. 91°–93°. The oil leaves a resinous evaporation residue and increases the congealing point of the rose oil. For detection, see also Vol. I of the present work, p. 328.

Oil of Geranium macrorrhizum L., the so-called "Zdravetz Oil," produced in Bulgaria, has occasionally been used for adulteration of rose oil, particularly by small producers who distilled the "Zdravetz Oil" specifically for this purpose. [59] For physicochemical properties of this oil, see Vol. IV of the present work, p. 734.

There are a number of other compounds, small quantities of which occur naturally in rose oil, but which are readily available from other sources and can, therefore, be used as adulterants in conjunction with the above-named products. Among these are *l*-linaloöl, nerol, citral, eugenol, methyl-

[56] *Compt. rend.* **122** (1896), 530.
[57] Mme. de Dortan-Sonntag. Thèse de Doctorat, Faculté des Sciences, Paris (1933). Cf. Sabetay, *Bull. soc. chim.* **45** (1929), 69. Fourneau and Puyal, *ibid.* **31** (1922), 424.
[58] *Ann. fals.* **27** (1934), 264. Cf. *Perfumery Essential Oil Record* **25** (1934), 347.
[59] A. Zlatarov, "La Rose et l'Industrie de l'Essence de Roses en Bulgarie," Sofia (1926).

eugenol, nonyl aldehyde, and farnesol. If added in traces only, these can hardly be detected.

All of the above proves that oil of rose is strictly an article of confidence and should be purchased only from the most reliable dealers.

Chemical Composition.—The early examinations of rose oil by de Saussure,[60] Blanchet,[61] Baur [62] and Gladstone [63] dealt chiefly with stearoptene and "Elaeoptene"; since the results were vague and inconclusive, they require no discussion in these pages.

The first thorough and fruitful investigation of the main constituent of rose oil is that of Eckart,[64] who in 1890 isolated an alcohol (similar to geraniol) which he named "Rhodinol." (The compound described by Eckart appears actually to have been a mixture of geraniol and *l*-citronellol.) In 1893, Markovnikov and Reformatski [65] isolated a compound "Roseol," to which they assigned the empirical molecular formula $C_{10}H_{20}O$, at the same time claiming it to be the chief constituent of rose oil. Simultaneously, Barbier [66] concluded that the formula $C_{10}H_{18}O$ originally proposed by Eckart was correct. The discrepancies in the results obtained by the above-named researchers induced Bertram and Gildemeister [67] to investigate the chemical composition of rose oil more closely. Examining oils distilled in Bulgaria and Germany, they arrived at the conclusion that the principal component is geraniol, an alcohol $C_{10}H_{18}O$ discovered by Jacobsen [68] in palmarosa oil as far back as 1870, and that the "Rhodinol" of Eckart was only an impure geraniol. In the same year in which Bertram and Gildemeister published their findings, Hesse [69] expressed the opinion that his "Reuniol" (an alcohol $C_{10}H_{20}O$ he had found in geranium oils) occurs probably also in rose oil. Two years later, Tiemann and Schmidt [70] proved the identity of Hesse's "Reuniol" with citronellol obtained by reduction of citronellal. The problem of the identity of "Rhodinol" led to a lively controversy among some of the most brilliant terpene chemists for several decades after 1890. (For details see the original literature,[71] and Vol. II of the

[60] *Ann. chim. phys.* [2], **13** (1820), 337.
[61] *Liebigs Ann.* **7** (1833), 154.
[62] *Neues Jahrb. Pharm.* **27** (1867), 1; **28** (1867), 193. *Jahresber. Pharm.* (1867), 350. *Dinglers polytech. J.* **204** (1872), 253.
[63] *J. Chem. Soc.* **25** (1872), 12. *Pharm. J.* [3], **2** (1872), 747.
[64] *Arch. Pharm.* **229** (1891), 355. *Ber.* **24** (1891), 4205. Cf. Poleck, *Ber.* **23** (1890), 3554.
[65] *J. prakt. Chem.* [2], **48** (1893), 293. *Ber.* **23** (1890), 3191; **27** (1894), 625 Ref.
[66] *Compt. rend.* **117** (1893), 177. Cf. Tiemann and Semmler, *Ber.* **26** (1893), 2708.
[67] *J. prakt. Chem.* [2], **49** (1894), 185.
[68] *Liebigs Ann.* **157** (1871), 232.
[69] *J. prakt. chem.* [2], **50** (1894), 472.
[70] *Ber.* **29** (1896), 903, 922; **30** (1897), 33.
[71] Erdmann and Huth, *J. prakt. Chem.* [2], **53** (1896), 42. Bertram and Gildemeister, *ibid.*, 225. Hesse, *ibid.*, 238. Erdmann and Erdmann, *ibid.* [2], **56** (1897), 1. Bertram and Gildemeister, *ibid.*, 506. Poleck, *ibid.*, 515. *Ber.* **31** (1898), 29. Bertram

present work, p. 178.) Today rhodinol [72] is generally considered to be *l*-citronellol, the latter consisting of a mixture of two isomeric forms in which, according to Doeuvre,[73] the terpinolene or "*β*"-form (isopropylidene form) largely predominates. Studying the infrared spectra of 25 purified acyclic monoterpene compounds, with particular reference to the position and configuration of the double bonds, Barnard, Bateman, Harding, Koch, Sheppard and Sutherland [74] recently obtained very strong evidence that naturally occurring acyclic monoterpenes as a class possess the homogeneous isopropylidene end-group structure. In every purified sample, citronellol (rhodinol) among them, examined by these authors, the proportion of the compound in the isopropenyl form never exceeded 3 per cent and was usually very much less or possibly nil.

In addition to rhodinol (*l*-citronellol) and geraniol, the two principal constituents of rose oil obtained by hydrodistillation of roses, a number of other components have been identified; all constituents are listed below, according to approximate importance:

l-Citronellol. The quantity of *l*-citronellol occurring in rose oil depends chiefly˘upon the method of distillation (cf. above and "Physicochemical Properties"). Some experts evaluate a rose oil according to its rhodinol (*l*-citronellol) content, claiming that a high rhodinol percentage indicates a very good quality.

Geraniol. The next most important constituent of the oil.

Nerol. According to von Soden and Treff,[75] oil of rose contains from 5 to 10 per cent of nerol, which has a marked influence upon its odor. Von Soden and Treff identified nerol by means of its diphenylurethane m. 52°–53°.

l-Linaloöl. Schimmel & Co.[76] characterized this alcohol by oxidation to citral (naphthocinchoninic acid compound m. 197°–199°).

Phenylethyl Alcohol. Identified simultaneously, but independently, by von Soden and Rojahn,[77] and by Walbaum.[78] Because of its relatively good solubility in water,

and Gildemeister, *ibid.*, 749. See also Barbier, *Compt. rend.* **117** (1893), 177; Monnet and Barbier, *ibid.*, 1092; Barbier and Bouveault, *ibid.* **118** (1894), 1154; **119** (1894), 281, 334; **122** (1896), 530, 673. Bouveault, *Bull. soc. chim.* [3], **23** (1900), 458. Bouveault and Gourmand, *Compt. rend.* **138** (1904), 1699. Cf. Barbier and Bouveault, *ibid.* **122** (1896), 737. Wallach and Naschold, *Nachr. Ges. Wiss. Göttingen* (1896), Meeting February 8th. *Chem. Zentr.* (1896), I, 809. Bertram and Gildemeister, *J. prakt. Chem.* [2], **49** (1894), 185. Cf. *Ber. Schimmel & Co.*, October (1894), 23; April (1895), 37. Barbier and Locquin, *Compt. rend.* **157** (1913), 1114.

[72] The term rhodinol used in its scientific sense, and not to be confused with commercial rhodinol. The latter is usually a mixture of *l*-citronellol, geraniol, and small quantity of other alcohols. Cf. Vol. IV of the present work, p. 701.

[73] *Parfums France* **12** (1934), 197.

[74] *J. Chem. Soc.* (1950), 915. Cf. Naves, *Perf. Ess. Oil Rec.* **43** (1951), 294.

[75] *Ber.* **37** (1904), 1094.

[76] *Ber. Schimmel & Co.*, October (1900), 57.

[77] *Ber.* **33** (1900), 1720.

[78] *Ibid.*, 1903, 2299. Walbaum and Stephan, *ibid.*, 2302.

phenylethyl alcohol occurs in *distilled* rose oil only in small quantities (about 1 per cent). However, the extracted products—concretes and absolutes of rose—contain large quantities of phenylethyl alcohol (see below). In the oil extracted from (distilled) rose water von Soden and Rojahn [79] found 35 per cent of phenylethyl alcohol, and in the volatile oil distilled from rose pomades about 46 per cent. The same authors noted the presence of phenylethyl alcohol also in the residual waters remaining after distillation of the flowers. That phenylethyl alcohol (and eugenol—see below) are actually present in the residual waters, was later confirmed by Guenther and Garnier, [80] in collaboration with Walbaum and Rosenthal.

Farnesol. Reported a constituent of the distilled oil by von Soden and Treff. [81]

Esters. The above-named alcohols occur in the oil chiefly in free form, and to a small degree in esterified form, the ester content (calculated as geranyl acetate) ranging from about 2.5 to 6.0 per cent. These esters are very important as regards the odor of the oil.

Nonyl Aldehyde and Other Higher Aldehydes(?). Schimmel & Co. [82] identified nonyl aldehyde in the oil by oxidation to pelargonic acid m. 252°–253°. This aldehyde, which is probably accompanied by lower and higher homologues, exerts a considerable influence upon the odor of the oil.

Citral. In the course of the same investigation, Schimmel & Co. [83] also noted the presence of citral in rose oil. It was identified by preparation of the naphthocinchoninic acid compound m. 197°–199°.

Eugenol. Von Soden and Treff [84] first found that rose oil contains about 1 per cent of eugenol. This was later confirmed by Naves. [85] Guenther and Garnier [86] observed the presence of eugenol in the residual waters remaining after distillation of the flowers (see above—"Phenylethyl Alcohol").

Eugenol Methyl Ether. Naves [87] recently noted that rose oil contains from 1 to 1.2 per cent of methyleugenol.

Carvone. The same author [88] also reported traces of carvone in the oil.

Sesquiterpenes(?). Investigating genuine Bulgarian rose oils, Garnier and Sabetay [89] observed that they contained small quantities of azulenogenic sesquiterpenes. Like the stearoptene, these sesquiterpenes are insoluble in 75 per cent alcohol, hence they appear as "Stearoptene" in the isolation and assay of the latter by means of 75 per cent alcohol.

[79] *Ibid.* **33** (1900), 1720, 3063; **34** (1901), 2803.
[80] *Am. Perfumer* **25** (December 1930), 621.
[81] *Ber.* **37** (1904), 1094.
[82] *Ber. Schimmel & Co.,* October (1900), 57.
[83] *Ibid.*
[84] *Ber.* **37** (1904), 1094.
[85] *Helv. Chim. Acta* **32** (1949), 967.
[86] *Am. Perfumer* **25** (December 1930), 621.
[87] *Helv. Chim. Acta* **32** (1949), 967.
[88] *Ibid.*
[89] *Ann. fals.* **28** (1935), 585.

Stearoptene: According to Flückiger,[90] the stearoptene present in rose oil consists not of one, but of at least two homologues of the paraffin series C_nH_{2n+2}, because it can be separated into two fractions m. 22°, and m. 40°–41° (cf. Vol. I of the present work, p. 328).

Aside from the constituents described above, oil of rose undoubtedly contains still other substances which play a most important role in the total odor of the oil. Any attempt to compound a synthetic rose oil by using only the constituents just enumerated, and according to the proportions established in natural rose oil, results only in mixtures which exhibit some similarity to the natural product, but which as regards odor are still unsatisfactory. The so-called trace compounds, to which natural rose oil owes its characteristic and powerful odor, occur in it only in minute quantities. They have not yet been identified; or at least nothing definite has been published about their nature. Some years ago, Garnier and Sabetay [91] reported that a small, strongly dextrorotatory fraction occurring between the ethyl alcohol fraction and the rhodinol-geraniol fraction appears to be particularly important as regards the odor of the oil.

In this connection it should also be mentioned that the odor of the distilled rose oil differs greatly from that of the extracted rose products (concretes and absolutes—see below). This difference is caused not only by the large quantity of natural waxes present in the latter products, but also by the occurrence, in the concretes and absolutes, of compounds—particularly phenylethyl alcohol and eugenol—which on distillation of the roses remain mostly in the residual waters, and are thrown away—hence are present in the distilled oil in very small quantities only.

B. Extracted Rose Oil
(Concrete and Absolute of Rose)

The odorous principles present in rose flowers can be isolated more efficiently and completely by the modern method of extraction with volatile solvents (usually petroleum ether) than by hydrodistillation. The so-called concretes and absolutes of rose thus obtained have come to play a very important role in the art of the perfumer, and today the quantity of rose concretes produced largely exceeds that of the distilled oil. Principal areas of production of the concrete are now the Grasse region of Southern France and, to a much smaller extent, Morocco. In these countries *Rosa centifolia* L. is used exclusively; it does not lend itself advantageously to hydrodistillation because it gives only a small yield of distilled oil. The Bul-

[90] *Pharm. J.* [2], **10** (1869), 147. *Z. Chem.* **13** (1870), 126. *Jahresber. Chem.* (1870), 863. "Pharmakognosie," 3d Ed., 170.
[91] *Compt. rend.* **197** (1933), 1748.

garian roses (*Rosa damascena* Mill. and *R. alba* L.), on the other hand, can be processed either by hydrodistillation or by extraction with solvents. For details of the extraction process itself the reader is referred to Vol. I of the present work, pp. 200 ff.

The solvent extraction process was introduced to Bulgaria by Charles Garnier, who in 1904 erected a plant in Kara Sarli, equipped with a battery of six rotary extractors. In the next thirty years other large-scale rose oil producers followed his example, and in 1938 there were in Bulgaria altogether 32 apparatus [92] for extraction of roses with solvents.

According to trade reports,[93] the following quantities of distilled rose oil and rose concrete were produced in Bulgaria between 1938 and 1945:

Year	Distilled Rose Oil (in kg.)	Concrete of Rose (in kg.)
1938	1,720	2,313
1939	3,779	86
1940	1,848	2,104
1941	836	1,205
1942	271	1,268
1943	1,133	2,016
1944	734	...
1945	351	...
1946 [94]	794	...

From these figures it appears that at times the quantities of rose concrete supplied by Bulgaria exceeded those of the distilled oil. In recent years, however, production of the concrete seems to have been curtailed, owing perhaps to political and economic developments in that country.

Yield and Physicochemical Properties.—According to the author's experience in Bulgaria, from 400 to 450 kg. of *Rosa damascena* Mill. are required to yield 1 kg. of rose concrete which, in turn, gives about 520 g. of alcohol-soluble absolute. Naves and Mazuyer [95] reported yields of concrete (petroleum ether as solvent) ranging from 0.22 to 0.25 per cent. The concrete gives from 50 to 60 per cent of absolute, and contains from 35 to 41 per cent of steam-volatile oil. (The latter is not produced commercially, but only in the laboratory to test the purity of a concrete or absolute—cf. Vol. I of the present work, p. 215.)

[92] Karaivanoff, *Drug Cosmetic Ind.* **42** (1938), 582.

[93] *Annual Rept. Schimmel & Co., Inc., New York* (1950), 50.

[94] Private communication from Mr. T. W. Delahanty, Chief, Consumers Merchandise Branch, U. S. Department of Commerce, Washington, D. C. According to the same source, Bulgaria's rose acreage in 1946 was 6,943 acres, a slight increase over the acreage in 1945 (5,700 acres)—see above. In 1948, production of rose oil in Bulgaria was less than 800 kg.

[95] "Les Parfums Naturels," Paris (1939), 269.

Concrete of rose produced in Bulgaria by extraction with petroleum ether is a light to dark brown waxy mass with a characteristic rose odor, only partly soluble in high-proof alcohol. Walbaum and Rosenthal [96] indicated these properties for Bulgarian rose concrete:

$$\text{Congealing Point} \ldots \ldots \ +41° \text{ to } +46.5°, \text{ average } +44.4°$$
$$\text{Acid Number} \ldots \ldots \ldots \ 31 \text{ to } 56$$

These authors pointed out that the commercial products frequently contain petroleum ether, which has not been completely removed by careful purification and which lowers the congealing point of the concrete. Thus a congealing point of 44.5° in a pure concrete is lowered to 41.5° by the presence of 3 per cent of petroleum ether, to 37° by 15 per cent, and to 34° by 25 per cent of petroleum ether.

As regards the Bulgarian *absolute* of rose, it is a viscous greenish-brown liquid, soluble in high-proof alcohol. Its odor is strong and characteristic of rose, softer, more mellow, and much more lasting than that of the distilled oil.

Schimmel & Co.[97] (I to III), and Naves [98] (IV to VI) reported the following properties for Bulgarian rose absolutes:

	Specific Gravity at 15°	Optical Rotation *	Refractive Index at 20°	Acid Number	Ester Number	Ester Number after Acetylation	Total Alcohol Content, Calculated as Geraniol
I.....	0.9729	+13° 10'	1.50808	9.3	29.9	217.5	71.5%
II.....	0.9682	+14° 25'	1.50633	9.3	30.9	211.9	69.3%
III.....	0.9916	+14° 20'	1.51566	11.2	27.1	218.4	71.8%
IV......	0.983	+9° 40'	1.5078	4.8	22.4	216.1	70.9%
V......	0.991	+10° 24'	1.5072	3.5	19.6	243.2	81.8%
VI......	0.973	+11° 10'	1.5081	4.34	23.1	231.7	77.0%

* Because of their dark color, Schimmel & Co. determined the optical rotation of the absolutes in alcoholic solution.

Note that the Bulgarian rose absolutes exhibit a pronounced dextrorotation, whereas all distilled rose oils are laevorotatory. The compounds causing the dextrorotation in the absolutes are nonvolatile, as is proved by the fact that the volatile oils distilled in the laboratory from the absolutes are laevorotatory (see below).

As was mentioned above, the concrete and absolute contain a certain per-

[96] *Ber. Schimmel & Co., Jubiläums-Ausgabe* (1929), 192.
[97] *Ber. Schimmel & Co.* (1926), 104.
[98] "Les Parfums Naturels," Paris (1939), 270.

centage of volatile oil which is occasionally isolated in order to determine its physicochemical properties and thereby check the purity of the concretes and absolutes. The properties of seven volatile rose oils thus obtained, and described by Naves and Mazuyer [99] varied within these limits: [100]

Specific Gravity at 15°...........	0.948 to 0.992
Optical Rotation.................	−0° 54′ to −2° 42′
Refractive Index at 20°..........	1.5046 to 1.5190
Acid Number....................	2.1 to 5.1
Ester Number...................	5.6 to 10.4
Ester Number after Acetylation....	278.6 to 320.6

Chemical Composition.—The volatile oil present in the concrete and absolute of rose probably contains the same constituents as the distilled oil of rose, but in different proportions. Exact data on the Bulgarian concrete and absolute of rose are scarce, because they have not been investigated as thoroughly as the distilled oil (see above). Phenylethyl alcohol, a minor constituent of the oil obtained by hydrodistillation of the flowers, is a very important component of the extracted products. Glichitch and Naves [101] submitted a Bulgarian concrete of rose to steam distillation at reduced pressure, and obtained a volatile oil which consisted of:

Phenylethyl Alcohol.......	63.7%
Citronellol...............	22.1%
Geraniol + Nerol.........	13.7%
	99.5%

Naves [102] reported that concrete of rose also contains small quantities of *eugenol, eugenol acetate,* and traces of a ketone with a minty odor.

II. THE TURKISH ROSE OIL INDUSTRY

Roses have been cultivated in Asia Minor for a long time. Before the turn of the century they were used for the distillation of fragrant rose water popular in the Near East as a flavorant in confectioneries, as a medicine, and as a cosmetic. The Turkish rose *oil* industry is said to have had its beginning [103] in 1894 when a Bulgarian peasant of Turkish descent came to Turkey and planted cuttings of *Rosa damascena* Mill., which he had pro-

---

[99] *Ibid.* Cf. Naves, Sabetay and Palfray, *Perfumery Essential Oil Record* **28** (1937), 336. Glichitch and Naves, *Parfums France* **11** (1933), 163.
[100] The concretes and absolutes investigated had been obtained by extraction with petroleum ether (not benzene).
[101] *Parfums France* **11** (1933), 163.
[102] "Les Parfums Naturels," Paris (1939), 271.
[103] *Ber. Schimmel & Co.* (1917), 45; (1926), 267. Cf. Jeancard, *Perfumery Essential Oil Record* **11** (1920), 210.

cured in Kazanlik and smuggled out of Bulgaria. Encouraged by the Turkish Government he showed the peasants of Anatolia how to produce rose oil in the type of primitive small field stills used at that time in Bulgaria. From this modest start there developed the Turkish rose oil industry which, in the years prior to World War I, attained considerable importance, up to 800 kg. of rose oil being produced yearly at that time. World War I, the ensuing Greco-Turkish War, and the expulsion of the Greeks and Armenians from Anatolia caused a temporary decline in the agricultural and horticultural activities of that part of Turkey, and the rose fields in particular were seriously neglected. Gradually, however, and with the aid of the Turkish Government, the rose industry recuperated, and in the period between the two World Wars, Turkey came to produce from about 100 to 300 kg. of rose oil per year.

The producing regions are located in the southwestern part of Anatolia, particularly in the Vilayets of Burdur and Isparta, center of production being the village of Atabey, most of whose 3,000 inhabitants participate in the cultivation of the flowers.[104] Other, but less important, rose producing regions include Biledjik, Antalya, Konia, Brussa, Aidin, Kütahia, Afion, Karahissar.[105] In 1933 about 450 hectares were under rose in these sections of Anatolia. According to a report from Ankara,[106] the yearly production of rose flowers in the region of Isparta averages 250 metric tons in Isparta, 250 tons in Atabey and Barla, and 175 tons in Keçibarlu. Individual rose plantings are small; in 1937 there were about 1,100 rose field proprietors in Isparta and Burdur alone.[107]

The roses cultivated in Anatolia are chiefly the pink rose (*Rosa damascena* Mill.), and—to a much smaller extent—the white rose (*Rosa alba* L.), and the so-called "Hafis," a small dark red, thornless rose. The latter is cultivated mostly in Brussa; it gives a high yield of oil which, however, is inferior in quality to that of *Rosa damascena*. Rose oil dealers often use the oil from the "Hafis" for admixture with the high-grade oil from *Rosa damascena*.

The soil in which roses thrive best in Anatolia is clayey, rich in humus, slightly sandy and contains traces of ferrous salts. It should be slightly moist, not too dry. Most suitable are the slopes of hills, at 2,500 to 3,250 ft. altitude, well protected against cold north winds and hot south winds. In periods of drought the rose fields must be irrigated. The climate should be humid rather than dry; in years with ample rains the roses yield more oil than in dry years.

[104] Erhardt, *Seifensieder-Ztg.* **65** (1938), 573, 594.
[105] *Ber. Schimmel & Co.* (1933), 50. Cf. Hackforth-Jones, *Am. Perf.* **58** (1951), 257.
[106] *Öle, Fette, Wachse, Seife, Kosmetik* (1936), No. 11, p. 13.
[107] *Perfumery Essential Oil Record* **28** (1937), 384.

New fields are started in the fall. The ground is plowed to a depth of 50 to 60 cm.; ditches, 35 to 40 cm. deep, 60 to 70 cm. wide, and 1.5 m. apart, are dug and well fertilized. In the spring the cuttings are placed into the ditches at the rate of 13,000 to 15,000 cuttings per hectare. The first worthwhile harvest can be obtained in the third year after planting, a normal harvest in the fifth year. The bushes remain productive for about twenty or twenty-five years at the most. The harvest takes place from May to June.

Distillation of the roses is carried out in numerous small open-fire stills resembling those used in Bulgaria. Some stills are fairly large—up to 1,000 liters capacity. For distillation of the oil, the flowers are boiled in ten times their weight of water. The yield of oil averages about 1 kg. per 3,000 kg. of roses.

Until 1934 Turkey's rose oil industry was a primitive house-and-village activity, entirely in the hands of numerous small producers. In that year, however, the Turkish Government, through the agency of the Sumer Bank, erected a large modern distillery in Isparta, in order to produce a rose oil of high quality which could compete with the Bulgarian product. The stills in this plant are heated with steam; 30 metric tons of roses can be processed daily. In 1938 the distillery produced about 20 per cent of Turkey's total output of rose oil.

As regards the quality of the Turkish rose oil in general, unfortunately it has not enjoyed too good a reputation on the world market. Even today it is considered inferior to the Bulgarian product. The causes are not quite clear, but several factors are probably responsible:

(a) The primitive equipment used by the peasant producers.

(b) Lack of "know-how," and faulty methods of distillation.

(c) Poor condition of the flower material (stale, slightly fermented roses, instead of fresh ones).

(d) Admixture of roses other than *Rosa damascena* Mill.

(e) Wide-spread adulteration of the oil by unscrupulous dealers and exporters. Oil of palmarosa, the so-called "Turkish Geranium Oil" for a long time has been a favored adulterant of Turkish rose oil.

That Anatolia could produce a rose oil of normal quality was demonstrated some years ago when the Turkish Government submitted samples of genuine rose oil to the Pharmazeutisches Institut of the University of Berlin for analysis.[108] The oils were found to be normal in physicochemical properties, and in odor.

Most commercial shipments, however, exhibit somewhat abnormal properties and an odor decidedly inferior to that of the Bulgarian oils. According

[108] *Ber. Schimmel & Co.* (1933), 51.

to Gildemeister and Hoffmann,[109] a sample of Anatolian rose oil, distilled from red and white roses of the Bulgarian type, had the following properties:

Specific Gravity at 30°	0.8589
Optical Rotation	−2° 20′
Refractive Index at 25°	1.46486
Congealing Point	+19.5°
Acid Number	1.4
Ester Number	9.8
Total Geraniol Content	74%

In view of the political events and the economic developments that have taken place in Bulgaria since the end of World War II, it appears highly desirable that the Anatolian rose oil industry be developed under the guidance of experts and, more important still, that the Government of Turkey take energetic steps to stop the wide-spread practice of adulteration, from which in the past the Anatolian oil has suffered most.

III. The French Rose Oil Industry

The rose oil industry of the Grasse region in Southern France differs from the Bulgarian chiefly in two respects:

(a) The rose grown in the Grasse region for industrial purposes is *Rosa centifolia* L., locally called "Rose de Mai," a light pink-colored rose related to the Bulgarian (and Turkish) *Rosa damascena* Mill.

(b) On hydrodistillation, *Rosa centifolia* gives a smaller yield of oil than *Rosa damascena;* hence in the Grasse region the roses are extracted with volatile solvents, yielding the concrete and absolute of rose (cf. Vol. I of the present work, pp. 200 ff.). Before the introduction of this modern process, the flowers were treated by maceration with hot fat (cf. Vol. I, p. 198), which gave the so-called *"Pommades"* and *"Extraits de Rose."* A small quantity of the roses in the Grasse region has always been processed by distillation—not to isolate the volatile oil, but to prepare the fragrant rose water so popular in Mediterranean and Latin countries for flavoring, cosmetic, and pharmaceutical purposes.

Producing Regions, Planting, and Cultivation.—The rose plantations— all relatively small and mostly operated by individual farmers and their families—are located near Grasse, Pégomas, Mougin, Montauroux, and other parts of that picturesque and beautiful country; two-thirds of the crop comes from La Colle, Grasse, and Saint-Paul de Vence. In 1939 the extraction plants of the Grasse region processed altogether 750,000 kg. of roses; in 1946 the quantity declined to about 400,000 kg. (Girard [110]). In 1949 the

---

[109] "Die Ätherischen Öle." 3d Ed., Vol. II, 822.
[110] *Ind. parfum.* **2** (1947), 185.

Grasse region produced 500,000 kg. of roses, in 1950 only 300,000 kg.[111] According to Ellmer,[112] roses cultivated in the higher altitudes (about 1,150 m.) yield from 30 to 35 per cent more concrete than the flowers from the plains (100 to 250 m. altitude).

In the Grasse region the roses are usually grown on nonirrigated land for the simple reason that they withstand spells of drought better than other flowers—jasmine, e.g. As a rule, however, rose bushes require water. Plantings are started in the fall, after deep plowing of the soil. According to Arnaud,[113] *Rosa centifolia* is propagated not by grafting, but by means of carefully selected, vigorous, and healthy suckers, planted 0.7 by 1.5 m. apart. To keep the ground in good condition, it should be frequently plowed, and dressed with organic and inorganic fertilizers. To obtain a high yield of flowers per branch, the latter should be allowed to grow as long as possible and trained into a horizontal position. This can be effected by preserving, in the pruning operation, two or three long stems which are intertwined with the branches of the neighboring rose bush. In this operation old woody stems that have already produced their share of flowers are cut out, and the young shoots are properly dressed for new flower growth.

Harvesting starts in the second year after planting, reaching its maximum in the fifth year. After twelve years the productivity of a planting declines rapidly. On the average, a single rose bush produces 250 g. of flowers per season. A yield of 3,000 kg. of roses per hectare is considered very satisfactory. The harvest takes place in May, very early every morning (from 4 to 8 A.M.), before the dew has disappeared from the flowers. One harvester can pick from 4 to 10 kg. of roses per hour. After collection, the roses should be processed as soon as possible; otherwise they will wither and give a poor yield on extraction.

A. Distilled Rose Oil

Years ago relatively large quantities of "Rose de Mai" were processed in the Grasse region by hydrodistillation for the production of fragrant rose water. To a limited extent this is still practiced. For this purpose 1,000 kg. of *Rosa centifolia* L. are charged into a still and boiled in water until 1,000 liters of water have distilled over. This water is not redistilled (cohobated) and represents the commercial "Rose Water." In the process of distillation about 100 g. of (direct) rose oil separate in the Florentine flask. This quantity corresponds to a yield of 1 kg. of rose oil per 10,000 kg. of *Rosa centifolia* L., as compared with 1 kg. of rose oil per 4,000 kg. of *Rosa damascena*

[111] *Soap, Perfumery & Cosmetics* **23** (1950), 913.
[112] *Riechstoff Ind.* **5** (1930), 193.
[113] *Ind. parfum.* **4** (1949), 367.

Mill., obtained in Bulgaria. (It must, however, be kept in mind that in Bulgaria the rose water is repeatedly cohobated in order to recover the relatively large quantity of rose oil suspended and dissolved therein. The yields indicated above, therefore, do not represent the actual oil contents of the two rose species.) The rose oil thus obtained in the Grasse region is only a by-product in the preparation of rose water; it is usually not sold on the market as a commercial product, but used by the essential oil houses of Grasse (mostly in perfume compositions).

A French rose oil obtained in the manner described above (without cohobation of the distillation water) was examined by Jeancard and Satie [114] who reported these properties:

Congealing Point	+25.9°
Acid Number	2.24
Saponification Number	14.70
Total Alcohol Content	32%
Citronellol Content	15.10%
Stearoptene Content	58.88%

As regards the rose water—it must meet the standards of the various Pharmacopoeias. To protect it from infestation by microorganisms, it should be stored in very clean containers, in a dark and cool place.

B. *Extracted Rose Oil*
(Concrete and Absolute of Rose)

As was said above, by far the greater part of the roses (*Rosa centifolia* L., or "Rose de Mai") grown in the Grasse region of Southern France are now worked up by the modern process of extraction with volatile solvents, thereby yielding concrete and absolute of rose. The solvent most generally used is petroleum ether which, according to Naves and Mazuyer,[115] yields from 0.24 to 0.265 per cent of concrete, the wider limits being 0.17 to 0.27 per cent. The concrete gives from 55 to 65 per cent of alcohol-soluble absolute, and contains from 25 to 32 per cent of steam-volatile oil. In the experience of the author of the present work, from 400 to 500 kg. of *Rosa centifolia* L. are required to produce 1 kg. of concrete; the latter yields from 500 to 600 g. of absolute.

There is another type of rose cultivated on the French and Italian Riviera, principally for use as an ornamental flower. This is the so-called "Rose Brunner"; flowers that cannot be sold to florists are occasionally purchased by the extraction plants in Grasse and processed with volatile solvents.

[114] *Bull. soc. chim.* [3], **31** (1904), 934.
[115] "Les Parfums Naturels," Paris (1939), 268.

According to Naves and Mazuyer,[116] the yield of concrete from "Rose Brunner" ranges from 0.10 to 0.14, occasionally to 0.19 per cent; the concrete gives from 30 to 35 per cent of absolute. Because of the poor yield and odor, only insignificant quantities of concrete and absolute are produced from "Rose Brunner."

Physicochemical Properties.—Concrete of rose extracted from *Rosa centifolia* L. ("Rose de Mai") with petroleum ether (the usual solvent) is a waxy, light to dark brown mass with a characteristic and strong rose odor. Compared with that of the Bulgarian concrete (from *Rosa damascena* Mill.), the odor of the French product is softer and more spicy, less reminiscent of distilled rose oil.

Girard [117] (I), and Naves and Mazuyer [118] (II) reported these properties for rose concrete from "Rose de Mai":

	I	II
Congealing Point..............	+52° to +53°	+43° to +48°
Melting Point..................	...	+49° to +54°
Optical Rotation...............	+10° 20′	...
Refractive Index at 60°.........	1.4811	...
Acid Number..................	14.28	9.8 to 14.4
Ester Number.................	23.1	19.6 to 25.2

The corresponding *absolute* is a yellowish to light brown viscous liquid with a strong and very lasting rose odor.

The following properties are those noted by Schimmel & Co.[119] for a French absolute of rose (I), by Naves [120] for three absolutes from the Département of Var in Southern France (II), and by Girard [121] for absolutes from Grasse (III). All oils were extracted from "Rose de Mai":

	I	II	III
Specific Gravity at 15°....	0.9640	0.981 to 0.993	0.964 to 0.988
Optical Rotation.........	+13° 55′	+10° 41′ to +13° 10′	+10° 41′ to +13′ 55′
Refractive Index at 20°...	1.50873	1.5096 to 1.5122	1.5087 to 1.5104
Acid Number...........	9.3	3.4 to 7.56	3.4 to 9.3
Ester Number..........	29.0	17.5 to 26.6	17.5 to 29
Ester Number after Acetylation...........	220.3	229.9 to 236.6	...

As was pointed out above, the concrete and absolute of rose contain volatile oils which are occasionally isolated in the laboratory, by steam distilla-

[116] *Ibid.*, 269.
[117] *Ind. parfum.* **2** (1947), 185.
[118] "Les Parfums Naturels," Paris (1939), 269.
[119] *Ber. Schimmel & Co.* (1926), 104.
[120] "Les Parfums Naturels," Paris (1939), 269.
[121] *Ind. parfum.* **2** (1947), 185.

tion at reduced pressure, in order to check the purity of the corresponding concrete or absolute. These volatile oils are not commercial products, however. Naves, Sabetay and Palfray [122] examined a volatile oil which they had obtained from an extraction product of "Rose de Mai":

Specific Gravity at 15°..............	0.976
Optical Rotation...................	−2° 24′
Refractive Index at 20°.............	1.5038
Acid Number......................	0.8
Ester Number.....................	7.2
Ester Number after Acetylation......	292.6

Chemical Composition.—The chemical composition of the volatile constituents present in the concrete and absolute of *Rosa centifolia* L. was investigated by von Soden,[123] and Elze [124] who identified the following compounds:

Phenylethyl Alcohol. The principal constituent. Treating the volatile oil (derived from French rose extracts) with phthalic anhydride, von Soden isolated 75 to 80 per cent of alcohols; these consisted of 75 per cent of phenylethyl alcohol and 25 per cent of other primary alcohols—the latter identical with those occurring in distilled rose oil.

Geraniol. Reported by Elze. Diphenylurethane m. 81°–81.5°.

Citronellol. Characterized by oxidation to citronellal, which was identified by means of the naphthocinchoninic acid compound m. 225° (Elze).

Nerol. Preparation of the diphenylurethane m. 50°–50.5° (Elze).

Farnesol. Reported by Elze.

According to Naves,[125] concrete of rose also contains small quantities of *eugenol, eugenol acetate,* and traces of a ketone with a minty odor. However, Naves does not specify whether the rose concrete from which he isolated these compounds was of French or Bulgarian origin.

IV. THE MOROCCAN ROSE OIL INDUSTRY

Most probably the rose plant was introduced to Morocco by the Moslem Arabs in the seventh century A.D., when they swept over the whole of North Africa and founded the Ommiad Caliphate. Since then the rose has remained the favorite flower of the Moors, who grow it in cities and villages for ornamental purposes and for distillation of rose water. Next to gera-

122 *Perfumery Essential Oil Record* **28** (1937), 336.
123 *J. prakt. Chem.* [2], **69** (1904), 265.
124 *Chem. Ztg.* **43** (1919), 747.
125 Naves and Mazuyer, "Les Parfums Naturels," Paris (1939), 271.

nium, the rose is today perhaps the most widely cultivated of all North African perfume plants. For centuries the production of rose water has been a picturesque home industry practiced by the families of wealthier natives, particularly in Morocco. The perfumed water thus prepared serves for toilet requirements and for the flavoring of beverages and confectionery. In the primitive process the essential oil is seldom collected because the distillation water would have to be cohobated to obtain an appreciable yield of oil.

The rose cultivated in Morocco was formerly thought to be *Rosa damascena* Mill. In 1930, however, the botanist Jean Gattefossé made an expedition to various oases in the southern part of Morocco, where roses grow abundantly at altitudes ranging from 3,000 to 4,800 ft.; he identified the plant as belonging to the species cultivated in Southern France, viz., *Rosa centifolia* L.[126] The observations of Gattefossé have shown that in Morocco the rose bushes thrive at much higher altitudes than in the Grasse region of Southern France; they are capable of resisting very wide ranges of temperature and particularly drought—provided the soil is properly taken care of. As a matter of fact, "rust," that dreaded disease which frequently attacks rose bushes growing in humid or irrigated soil, is almost unknown in certain parts of southern Morocco.

While the rose bush of Marrakesh, the most popular of all Moroccan roses, is of the species *Rosa centifolia* L., there seem to be other types in southern Morocco—e.g., in Mesguita, south of Quarzazat—which belong to the species *Rosa damascena* Mill. It is quite possible that the original rose stock, introduced by the Arabs (perhaps from Persia) was *Rosa damascena*, or even a more primitive parent species, but that as a result of ecological and cultural factors the plant underwent morphological changes and finally developed into *Rosa centifolia* and closely related species. It is a well-known fact that species of *Rosa* hybridize very easily.

The rose producing regions of Morocco are located chiefly between the High Atlas in the north and the Djebel Sarro in the south. Here the bushes are grown not on special plantations, but in rows between trees, and in the form of dense hedges surrounding gardens, wheat, and leguminous fields. The bushes may attain a height of 6 ft., and each bush produces thousands of blossoms. The roses are clipped off in March and April before the buds have opened and are then dried in the sun—a simple process in the warm and dry climate of Morocco. Five kilograms of fresh buds yield about 1 kg. of dried buds. According to Igolen,[127] native producers and dealers distinguish several qualities, the lowest being the so-called "Entifa" and "Glaoua," the next "Skoura," and the best "Daddès." Center of the trade

[126] Trabaud, *Drug Cosmetic Ind.* **40** (1937), 782.
[127] *Rev. marques parfums France* **17** (1939), 175.

in dried roses is the old and picturesque city of Marrakesh; here large quantities of dried rose buds arrive by caravan from distant oases in the south and southwest. Before World War II Morocco produced from 400 to 750 metric tons of dried rose buds per year; one-third of this quantity was exported to Marseilles and New York, two-thirds was used locally for the preparation of rose water by small-scale native distillers.

Hydrodistilling dried Moroccan rose buds, Igolen [128] obtained 0.03 per cent of a rose oil, which at room temperature was a greenish, solid mass with these properties:

Congealing Point..................... +27°
Melting Point........................ +28.7°
Specific Gravity at 15°............... 0.8628
Optical Rotation at 25°.............. ±0°
Refractive Index at 25°.............. 1.4627
Acid Number........................ 19.6
Ester Number....................... 35.07
Ester Number after Formylation....... 68.74
Rhodinol Content................... 19.8%

The oil exhibited the characteristic, slightly unpleasant, hay-like odor of the dried buds.

Extracting dried Moroccan rose buds with petroleum ether, the same author [129] obtained a dark green, semisolid concrete (melting point 46°–47°, acid number 18.2, ester number 64.4) which on steam distillation and cohobation of the distillation waters yielded 1.05 per cent of a volatile oil with the following properties:

Congealing Point.................... +11°
Specific Gravity at 15°.............. 0.8666
Optical Rotation.................... −0° 20′
Refractive Index at 20°............. 1.4637
Acid Number....................... 16.8
Ester Number...................... 39.2
Ester Number after Acetylation....... 129.06
Ester Number after Formylation...... 105.2
Rhodinol Content.................. 30.9%

Investigating this oil more closely, Igolen found that it contained relatively large quantities of *phenols, aldehydes* (*benzaldehyde* and perhaps *nonylaldehyde*) and free *acids*. The benzaldehyde probably originated during the drying process of the rose buds, by oxidation of certain degradation products.

In the course of the past thirty years several attempts have been made by French concerns to improve the rose industry in Morocco and to produce

[128] *Ibid.*   [129] *Ibid.*

distilled rose oil as well as concretes in modern equipment. For example, at Sebaa-Aioun [130] a factory has been equipped not only for the distillation of roses, but also for extraction by the volatile solvent process. Because dried rose buds are used, the products obtained in this plant differ from those of the Grasse region, but certain perfumers have been using them for some time with quite satisfactory results.

According to Trabaud,[131] another attempt along similar lines was made near Meknès, at 1,600 ft. altitude, where many acres of a thornless variety of *Rosa centifolia* L. were planted under ideal conditions by the same French concern. This variety is a hybrid produced from a crossing of *Rosa centifolia* and *Rosa indica*. It produces more flowers, but their perfume is not quite so fine as that of the thorny variety. Yield of distilled oil approximates that obtained in Bulgaria, viz., 1 kg. of oil per 3,000 to 4,000 kg. of flowers. The oil has a yellow, slightly greenish color, and a delicate odor differing quite markedly from that of the Bulgarian distilled oil (*Rosa damascena*). Its aroma resembles that of the *Rosa centifolia* grown in the Grasse region of Southern France. For oils distilled in 1933, 1935, and 1936 Trabaud reported these properties:

Complete Oil	*1933*	*1935*	*1936*
Stearoptene Content............	13%	15%	17.5%
Congealing Point..............	+22°	+20°	+20.5°
Melting Point.................	+23.5°	+20.5°	+21°
Specific Gravity at 15°.........	0.873	0.871	0.8706
"Elaeoptene" (Liquid Portion)			
Congealing Point..............	−9°	−10°	−15°
Specific Gravity at 15°.........	0.8782	0.885	0.8805
Optical Rotation..............	−2° 0′	−2° 0′	−2° 10′
Refractive Index at 20°.........	...	1.4702	1.4523
Acid Number.................	2.5	0.8	0.9
Ester Number.................	13.75	14	7.4
Ester Number after Acetylation..	263.2	253.87	261.3
Total Alcohol Content.........	90%	86.23%	89.4%
Geraniol Content..............	37%	50%	55.5%
Citronellol Content (By Formylation)............	50%	32%	31.8%

Compared with that of the Bulgarian oils, the stearoptene content of the Moroccan oils is very low; the percentage of total alcohols approximates the upper limits found in Bulgarian oils.

Because of its small production, the Moroccan distilled rose oil is not yet well known in the trade. According to Trabaud,[132] however, the oil may well serve as a replacement for the very expensive (distilled) "Rose de Mai"

[130] Trabaud, *Drug Cosmetic Ind.* **40** (1937), 816.
[131] *Ibid.*   [132] *Ibid.*

oil from Grasse, production of which has been practically abandoned in favor of the (extracted) concrete.

An entirely new chapter in the rose industry of Morocco was opened when in 1941 an experienced and progressive essential oil producer from Southern France started large-scale plantations of aromatic plants near Khemisset, about halfway between Rabat and Meknès. Details of this development are given in the monographs on "Oil of Geranium Moroccan" [Vol. IV of the present work, p. 709, and "Concrete and Absolute of Jasmine" (Morocco)—this volume, p. 323]. After selection of the most suitable locality, plantations were started with modern farming equipment; and carefully selected, vigorous stocks of *Rosa centifolia* L. imported from the Grasse region of Southern France were planted. Great difficulties were encountered at first; indeed, it was only in 1950 that the plantations began to produce.[133] Planting was done as with grape vines in Algeria, and with similar spacing between plants—i.e., 1.5 m. between rows and 1 m. between individual plants in the same row. This method permitted setting out only 4,000 cuttings per hectare. In the pruning operation, the time-honored method practiced in the Grasse region was followed: the rose shoots were trained to grow horizontally (by intertwining branches of adjacent bushes). In this position, they bear more flowers. Use of special cultural techniques along the principles of dry farming (frequent tillage, dressing, etc.) obviated irrigation of the rose fields, which may cause the worst of all rose diseases—rust. In spite of the fact that the number of plants per hectare is much smaller than is usual in Southern France, the yield of roses per hectare in Morocco approaches that of the Grasse region (2 to 3 metric tons). However, yield of concrete from the flowers is smaller than that obtained in France, 1,000 kg. of roses yielding less than 2 kg. of concrete. In France the rose harvest lasts only about two weeks; in Morocco flowers can be gathered for almost two months. The picking is done by hundreds of native children, in the very early morning hours, starting at dawn. The freshly harvested roses are brought to the modern extraction plant located in the center of the flower fields, and extracted with petroleum ether immediately upon arrival. Concretes and absolutes of rose produced in Khemisset have been favorably received by perfumers. Provided no unforeseen events interfere with this new development, it may well become a serious competitor of the Grasse industry.

V. THE RUSSIAN ROSE OIL INDUSTRY

Comparatively little is known about the production of rose oil in the U.S.S.R.

[133] Private communication from Mr. Pierre Chauvet, Seillans (Var), France.

(*Top*) Rose distillation in Bulgaria. A battery of old-fashioned direct fire stills. (*Left*) A modern rose distillery in Bulgaria equipped with direct fire stills. (*Right*) Separation of the condensate into rose oil and distillation water. *Photos Fritzsche Brothers, Inc., New York.*

(*Top*) East Indian nutmeg. The whole nutmeg fruit, and cut fruit showing the mace surrounding the shell which encloses the nutmeg. (*Bottom*) Nutmeg production in Grenada, B.W.I. Sun-drying of nutmegs on large wooden trays (barbecues). The nutmegs have to be turned over from time to time. *Photos Fritzsche Brothers, Inc., New York.*

In 1937 Kitchounow [134] reported that roses of the Kazanlik type (one variety with pink flowers and another variety with red flowers) had been planted in the Crimean Peninsula, the total acreage covering 350 hectares, and one hectare comprising 4,000 rose bushes. The yield of flowers can be substantially increased by irrigation from May to August. The harvest takes place from the end of May to the beginning of June. If the flowers are picked after 10 A.M., the yield of oil is as much as 59 per cent lower than when they are collected in the early morning. If possible, the plants are propagated by means of suckers, which give better results than young plants.

According to Schimmel & Co. [135] the first experiments in growing roses in the Crimea were undertaken by the Soviet Government about 1930, in the Nikita Gardens near Yalta. The first plantings in the field were laid out on the southern slopes of the Jaila Mountains and later were extended into the plains to the north of these mountains. Distillation of the flowers was carried out in several distilleries. The plantations suffered during World War II, but some rose oil was produced under the direction of German experts.

According to another trade report,[136] the so-called "Valley of the Roses" ("Dolina Ros") on the Crimean Peninsula in 1941 contained about 70 hectares of Kazanlik roses. The flowers were distilled by the same government corporation that owned the fields. The districts of Lagodechski and Telavski in the Georgian Soviet Republic also started cultivation of roses on a rapidly increasing scale. To what extent these developments were affected by World War II is difficult to ascertain. Quite possibly the Soviet Government is now attempting to produce rose oil on a larger scale than before the war.

Use of Oil, Concrete, and Absolute of Rose

Oil of rose is one of the oldest and most valuable perfumers' materials. It would not be an exaggeration to say that no high-grade perfume is complete that does not contain at least small quantities of rose oil, which lends beauty and depth to odor blends.

The distilled oil of rose imparts characteristic flowery top notes to perfumes; the extracted absolute adds lasting tonalities and increases fixation of odors. A mixture of distilled oil and extracted absolute combines the advantages of the two products.

Where solubility in dilute alcohol is important, the distilled oil will have to be employed; the absolute is soluble only in high-proof alcohol and can

---

[134] *Parfumerie moderne* **31** (1937), 175. [136] *Chem. Ind.* **64** (1941), 366.
[135] *Ber. Schimmel & Co.* (1942/43), 59.

therefore be used only in handkerchief perfumes, or in cosmetics where solubility plays no role.   In powders and creams even the much lower priced concretes (which are only partly soluble in high-proof alcohol) may give excellent results.

For many years, large quantities of distilled rose oil have been used for the flavoring of certain types of tobacco, particularly snuff and chewing tobacco.   Limited quantities of the oil are employed also in the flavoring of soft drinks and alcoholic liqueurs.

SUGGESTED ADDITIONAL LITERATURE

G. Maranca, "Le Parfum des Feuilles de Rosier de Mai," *Ind. parfum.* 3 (1948), 360.

Paul Langlais et Jean-Louis Bollinger, "Recherches dans le Domaine des Parfums Naturels.   L'Hydrolyse des Glucosides de la Rose," *Ind. parfum.* 2 (1947), 13.

A. Janistyn, "Rosen von Dadès," *Parfümerie Kosmetik* 32 (1951), 248.

OIL OF BITTER ALMOND

Essence d'Amandes Amères   *Bittermandelöl*
Aceite Esencial Almendras Amargas   *Oleum Amygdalarum Amararum*

Bitter almond oil is the volatile oil derived, by steam distillation, from the dried, ripe kernels of *Prunus amygdalus* Batsch var. *amara* (DC.) Focke (fam. *Rosaceae*), or from other kernels containing amygdalin.   Prior to distillation, the kernels must be freed of their fatty (fixed) oil by expression, and the powdered pressed cake must be macerated in warm water in order to split the glucoside amygdalin (see below).

The bitter almond tree is widely cultivated in Southern Europe (particularly Spain), North Africa (Morocco and Algeria), Asia Minor, and lately also in California.   However, in modern industrial production of bitter almond oil the kernels of the bitter almond tree have for quite some time been replaced by lower priced and higher yielding kernels of other fruit trees, *which contain amygdalin* and yield an essential oil that cannot be distinguished from that of the bitter almonds.   Obviously only bitter kernels can be used in the production of the volatile oil.   Most important among them are apricot kernels (commercially often called "peach kernels"), large quantities of which have been imported to Europe and the United

States from Morocco, Syria, Asia Minor, and the Far East (China and Japan).

A new supply has been opened in the last twenty years with the availability of substantial quantities of stones and kernels obtained as by-products in fruit canning. Tilgner [1] reported that the stones of apricots yield from 22 to 25 per cent of kernels, those of peaches 7 per cent, cherries 28 per cent, and plums 12 per cent of kernels. The kernels are heated to 85° C. and pressed, first in a continuous conical press and then in a hydraulic press at a pressure of at least 350 kg. per sq.cm. Apricots thus give 35 per cent of fatty (fixed) oil, peaches 25 per cent, cherries and plums 30 per cent. After filtration the fatty oil is neutralized with sodium carbonate and decolorized with Fuller's earth or charcoal. According to Gildemeister and Hoffmann,[2] bitter almonds, on *cold* expression, yield about 50 per cent, apricot kernels from 35 to 38 per cent, of fatty oil.

The essential (volatile) oil is not present as such in the kernels, but in the form of a cyanogenetic glucoside, viz., amygdalin, which has this constitution:

$$C_6H_5C \overset{\displaystyle H}{\underset{\displaystyle CN}{{-}O \cdot C_{12}H_{21}O_{10}}}$$

(Mandelonitrile-gentiobioside)

Under the influence of the enzyme emulsin, the amygdalin is decomposed into a molecule of benzaldehyde, a molecule of hydrocyanic acid, and two molecules of glucose, according to the following reaction:

$$\underset{\text{Amygdalin}}{C_{20}H_{27}NO_{11}} + \underset{\text{Water}}{2H_2O} \rightarrow \underset{\text{Benzaldehyde}}{C_6H_5CHO} + \underset{\substack{\text{Hydro-}\\\text{cyanic}\\\text{Acid}}}{HCN} + \underset{\text{Glucose}}{2C_6H_{12}O_6}$$

Distillation.—For commercial production of the essential oil the pressed cake is powdered, mixed with ten parts of water, and permitted to macerate for about 12 hr. at a temperature ranging from 50° to 60° C. During this period the emulsin, which is present in the kernels and freed by crushing of the latter, reacts upon the amygdalin. To induce and accelerate the process it may be advisable to add to the mass a small quantity of warm water thoroughly mixed with finely powdered bitter almond kernels. After complete hydrolysis of the amygdalin by means of emulsin, the mass is steam distilled. Because of the toxicity of the hydrocyanic acid originating

[1] *Konserven Ind.* **18** (1931), 257. *Chem. Abstracts* **25** (1931), 4322.
[2] "Die Ätherischen Öle," 3d Ed., Vol. II, 844.

during the process, utmost caution must be exercised in the course of distillation. Since poisonous hydrocyanic acid vapors may escape into the room, the oil separator and receiver should be tightly enclosed. Any vapors that cannot be condensed should be conducted into the open air where they can do no harm. Moreover, the distillation waters, in which substantial quantities of hydrocyanic acid are dissolved, should also be properly disposed of. In some countries, the distillation waters discarded from a rural distillery serve for irrigation or for watering of cattle; in the case of bitter almond distillation this must be strictly avoided.

The residual powdered material can be pressed into cakes and used as cattle feed.

Yield of Oil.—Benzaldehyde, the chief constituent of bitter almond oil, is quite soluble in water. In order to obtain the bulk of the oil, the distillation water must therefore be redistilled (cohobated).

Gildemeister and Hoffmann [3] reported that bitter almonds yield from 0.5 to 0.7 per cent, apricot kernels from 0.6 to 1.8 per cent of volatile oil. According to Tilgner [4] peach kernels give about 0.7 per cent, apricot kernels 1.2 per cent, plum kernels 0.4 per cent, and cherry kernels 0.6 per cent of volatile oil. Apricot kernels, therefore, represent the best starting material for the production of bitter almond oil.

Oil of bitter almond contains from 2 to 4 per cent of hydrocyanic acid. To remove the latter, if so required, the oil is shaken with milk of lime and iron vitriol (hydrous ferrous sulfate), whereby the hydrocyanic acid is precipitated as insoluble calcium ferrocyanide. The remaining benzaldehyde is then steam distilled. Provided the operation has been properly conducted the rectified oil contains no trace of hydrocyanic acid. (Re Prussian Blue Test, see Vol. I of this work, p. 304.) Tilgner [5] recommended adding sodium sulfite to the oil and distilling the mass, in order to remove the hydrocyanic acid.

Physicochemical Properties.

(a) *Oil Containing Hydrocyanic Acid.*—Oil of bitter almond, which contains hydrocyanic acid, is a colorless, highly refractive liquid, exhibiting the typical odor of chewed bitter almonds. On aging of the oil its color turns yellow. Because of its toxicity the oil should be smelled with great caution.

Gildemeister and Hoffmann [6] reported these properties for bitter almond oil from which the hydrocyanic acid has *not* been removed:

[3] *Ibid.*, 847.
[4] *Konserven Ind.* **18** (1931), 257. *Chem. Abstracts* **25** (1931), 4322.
[5] *Ibid.*
[6] "Die Ätherischen Öle," 3d Ed., Vol. II, 847.

Specific Gravity at 15°.......... 1.045 to 1.070 (a higher specific gravity indicates an abnormally high content of hydrocyanic acid or, more correctly, of benzaldehyde cyanohydrin. For further details see the section on "Chemical Composition," below)

Optical Rotation............... In most cases optically inactive; in a few instances very slightly dextrorotatory, up to $+0°$ 9'

Refractive Index at 20°......... 1.532 to 1.544 (the magnitude of the refractive index is inversely proportional to the hydrocyanic acid content)

Hydrocyanic Acid Content...... Varies greatly and is much higher in crude oils than in rectified oils. In a crude oil (d_{15} 1.090, n_D^{20} 1.52986) as much as 11% of hydrocyanic acid was observed; in a rectified oil (d_{15} 1.053, n_D^{20} 1.54497) only 0.56% was found (see also below)

Benzaldehyde Content.......... Must conform to the requirements of the various pharmacopoeias (see also below)

Solubility..................... The oil is quite soluble in water, 1 vol. of oil being soluble in about 300 vol. of *pure* water. Water containing hydrocyanic acid dissolves the oil even more readily. The oil is soluble in 1 to 2 vol. and more of 70% alcohol; soluble in 2.5 vol. and more of 60% alcohol; on aging, the oil loses its solubility in 60% alcohol

Pure oils of bitter almond (containing hydrocyanic acid) examined by Fritzsche Brothers, Inc., New York, had properties varying within the following limits:

Specific Gravity at 25°/25°........... 1.049 to 1.058
Optical Rotation................... Inactive to $+0°$ 11'
Refractive Index at 20°.............. 1.5403 to 1.5435
Aldehyde Content, Calculated as Benz-
  aldehyde (Hydroxylamine Hydrochlo-
  ride Method)...................... 81.3 to 93.2%
Total Hydrocyanic Acid Content (In-
  cluding Benzaldehyde Cyanohydrin). 2.1 to 4.0%
Halogen Test....................... Negative
Heavy Metals Test.................. Negative
Solubility.......................... Soluble in 1 to 2 vol. of 70% alcohol; sometimes slightly hazy upon further dilution. Usually soluble in 5 to 6 vol. of 50% alcohol; sometimes with slight opalescence or haze
Solubility in Bisulfite Solution........ Slight oily separation

As regards the content of benzaldehyde, it should not be less than 80 per cent. In most oils the aldehyde content, calculated as benzaldehyde, ranges from 84 to 87 per cent. For method of determination, see Vol. I of the present work, p. 286.

Regarding the content of hydrocyanic acid, it should not be less than 2 per cent and not more than 4 per cent. For quantitative determination, see Vol. I of the present work, pp. 304, 305.

Freshly distilled bitter almond oil is neutral, but on aging the oil becomes acidic, because of oxidation of benzaldehyde to benzoic acid.

When oil of bitter almond is distilled over direct fire (caution: development of hydrogen cyanide vapors) the first fractions will contain substantial quantities of hydrogen cyanide, the later fractions much less. The distillation residue contains benzoin, a product originating by polymerization of two molecules of benzaldehyde under the influence of hydrocyanic acid:

$$C_6H_5CHO + OHCC_6H_5 \rightarrow C_6H_5CH(OH)COC_6H_5$$
$$HCN$$

Benzaldehyde   Benzaldehyde   Benzoin

(b) *Oil Free of Hydrocyanic Acid.*—Oil of bitter almond from which the hydrogen cyanide has been removed consists almost entirely of benzaldehyde and may justly be called natural benzaldehyde.

According to Gildemeister and Hoffmann [7] the oil has these properties:

Boiling Point............... 179°
Specific Gravity at 15°....... 1.050 to 1.055 (a higher specific gravity may be caused by a greater content of benzoic acid)
Optical Rotation............ Optically inactive
Refractive Index at 20°...... 1.542 to 1.546
Solubility.................. Soluble in 1 to 2 vol. and more of 70% alcohol

This type of oil is much more sensitive to oxidation by air than the oil which contains hydrogen cyanide. The latter acts as a preservative. Pure benzaldehyde is readily oxidized to benzoic acid by exposure to air.

Adulteration.—At one time bitter almond oil was occasionally adulterated with nitrobenzene; but this crude form of adulteration is no longer encountered.

A much more annoying adulterant is synthetic benzaldehyde because it cannot be detected readily. In past years the synthetic benzaldehyde used for this purpose sometimes contained chlorine, the result of improper purification. Such benzaldehyde could then be detected quite readily by the Halogen Test, details of which will be found in Vol. I of this work, pp. 307,

[7] *Ibid.*, 848.

308. Today, however, the Halogen Test has lost much of its importance because synthetic benzaldehyde is now produced in a high state of purity.

Storage.—Oil of bitter almond has to be stored in well-filled and well-stoppered bottles in order to prevent any access of air. Otherwise, part of the benzaldehyde will be oxidized to benzoic acid. As was mentioned above, oils containing hydrogen cyanide are much more stable than oils from which the hydrogen cyanide has been removed.

Schimmel & Co.[8] found that alcohol added in the amount of 10 per cent acts as a preservative of the oil. On the contrary, if only 5 per cent of alcohol is added, oxidation takes place even more rapidly than in the undiluted oil. This applies to both types of oil.

Chemical Composition.—As was explained above, splitting of the cyanogenetic glucoside amygdalin under the influence of the enzyme emulsin results in the formation of benzaldehyde, hydrocyanic acid, and glucose. Distillation of the reaction products yields a volatile oil composed chiefly of benzaldehyde and a small quantity of hydrocyanic acid, the greater part of the latter being dissolved in the distillation water. On prolonged contact of benzaldehyde with hydrocyanic acid the following reaction takes place:

$$C_6H_5CHO + HCN \rightleftarrows C_6H_5C\diagup^{H}_{\diagdown^{OH}}_{CN}$$

Benzaldehyde	Hydro-cyanic Acid	Benzaldehyde Cyanohydrin

In other words, benzaldehyde and hydrocyanic acid combine to form benzaldehyde cyanohydrin.[9] However, the latter compound is readily decomposed into its components, even on steam distillation or on distillation *in vacuo*. Therefore, benzaldehyde cyanohydrin originates only *after* distillation of the reaction mass, particularly when the oil is left standing for some time in contact with distillation water that contains substantial quantities of hydrocyanic acid.

That bitter almond oil actually contains benzaldehyde cyanohydrin was proved more than seventy years ago by Fileti,[10] who reduced bitter almond oil and obtained phenylethylamine. Reduction of a freshly prepared mixture of benzaldehyde and hydrocyanic acid, on the other hand, yielded only methylamine.

Benzaldehyde cyanohydrin possesses a relatively high specific gravity

[8] *Ber. Schimmel & Co.*, April (1895), 47.
[9] Cf. Wirth, *Arch. Pharm.* **249** (1911), 382.
[10] *Gazz. chim. ital.* **8** (1878), 446. Cf. *Ber.* **12** (1879), 297.

(d 1.124). Therefore, the more benzaldehyde cyanohydrin a bitter almond oil contains, the higher will be its specific gravity and the greater will be its content of hydrocyanic acid. According to Gildemeister and Hoffmann [11] normal bitter almond oils with a specific gravity of 1.052 to 1.058 contain from 1.6 to 4 per cent of hydrocyanic acid, while oils with a specific gravity of 1.086 to 1.096 contain from 9 to 11.4 per cent of hydrocyanic acid. Experiments carried out by Schimmel & Co.[12] showed that the specific gravity of pure benzaldehyde increased from 1.054 to 1.074, after the aldehyde had been left in contact with a 20 per cent aqueous solution of hydrocyanic acid for two days.

Aside from benzaldehyde, hydrocyanic acid, and benzaldehyde cyanohydrin, oil of bitter almond most likely contains traces of other substances which are responsible for the finer odor and flavor of natural benzaldehyde isolated from the oil, as compared with the synthetic product. These trace substances have not yet been identified (cf. Dodge, Eighth International Congress of Applied Chemistry, Washington and New York, Vol. 17 [1912], 15).

Use.—Oil of bitter almond is a very important flavoring agent, used particularly in baked goods, cakes, confectioneries, and candies. However, for the scenting of cosmetics, soaps, and technical preparations, the much lower priced synthetic benzaldehyde is now employed almost exclusively.

OIL OF CHERRY LAUREL

Essence de Laurier Cerise   *Kirschloorbeeröl*
Aceite Esencial Laurel Cerezo   *Oleum Laurocerasi*

Prunus laurocerasus L. (fam. *Rosaceae*), our common cherry laurel, is an evergreen bush or small tree, native of southeastern Europe and Asia Minor. It is cultivated as an ornamental plant in many temperate countries. In Southern France, Italy, and other Mediterranean countries the leaves are distilled, whereby an aromatic water, the so-called cherry laurel water, is obtained.

The leaves contain a monoglucoside, viz., *prulaurasin* $C_{14}H_{17}O_6N$, which, according to the recent work of Rosenthaler [1] is the nitrile glucoside of

[11] "Die Ätherischen Öle," 3d Ed., Vol. II, 849.
[12] *Ber. Schimmel & Co.*, April (1893), 42.
[1] *Pharm. Acta Helv.* **19** (1944), 100.

l-mandelic acid. Splitting of the glucoside prulaurasin by means of emulsin yields 61.24 per cent of β-glucose, 8.58 per cent of hydrocyanic acid, and benzaldehyde. In the industrial production of cherry laurel water it is, therefore, necessary to macerate the well triturated leaves in warm (but not hot) water for some time, prior to actual distillation. To induce and accelerate splitting of the glucoside, it is advisable to add to the macerating mass a tepid infusion of powdered cherry laurel leaves. Great care has to be exercised in the industrial production of this water, as it contains substantial quantities of hydrocyanic acid which is highly poisonous (cf. Vol. II of the present work, p. 721).

As has been mentioned, distillation of the macerated leaves yields aromatic cherry laurel water. To isolate the (concentrated) essential oil, the distillation water may have to be redistilled (cohobated). The yield of oil, calculated upon the leaves, averages 0.05 per cent.

Physicochemical Properties.—Oil of cherry laurel differs from bitter almond oil chiefly in regard to odor. The physicochemical properties of the two oils are similar. Gildemeister and Hoffmann [2] reported these properties for the volatile oil of cherry laurel:

Specific Gravity at 15°......... 1.046 to 1.067
Optical Rotation.............. Mostly inactive, occasionally with a very slight rotation, $+0°\ 12'$ to $-0°\ 46'$
Refractive Index at 20°........ 1.539 to 1.543
Acid Number................. 1.6 to 2.8 (Also reported up to 10 in the same source)
Hydrocyanic Acid Content...... 0.4 to 4.1%, seldom higher (up to more than 8%!)
Solubility.................... Usually soluble in about 7 vol. of 50% alcohol; soluble in 2.5 to 4 vol. of 60% alcohol (the solubility in 60% alcohol decreases on aging of the oil); soluble in 1 to 2 vol. of 70% alcohol

Genuine oils of cherry laurel imported by Fritzsche Brothers, Inc., New York, from Southern Europe had properties varying within the following limits:

Specific Gravity at 15°/15°... 1.044 to 1.067
Optical Rotation............ Mostly inactive, occasionally with a slight rotation up to $+0°\ 10'$
Refractive Index at 20°...... 1.5417 to 1.5440
Aldehyde Content, Calculated as Benzaldehyde.......... 82.7 to 91.2%
Hydrocyanic Acid Content... 1.4 to 2.8%
Solubility................. Soluble in 2.5 to 3 vol. of 60% alcohol and more

[2] "Die Ätherischen Öle," 3d Ed., Vol. II, 853.

Because of its content of hydrocyanic acid oil of cherry laurel is very poisonous and must be handled with the greatest of care.

Chemical Composition.—Tilden,[3] Fileti,[4] and Rosenthaler[5] reported the presence of the following compounds in the volatile oil derived from cherry laurel leaves:

Benzaldehyde. The chief constituent.

Hydrocyanic Acid. The poisonous component of the oil.

Nitrile of *l*-Mandelic Acid. Formerly reported to be benzaldehyde cyanohydrin, but more recently identified by Rosenthaler as the nitrile of *laevo*rotatory mandelic acid.

Benzyl Alcohol. Presence possible (Tilden).

Use.—Small quantities of oil of cherry laurel are used in certain medicinal preparations.

OIL OF *SPIRAEA ULMARIA* L.

According to Schneegans and Gerock,[1] the volatile oil distilled from the *flowers* of *Spiraea ulmaria* L. (fam. *Rosaceae*), an attractive deciduous shrub, contains *salicylaldehyde* as chief constituent, in addition to *methyl salicylate,* free *salicylic acid, piperonal* (heliotropin), and *vanillin.*

As regards the volatile oil distilled from the *herb,* Wicke[2] identified *salicylaldehyde.*

The oils are not produced on a commercial scale, synthetic salicylaldehyde being used instead.

[3] *Pharm. J.* [3], **5** (1875), 761.
[4] *Gazz. chim. ital.* **8** (1878), 446. Cf. *Ber.* **12** (1879), 297.
[5] *Pharm. Acta Helv.* **19** (1944), 100.
[1] *J. pharm. Elsass-Lothr.* **19** (1892), 3, 55. *Jahresber. Pharm.* (1892), 164.
[2] *Liebigs Ann.* **83** (1852), 175.

CHAPTER II

ESSENTIAL OILS OF THE PLANT FAMILY *MYRISTICACEAE*

OILS OF MYRISTICA
Oil of Nutmeg and Oil of Mace
Essence de Muscade and *Essence de Macis*
Aceite Esencial Nuez Moscada and *Aceite Esencial Macis*
Muskat- and *Macisöl*   *Oleum Nucis Moschati* and *Oleum Macidis*

Introduction and History.—Nutmeg (myristica) and mace are both derived from the fruit of *Myristica fragrans* Houtt. (fam. *Myristicaceae*), nutmeg being the dried ripe seed, and mace the dried arillode which envelops the shell containing the seed or nutmeg. Both are very important spices, which have been used for a long time in the flavoring of savory dishes, baked goods, and other food products. They owe their characteristic aroma chiefly to the presence of an essential oil which can be isolated by steam distillation. Oil of nutmeg and oil of mace are very similar in odor and flavor, and since the nutmeg suitable for distillation is usually lower priced than mace, the former is generally employed for commercial production of the volatile oil.

According to some authors, nutmeg and mace—the latter particularly— were highly valued by the ancient Romans, who sometimes used them as a form of currency. According to other, and more reliable, sources, however, nutmeg was unknown to the Romans, becoming familiar in Europe only during the twelfth century, after it had been introduced from the East perhaps by returning Crusaders or by Arabian physicians.[1] In 1158 *"Nuces muscatarum"* imported from Alexandria were traded in the harbor of Genoa.[2] The true origin of the spice was discovered only at the beginning of the sixteenth century (in 1512) by two travelers, Barthema and Pigafetta.[3] At that time the Portuguese conquered the Moluccas, those fabled spice islands, and declared the trade in nutmeg and mace a monopoly. In 1605 the Hollanders drove the Portuguese out of the Moluccas and, taking over the spice monopoly, restricted the production of nutmeg and mace to Banda, and that of clove to Amboyna, in order to keep up prices. The monopoly of the Hollanders lasted for almost two hundred years. In 1769 the French succeeded in introducing the nutmeg tree to Ile

---

[1] Cf. "Nutmeg and Mace.—A Note on Their History," *Perfumery Essential Oil Record* **7** (1916), 76. Cf. *Ber. Schimmel & Co.* (1942/43), 26.

[2] Cf. Gildemeister and Hoffmann, "Die Ätherischen Öle," 3d Ed., Vol. I, 134.

[3] G. B. Ramusio, "Delle Navigationi et Viaggi," Venice (1554), fol. 183 and fol. 389b.

de France (now Mauritius), east of Madagascar; in 1796 Christopher Smith collected spice bearing plants in the Moluccas for the powerful East India Company. These were brought to Penang, where the first commercial crop of nutmeg and mace was gathered about 1802. This marked the end of the Dutch monopoly (which included cloves; in fact, the history of the clove trade almost parallels that of the commerce in nutmeg and mace). Early in the nineteenth century, numerous nutmeg plantations were started in many parts of Malaya and the Indonesian Archipelago, and about 1843 the tree was introduced also to Grenada in the West Indies, a small island which today produces about 40 per cent of the world's output of nutmeg and mace.

Producing Regions and Qualities.—The trade recognizes today two principal types of nutmeg (and mace), the difference between the two being based not only upon geographical origin, but also upon quality of the spice:

(1) *East Indian Nutmeg and Mace.*—According to Nijholt,[4] *Myristica fragrans* Houtt., which yields the so-called "Banda Nutmeg," is grown extensively in Indonesia, chief producing areas being the Moluccas, Minahassa (northern Celebes) and the Sangih Islands, Benkulen (West Sumatra) and Achin (North Sumatra), including the island of Nias. Throughout Indonesia nutmeg plantings are owned by natives, or occasionally by Chinese.

According to Landes,[5] there are four grades of East Indian nutmegs:

(a) The "Banda Nutmegs," considered perhaps the finest for commercial use, and containing about 8 per cent of essential oil.

(b) The "Siauw Nutmegs," considered almost as good as the "Banda," but containing only about 6.5 per cent of essential oil.

(c) The "Penang Nutmegs." Prior to World War II, these were of good quality; but they are now wormy and moldy beyond the possibility of reconditioning, probably as a result of neglecting plantations during the war, and can be used only for distillation purposes.

(d) The "Papua Nutmegs." These are among those nutmegs not derived from *Myristica fragrans* Houtt., but from an allied species, viz., *Myristica argentea* Warb. They can easily be recognized by their shape and peculiar, unpleasant odor. Papua nutmegs have been imported into the United States only since the last war. They should not be used for distillation, the yield of oil being low, and the oil possessing an odor reminiscent of sassafras.

In addition to the four above-mentioned grades, there is, perhaps, a fifth

[4] Private communication from Dr. J. A. Nijholt, Director of the Laboratorium voor Scheikundig Onderzoek, Buitenzorg, Java.
[5] Private communication from Dr. Karl H. Landes, New York.

grade, the so-called "Java Estate Nutmegs," small quantities of which enter the United States market. They are of very good quality and may be classified as an intermediate between the "Banda" and the "Siauw" nutmegs.

As was pointed out above, the "Papua" nutmegs are derived from a species of *Myristica* other than *fragrans* Houtt. Whether this is the case also with some of the other grades lower than the "Banda" cannot easily be ascertained. *Myristica succedanea* Bl., in Indonesia called "Pala Maba," e.g., produces a grade of nutmeg and mace which is only slightly, if at all, inferior to the first grade "Banda" spice. (The oils derived from species of *Myristica* other than *fragrans* Houtt. will be described in separate monographs. The present monograph will deal exclusively with the oils of nutmeg and mace from *Myristica fragrans*, both East Indian and West Indian.)

As regards the grades of mace produced in East India, Landes [6] distinguishes between the following:

(a) The "Banda Mace," considered the finest. This is prepared for the market with more care than any other grade of mace, being artificially dried and fumigated in specially constructed dryers and carefully sifted. Any broken pieces are eliminated. The color is bright orange, the aroma very fine.

(b) The "Java Estate Mace," grown near Semarang, in Java. Like the "Banda," this is artificially dried, and quite free of insect infestation. It can easily be recognized by its round shape and its gold-yellow color, interspersed with brilliant crimson streaks.

(c) The "Siauw Mace," originating from the islands of Celebes, Ternati, Talanda, and Sangi. The color is lighter than that of the "Banda"; the brighter the color, the higher the price. Depending upon color and size, the Siauw mace is classified as of No. 1 and No. 2 quality. Broken pieces are classified as siftings. In general, Siauw mace contains about 10 per cent less volatile oil than the Banda mace.

(d) The "Papua Mace," derived from *Myristica argentea* Warb. This type is entirely unsuitable for the purposes of grinding or distilling, since it contains a high percentage of fatty oil and comparatively little essential oil. (The latter exhibits an undesirable turpentine-like aroma.)

The above-mentioned grades of East Indian mace contain, on the average, the following quantities of volatile oil:

	Per Cent
"Banda Mace"	13
"Java Estate Mace"	11 to 12
"Siauw Mace"	10 to 11
"Papua Mace"	4

[6] *Ibid.*

West Indian mace (which will be described below) yields from 8 to 9 per cent of essential oil.

Nijholt [7] reported the following figures for nutmeg and mace exports from Indonesia in the years 1936 to 1940 inclusive, and 1948, in metric tons:

	Nutmeg	*Mace*
1936.	4,192	777
1937.	4,415	823
1938.	3,977	834
1939.	4,215	810
1940.	3,612	847
1948.	2,676	494

Shelled nutmegs are usually shipped directly from the harbors near the producing regions to the consuming countries. Singapore does not handle much of this type. However, of the unshelled nutmegs, about 75 per cent of those produced first go to Singapore for transshipment abroad.

In 1938 a total of 435,867 kg. of nutmegs ("distillers grade") was exported from Macassar (Banda). These exports were distributed as follows: [8]

	Kilograms
United States.	389,451
Holland.	15,345
Holland, Option.	6,131
Great Britain.	11,528
Great Britain, Option.	5,120
Germany.	8,292
	435,867

According to Swing,[9] the nutmeg tree was introduced to Ceylon in 1804, and now grows on the island in deep loamy soil, up to an elevation of 2,500 ft. In 1946 Ceylon exported 1,063 hundredweights of nutmeg, in 1947 the exports amounted to 1,488 hundredweights.

(2) *West Indian Nutmeg and Mace.*—Until 1939 the West Indies produced only about one-sixth of the world's supply of nutmeg and mace, but since the outbreak of World War II (when the spice from the East Indies became unavailable) production in the West Indies has increased to about 40 per cent of the world's output. The bulk of the West Indian nutmeg and mace originates from the small British island of Grenada, one of the Windward Islands, lying just to the north of Trinidad. Before World War

---

[7] Private communication from Dr. J. A. Nijholt, Director of the Laboratorium voor Scheikundig Onderzoek, Buitenzorg, Java.

[8] *Ber. Schimmel & Co.* (1939), 53.

[9] *Spice Mill* **72**, No. 2 (1949), 48.

II the West Indian spice was shipped largely to Great Britain and continental Europe, but during the war the picture changed and the United States became the principal buyer of the West Indian nutmeg.

According to Whitaker,[10] the nutmeg tree was introduced into Grenada about 1843, when the captain of a visiting Dutch ship gave one of Grenada's planters a few of the many nutmeg seeds he was bringing to the Netherlands from the island of Banda in Indonesia. These seeds produced tall, sturdy trees, which flourished in the soil and climate of Grenada. They formed the core of a new industry, which in later years came to provide employment for half of the island's population. As practically all of Grenada's nutmegs and mace are exported, the following table of exports for the years 1940 to 1945 may also be considered as production or crop figures:

Year	Total (in tons)	United Kingdom (in tons)	Canada (in tons)	United States (in tons)	Other (in tons)
Nutmegs					
1940	2,095	523	128	1,331	113
1941	1,910	700	144	1,015	51
1942	1,876	780	102	921	73
1943	2,270	351	80	1,636	203
1944	2,574	224	110	1,935	305
1945	2,341	500	73	1,553	215
Mace					
1940	274	222	31	20	1
1941	324	302	18	3	1
1942	340	288	10	34	8
1943	334	71	14	223	26
1944	338	41	22	233	42
1945	338	92	24	158	64

According to Landes,[11] the quality of the West Indian nutmegs has been gradually but steadily improving. Shipments now usually meet standards, and lots are rarely rejected because of worminess or an excessive amount of moisture. Recent shipments of West Indian nutmegs have shown a content of volatile oil as high as 9 per cent, which is greater than that of Banda nutmegs from Indonesia. The reason may lie in the fact that many of the nutmeg trees of Grenada have now reached an age of thirty to thirty-five years, i.e., the prime of their productivity.

As regards West Indian mace, its quality is still below that of the East Indian spice, and only the price difference between the two products has

[10] *U. S. Dept. Commerce, Foreign Commerce Weekly,* Washington, D. C., **25** (November 23, 1946), 8.
[11] Private communication from Dr. Karl H. Landes, New York.

compelled some of the grinders to substitute the West Indian for the East Indian mace. The former possesses a nice yellow color, but its texture is brittle, and its volatile oil content (8 to 9 per cent) below that of the East Indian mace. Moreover, the volatile oil exhibits a turpentine-like aroma.

Botany.—*Myristica fragrans* Houtt., one of about 100 known *Myristica* species,[12] is a bushy tree with numerous spreading branches; it attains a height of 30 to 40 ft., in some cases even 60 ft. In its general appearance the nutmeg tree resembles an orange tree. According to Ridley,[13] the trees are in most cases unisexual, bearing male flowers or female flowers only, but it is not uncommon to find a tree with flowers of both sexes upon it. It has long been known that a male tree after some years, usually about six, frequently commences to produce female flowers, and eventually becomes wholly female. The fruit appears on the tree mingled with flowers; in other words the nutmeg tree, like the lime tree, bears flowers and fruit at the same time. The orange-yellow, smooth fruit, when ripe, is one of the most beautiful in nature. A pendulous, fleshy drupe of quite variable form, it is globular on some trees, oval or pear-shaped on others. The size varies, but averages 2.5 in. In general, the fruit resembles a small peach. The husk (pericarp) is of fleshy, somewhat firm texture, and about 0.5 in. thick. It contains an acid, astringent juice with an aromatic flavor of nutmeg. When ripe, the fleshy husk opens by splitting into two halves, from the top nearly to the base, along the groove which runs down one side. On splitting of the husk, the nut appears. It has a deep brown, glossy seed coat or shell that contains the seed. The latter, when dried, is the nutmeg of commerce. The seed coat is partly covered by a peculiar, net-like and crimson, leathery arillode, growing from the base of the seed. The dried arillode is the mace of commerce.

The shell (*testa*) has a woody and brittle consistency; when dried it can easily be cracked to remove the seed (after the arillode has been peeled off the shell by hand). Within the shell a sound seed should measure about 1 in. in diameter. When fresh, the oval seed nearly fills the shell, but on drying, the seed (nutmeg) shrinks somewhat and then rattles within the shell. Seed used for sowing should not be dry enough to rattle, but for trade purposes the nutmegs should rattle in the shell when shaken. A dried, sound nutmeg, removed from the shell, is hard and woody in texture, ovoid or ellipsoidal in form, from 20 to 30 mm. in length and about 20 mm. in thickness, light brown to dark brown in color, reticulately furrowed. When cut, the inside shows a waxy and oily luster, with brown spots on a

---

[12] There are, however, a few others, the nuts of which are slightly aromatic, and occasionally collected by natives and exported to Europe, more as adulterants of true nutmeg rather than for separate use.

[13] "Spices," London, Macmillan & Company, Ltd. (1912), 94.

grayish ground. The dried seed (nutmeg) contains from 5 to 15 per cent of volatile oil, from 25 to 40 per cent of fixed oil, and from 5 to 15 per cent of ashes; the balance consists of moisture, fiber, and starch.

Mace (the arillode enlacing the shell of the seed), when freshly removed, exhibits a bright red color which, however, changes to orange-yellow on drying. It is marketed in the form of irregularly shaped pieces which swell when immersed in water. Odor and flavor are similar to, yet distinctly different from those of nutmeg. Mace contains from 4 to 14 per cent of a volatile oil closely resembling that derived from nutmegs. West Indian mace has a paler color than the East Indian and is less aromatic.

Production of Nutmeg and Mace in the East Indies.—The nutmeg tree flourishes on a variety of soils, except on bare clay slopes, sandy, or water-logged soils. The best altitude is from sea level to about 1,000 ft. The tree requires a hot, moist climate and appears to prefer proximity to the sea. Rainfall should be from 80 to 100 in. per year, well spread over the different months, with no absolutely dry spells of more than four to five days, and no continuous rains, without sun, of more than a fortnight. During wet spells parasitic fungi are most active.

According to Ridley,[14] carefully chosen, well-formed seed should be planted within 24 hr. of gathering, if possible, in beds of good soil, well dug over, manured and drained. The seed should be placed from 12 to 18 in. apart in rows, and at a depth of about 2.5 in. The beds need shading under a roof loosely made of leaves, admitting a certain amount of light. The beds should be watered every other day. The seeds require from four to six weeks to germinate, and sometimes longer. The young plants remain in the nursery until they are about 6 in. tall, i.e., for about six months. Then they have to be transplanted, during the rainy season, to their permanent position in the field, at a distance of 26 to 30 ft. apart, either in lines or in quincunx arrangement. The planting holes should be about 4 ft. wide, and at least 3 ft. deep. In hot places the young plants need a certain amount of shading until they have settled in the ground and begun to push out their roots. Being in its natural state a jungle plant, the nutmeg tree grows better and remains healthier when partly shaded (e.g., by the well-known "rain tree" *Pithecolobium saman*), than when fully exposed to the hot sun.

The tree does not flower until the eighth or ninth year, after which it bears flowers and fruit together for many years. The nutmeg tree is normally monoecious, i.e., either male and bearing only male flowers, or female and bearing only female ones. Obviously, only the female flowers produce fruit, but a small proportion of male trees (about one in ten) is required to

[14] *Ibid.*, 107.

provide the pollen necessary to fertilize the flowers on the female trees. Native growers, therefore, usually exterminate most of the male trees as soon as the trees have sufficiently developed to show their flowers and so become distinguishable. This, however, is possible only about seven years after planting. More progressive growers then head down any superfluous male trees and graft them with scions of female plants.

The trees commence to fruit usually in the eighth or ninth year, but reach their prime of productivity only when about twenty-five years old. They bear well up to sixty years, or even longer. The fruits ripen about six months after the flowering period. The fruit is ripe when it splits and shows the shell of the seed enlaced with its brilliant red mace. The fruits are sometimes allowed to drop to the ground; nuts and attached mace are then picked up daily and collected in baskets. The usual practice, in Indonesia, however, is to pull the fruit off the branches by hand, using hooked staffs to reach fruit on lofty branches. A good worker can collect from 1,000 to 1,500 nuts a day.

The yield of nutmeg and mace per tree varies considerably. A healthy tree should average 1,500 to 2,000 nuts per year. As regards weight, each tree should produce 10 lb. of nutmegs to 1 lb. of mace; some trees give much more than this (Ridley).

The tree bears fruit more or less all the year round, but in most places the heaviest crops are obtained in May and June, and again in August and September.

After arrival of the collected material at the working shed the mace (arillode) is detached from the shell of the seed with a knife, or simply by hand, by opening it from the top of the shell and reflexing it. The mace is attached to the seed only by its base, known as the heel of the mace. The freshly removed mace is flattened out by hand, spread on bamboo trays or on mats, and dried in the sun from 4 to 5 hr. a day for a fortnight. During dry weather drying may be accomplished in two or three days. Before nightfall the trays or mats must be brought into a drying-shed, to prevent wetting of the material by dew. Great care must be exercised to prevent the mace from getting moldy, which may easily happen. When fresh, the mace possesses a brilliant red color; on drying this changes to orange, and after a few months to yellow.

According to Ridley,[15] a perfect sample of mace should consist of entire double blades (not broken), flattened and of large size, horny in texture, not too brittle, and of good, clear and bright color.

After removal of the mace the nuts are placed on trays and exposed to the sun for several weeks until dry. Often drying is completed over a slow

[15] *Ibid.,* 146.

charcoal fire in a drying-house, care being exercised not to raise the temperature too high, as the seed will then shrivel and diminish in value. The seeds are dried while still in the shell (the thin, brittle outer coating, or *testa*) because otherwise they are attacked by beetles. Drying is completed when the seed rattles within the shell. When the seeds are sufficiently dry, the shells must be removed; this is usually accomplished by hitting them on the end (not on the side!) with a wooden club or mallet. Machines for cracking of nutmegs have been invented and are used on some of the large estates. Once the seed has been removed from the shell, it is exposed to the attacks of all kinds of beetles, especially while stored in the warehouse ("godown"). The beetles deposit eggs in the seed, and the larvae bore holes into it, destroying part of it. Most of these beetles are pests common in the godowns, attacking all kinds of food products. Therefore, the godowns should be thoroughly cleaned from time to time. To prevent damage by insects, the nutmegs are often limed at the place of production. In some parts of Indonesia this is done by sprinkling the dried nutmegs profusely with powdered lime, or by rubbing each nut individually with the chemical, or by dipping them into a mixture of lime and water. At one time, fumigation with sulfur dioxide was occasionally practiced, but this has now been largely abandoned. A more recent method is fumigation of the nutmegs with methyl bromide, in special chambers. However, at the present time, most of the Indonesian product reaches the world markets untreated.

On arrival at the warehouse, the nutmegs are tested by sounding. This is done simply by bouncing them individually on an iron plate. The assorter can thus detect worm-eaten and damaged specimens, which he removes. Sound nutmegs are assorted according to their size and number per pound. At the time of the author's visit to Malaya and Indonesia, just prior to the outbreak of World War II, nutmegs were classified as follows:

(a) Whole, sound nutmegs. These ranged from 60 to 125 nutmegs per pound, the greatest demand being for 85 to 95 nuts per pound. They were known as "Large" (60 to 80 per lb.), "Medium" (85 to 95 per lb.), and "Small" nutmegs (100 to 125 per lb.), and were sold to the spice trade.

(b) Sound shrivels. A quality much in demand for grinding.

(c) Rejections. Because of their low price, these are suitable for distillation.

(d) Broken and wormy. The quality most suitable for distillation because of the low content of fatty oil. Prior to World War II large quantities were shipped to Europe, particularly to Hamburg. The United States permits entry of this grade only under condition that the material be used exclusively for distillation.

Production of Nutmeg and Mace in the West Indies.—As has been mentioned, the nutmeg tree was introduced to Grenada about 1843, when a

Dutch ship bringing spices from Indonesia to Holland called on this small island in the West Indies.

According to Whitaker,[16] in the Windward Islands the trees appear to grow best at an elevation ranging from 500 to 1,500 ft. above sea level, in areas where the rainfall is fairly constant throughout the year. A large part of the fertile mountain slopes of Grenada above 800 ft. elevation is covered with dense groves of nutmeg trees. These trees grow very close together on the steep hillsides, forming an almost unbroken canopy above the ground. The hillsides slope so sharply that mechanical cultivation cannot be practiced; hence the trees do not have to be set out in lines. They are simply planted wherever there appears to be ample room for them to grow.

According to Noel,[17] the principal factor influencing the growth of the nutmeg tree is the amount of rainfall which should be well distributed and at least 80 in. per year. Optimum conditions in this respect prevail at elevations of from 600 to 800 ft., in the center of the island. The tree does not grow in the coastal plains.

Nutmeg trees bear the first fruit four years after planting, but the first commercial harvest takes place only after sixteen years. The trees continue bearing fruit for a hundred years and perhaps more.

It is estimated that there are about 10,000 acres devoted to the cultivation of nutmegs in Grenada. Large plantations comprise approximately 70 per cent of this acreage, the remaining 30 per cent being plots of from 1 to 5 acres in the hands of small farmers. Out of a population of 70,000, about 14,000 are such farmers.

Whereas in the Far East the fruits are usually picked from the tree before the husk splits, in Grenada the husk is allowed to split while the fruit is still on the tree, and the ripe fruit is collected from the ground. If the fruits are not picked up quickly, the nutmegs become waterlogged and ferment or start to grow, and weevils and worms enter the shell.

Nutmeg trees produce fruit all year round, but most heavily in August and September, and from February to April, inclusive. In June, November, and December collections are lightest. The productivity of a tree depends upon ecological (primarily soil) conditions and upon its vigor (plant selection). According to information gathered by the author during a visit to Grenada, the yield ranges from a few pounds of dried nutmegs to as many as 100 lb. per tree. Whitaker [18] reports that in Grenada the trees average

---

[16] *U. S. Dept. Commerce, Foreign Commerce Weekly,* Washington, D. C. **25** (November 23, 1946), 7. Cf. *Spice Mill* **71,** No. 5 (1948), 51.

[17] Author's conversation with Mr. Carlyle Noel, Grenville, Grenada, and Mr. William O'Brien Donovan, St. George, Grenada.

[18] *U. S. Dept. Commerce, Foreign Commerce Weekly,* Washington, D. C. **25** (November 23, 1946), 7. Cf. *Spice Mill* **71,** No. 5 (1948), 51.

about 1,000 nutmegs per year and that a tree in full bearing produces about 50 lb. of green nutmegs per year. An acre of nutmeg trees contains perhaps 90 or 100 trees. Many of these are young, not yet in full bearing or indeed not bearing at all, and some are past their prime. On a number of estates the less prolific young trees are cut out from time to time. The return per acre averages 1,500 lb. of green nutmegs a year—which, when dried and removed from the shells, yield about 720 lb. of sound nuts.

The yield of mace averages 150 lb. of green, or 30 to 40 lb. of cured, mace per acre. Ordinarily it takes about 100 lb. of green nutmegs to yield 8 lb. of fresh mace.

Total yearly production of nutmegs in Grenada averages 6,000,000 lb., that of mace 600,000 lb.

The Grenada Co-Operative Nutmeg Association advises its members never to collect or harvest the fruit prior to the natural opening of the "husk," considering it an absolutely unsound policy to "reap" fruit from the trees. Only fully ripened fruit fallen to the ground should be gathered, and this, under favorable weather conditions, within 24 hr. after their fall. Immature nutmegs result in a very low-grade spice. The work of collecting the fruit is done only once a day by women and children who go through the orchards. The gathered fruit (or the nutmegs for that matter) should never be heaped, as this may cause fermentation or sweating of the nutmegs and mold development. The soft husk is removed from the core, and the mace separated from the shell containing the nutmeg. Mace and nutmegs (the latter still within their shells) are then placed in separate baskets and brought to the "boucans" (curing houses) for drying. Here women flatten the pieces of mace and later spread them about 1 in. deep on large trays to dry in the sun for about 48 to 60 hr. When thoroughly dried, the mace—now reddish-brown and brittle—is packed away in large bins measuring about 6 ft. in each direction, where it is sealed tightly and kept from the ravages of the mace weevil by regular carbon bisulfide fumigation. A few teaspoonfuls of this chemical placed in a little saucer in the top of the bins every few days keep insects and mice away. After about five months the mace turns to a deep yellow color, about the shade of flint corn. At this stage it is ready to be prepared for export. For this purpose it is taken out of the storage bins and assorted by hand. In the cleaning process, mace is first sifted in a large mechanical oscillating sifter, then picked over by women experienced in this work. The large clear yellow pieces constitute the highest grade, or "pale whole mace." The darker colored pieces, somewhat smaller in size, are classified as "No. 1 broken grade," the small, dark-colored pieces as "No. 2 broken." Chaff and dark bits, stained from lying too long in the fields or in water, or which have some malformation, are put aside as an inferior type—"mace pickings"—

and seldom sold in the United States. After being sorted, mace is packed by hand according to grade in light plywood cases similar to tea chests, each containing 200 lb.

As regards nutmegs, they are dried (cured) while still enclosed in their shiny dark-brown shells, then removed (by cracking) from the shells, but only on orders for shipment of the spice. This helps to protect the kernels (nutmegs) against infestation by insects. As a general rule, the nuts are best cured in the air and away from sunshine, but such a procedure can be carried out only in buildings with an efficient system of aeration. The nuts are placed on shallow trays under cover, being spread no deeper than 3 in.; if spread more thickly they are liable to ferment or cure improperly. Some drying frames are 20 ft. long and 10 ft. wide; the drying sheds or "boucans" may be as large as 60 to 80 ft. long and 30 ft. wide. During the drying the nuts must be stirred two or three times a day with a long-handled hoe, to insure proper ventilation. A good product can be obtained in six weeks' time if only freshly fallen fruits are gathered within 24 hr. of their fall, and cured in the manner now usually employed, viz., aired for seven days; then exposed to early morning sunshine until about 9:30 A.M.—this for ten days; then to daily sunshine in the morning until not later than 11:30 A.M., and in the afternoon for about 1½ hr., until 5 P.M.—this for fourteen days; and finally ½ hr. of sunshine every morning—this for twelve days (with turning twice daily for each of these treatments). After about six weeks of curing, the dried nuts are shoveled into bags and stored in a clean, well-ventilated place to await shipment. In this stage the nutmeg is still enclosed in its shell, approximately a quarter of an inch longer and wider than the kernel (nutmeg) itself. Just prior to shipment, the nuts are taken out of the bags, one bag at a time, and dumped on a clean wooden floor. Women with small wooden hammers rapidly crack the shells surrounding the kernels. The nutmegs are then dumped into tanks filled with water, and slowly stirred. Those floating are considered "grinders" or "defectives," and employed for distillation purposes; nutmegs that sink to the bottom of the tank are considered "sound, unassorted" nutmegs, suitable as food. Both types are thoroughly dried in the air, carefully picked over, and finally exported from St. George, the only shipping port on the island of Grenada. Export of the spice is now a monopoly of the Grenada Co-Operative Nutmeg Association, with headquarters in St. George.

For shipment to England, Grenada nutmegs are graded in 60's, 65's, 80's, 110's and 130's—these numbers referring to the quantity of nuts per pound. When shipped to the United States they are not graded for size, but simply divided into two classes, "sound" nutmegs (used principally in food), and "defectives" (employed only for the distillation of nutmeg oil).

All nutmegs imported into the United States must meet standards set by the Pure Food and Drug Administration. The bulk of nutmeg oil produced in Europe and in the United States is derived from imported spice. However, a few years ago Grenada began local distillation of small quantities. Since nutmegs are available for the entire year, the oil is distilled at all seasons from both the shells and defective nuts. In 1946 there was only one still on the island for the distillation of nutmeg oil, but production could be largely increased by using rum stills or importing other stills. This would allow utilization of nutmeg shells, now a waste product. The bulk of nutmeg oil produced in Grenada goes to England. In 1945 about 3,400 lb. of oil were produced on the island; of this quantity 2,584 lb. went to the United Kingdom and 540 lb. to the United States. In 1946 the exports of nutmeg oil from Grenada amounted to 7,065 lb.[19]

The Fixed Oil.—Among other constituents nutmegs contain from 25 to 40 per cent of fixed oil, also known as concrete or expressed oil, or nutmeg butter (*Oleum myristicae expressum*). This is an orange-colored mass, very aromatic and of butter-like consistency at room temperature. The oil can be obtained by expressing crushed nutmegs between heated plates in the presence of steam, or by extracting with a volatile solvent.

Power and Salway[20] found that fixed oil of nutmeg contains 73 per cent of trimyristin (glyceryl myristate), 12.5 per cent of essential (volatile) oil, 3.5 per cent of fat (glyceryl oleate and linoleate), 2 per cent of resin, minute quantities of formic, acetic and cerotic acids, and 8.5 per cent of unsaponifiable residue.

The fixed oil is of no culinary interest, but may be used to a limited extent in medicinal preparations, particularly in mild stimulants for external application. According to Redgrove,[21] the fixed oil was at one time highly favored for the preparation of certain hair tonics.

It should be mentioned here that it is very difficult to remove the volatile oil from the fixed oil by distillation: the latter oil retains the volatile oil quite tenaciously through reduction of vapor pressure and vaporization rate (cf. Vol. I of the present work, p. 162).

The Volatile (Essential) Oil.—As was mentioned above, both nutmeg and mace contain a volatile oil which can be isolated by steam distillation. The yield of oil varies greatly with the geographical origin of the spice and with its quality; hence it is difficult to give average figures. Wormy nutmegs, e.g., in commercial distillation give a much better yield of volatile oil than do sound nutmegs, for the simple reason that in the former most of the fixed (fatty oil) has been devoured by the worms, while the strongly aro-

[19] *Bull. Imp. Inst.* **46** (1948), 167, 173.
[20] *J. Chem. Soc.* **93** (1908), 1653.
[21] "Spices and Condiments," London, Pitman and Sons, Ltd. (1933), 295.

matic volatile oil remains intact. Sound nutmegs, on the other hand, retain all their fixed oil, and the latter, on distillation, tends to retain the volatile oil, thus lowering its yield (see above).

In odor, flavor, physicochemical properties and chemical composition, the volatile oils of nutmeg and mace are so similar that the trade usually draws no distinction between them. Since mace is generally more expensive than nutmeg (particularly broken and wormy nutmeg), the essential oil is produced mostly from the latter, and only seldom from mace.

Distillation and Yield of Oil.—Prior to distillation, the nutmegs must be comminuted, and if *sound* nutmegs are to be used, most of the fixed oil must be removed by expression. In this latter event, however, the fixed oil will dissolve much of the volatile oil present in the nuts, and on distillation of the chopped and pressed nutmegs a low yield of volatile oil will be obtained. A more suitable, and incidentally much more economical, raw material is therefore the quality known in the trade as "broken and wormy," consisting of refuse and low-priced nutmegs, from which worms have removed much of the fixed oil.

Clevenger [22] found that shriveled East Indian nutmegs give a much larger percentage of volatile oil than do the mature sound ones. The same author also observed that the loss of volatile oil from *ground* mace or nutmegs is relatively rapid, amounting to approximately 80 per cent in two months. For this reason the material should be distilled immediately after grinding.

Gildemeister and Hoffmann [23] reported a yield of volatile oil ranging from 7 to 16 per cent for nutmegs, and from 4 to 15 per cent for mace. In the author's own experience, the yield of oil depends greatly upon the origin, condition, and age of the spice; in the case of nutmegs it varies between 6 and 16 per cent. Mace gives about 10 per cent of volatile oil.

Distillation of the comminuted material is best carried out with live steam; cohobation of the distillation waters may be necessary. With a low pressure steam, about 80 per cent of the oil distills during the first 2 hr.; the balance of the oil requires up to 10 hr. to distill over. High-pressure or superheated steam should not be employed as it carries over some of the fixed oil present in the spice.

PHYSICOCHEMICAL PROPERTIES OF OIL OF NUTMEG

Oil of nutmeg is a mobile, almost colorless or pale yellow liquid, possessing an odor and flavor characteristic of the spice, especially on dilution. With the passage of time the oil takes up oxygen and partly resinifies, becoming more viscous.

[22] *J. Assocn. Official Agr. Chem.* **18** (1935), 611.
[23] "Die Ätherischen Öle," 3d Ed., Vol. II, 596.

The physicochemical properties vary within quite wide limits, depending upon the origin and quality of the spice and the method of distillation. Thus the properties of the oils derived from East Indian nutmegs differ substantially from those of the oils distilled from West Indian nutmegs. Compared with the first, the West Indian type of oil exhibits a lower specific gravity, refractive index, and evaporation residue, and a higher optical rotation. This is the result chiefly of the predominance of terpenes in the West Indian oil. For the same reason the odor and flavor of the East Indian oil are much more pronounced and more characteristic of the spice than those of the West Indian type. The United States Pharmacopoeia, Fourteenth Revision, admits both oils, but requires that the label indicate whether the myristica (nutmeg) oil is of East Indian or West Indian origin.

A. East Indian Oil.—Gildemeister and Hoffmann [24] reported these properties for nutmeg oils, apparently of East Indian origin:

Specific Gravity at 15°.............. 0.865 to 0.925
Optical Rotation................... +8° 0′ to +30° 0′
Refractive Index at 20°............. 1.479 to 1.488
Acid Number...................... Up to 3.0
Ester Number..................... 2 to 9
Ester Number after Acetylation...... 25 to 31
Solubility......................... Soluble in 0.5 to 3 vol. of 90% alcohol
Evaporation Residue............... 1 to 1.5 per cent, if 5 g. of oil are slowly evaporated for 12 to 15 hr., until the weight of the evaporation residue is constant
Boiling Range..................... On distillation in a fractionation flask, about 60 per cent of the oil distills below 180°

Genuine East Indian nutmeg oils distilled by Fritzsche Brothers, Inc., New York, from imported nutmegs of various quality, had properties ranging within the following limits:

Specific Gravity at 25°/25°.......... 0.880 to 0.913
Optical Rotation................... +7° 53′ to +22° 10′, usually above +10°. Several lots of old nutmegs produced oils having abnormally low rotations, as low as +4° 46′
Refractive Index at 20°............. 1.4776 to 1.4861
Evaporation Residue............... 0.3 to 2.1%
Solubility at 20°.................. Soluble in 1 to 2.5 vol. and more of 90% alcohol

As regards the evaporation residue of the oil and method of determination, see Vol. I of the present work, pp. 259, 260.

[24] *Ibid.*

Clevenger [25] reported these properties for a number of oils distilled from Banda nutmegs (I), Padang nutmegs (II), and shriveled East Indian nutmegs (III):

	I	II	III
Yield of Oil (cc. per 100 g. of spice)	4 to 10	8 to 11.5	11.5 to 21.0
Specific Gravity at 20°/20°	0.919 to 0.956	0.878 to 0.909	0.897 to 0.916
Optical Rotation at 20°	+11° 42' to +20° 36'	+20° 42' to +27° 42'	+19° 18' to +21° 48'
Refractive Index at 20°	1.483 to 1.495	1.476 to 1.481	1.479 to 1.482
Acid Number	2.5 and 8.8	1.2 and 2.4	2.46
Ester Number	13.8 and 19.7	6.0 and 11.2	12.3

Clevenger [26] arrived at the conclusion that the volatile oils obtained from ground nutmegs and mace that have been exposed in the laboratory exhibit a definite increase in specific gravity, refractive index, acid and ester numbers, and a distinct decrease in optical rotation. These observations should prove valuable in determining the conditions under which these products are to be handled.

Distilling *fresh* (undried) East Indian nutmegs, de Jong [27] obtained an oil with the following properties:

Specific Gravity at 26°	0.940
Optical Rotation at 26°	+10° 20'
Boiling Range at atm. pr.	155° to 175°—9.5%
	175° to 200°—37%
	200° to 250°—22.0%
	250° to 285°—27.0%

An experimental distillation, on a commercial scale, of Ceylon nutmegs by Fritzsche Brothers, Inc., New York, gave an oil with the following properties:

Specific Gravity at 25°/25°	0.873
Optical Rotation	+28° 55'
Refractive Index at 20°	1.4765
Evaporation Residue	1.0%
Solubility	Soluble in 3 vol. of 90% alcohol and more

B. West Indian Oil.—Genuine West Indian nutmeg oils distilled by Fritzsche Brothers, Inc., New York, from imported nutmegs of various quality, had properties ranging within these limits:

[25] *J. Assocn. Official Agr. Chem.* **18** (1935), 614.
[26] *Ibid.*
[27] *Teysmannia* (1907), No. 8. Cf. *Ber. Schimmel & Co.*, October (1908), 91.

Specific Gravity at 25°/25°	0.859 to 0.865
Optical Rotation	+25° 45' to +38° 32'
Refractive Index at 20°	1.4729 to 1.4746
Evaporation Residue	0.2 to 0.8%
Solubility at 20°	Soluble in 2 to 3 vol. of 90% alcohol and more

Clevenger[28] reported the following properties for a number of oils distilled from West Indian nutmegs:

Yield of Oil (cc. per 100 g. of spice)	8.5 to 10.0
Specific Gravity at 20°/20°	0.859 to 0.868
Optical Rotation at 20°	+40° 48' to +49° 48'
Refractive Index at 20°	1.469 to 1.472
Acid Number	1.0 and 1.3
Ester Number	6.8 and 7.3

According to Clevenger,[29] West Indian nutmegs and mace yield volatile oils which may be distinguished from the corresponding East Indian oils by their low specific gravities and refractive indices, and high optical rotations.

Nutmegs imported from Grenada (B.W.I.) and distilled in the Imperial Institute, London,[30] yielded about 11 per cent of volatile oil with these properties:

	I	II
Specific Gravity at 15.5°/15.5°	0.8666	0.8682
Optical Rotation at 25°	+48° 42'	+48° 24' (at 24°)
Refractive Index at 20°	1.4728	1.4736
Solubility at 15.5°	Soluble in 4 vol. of 90% alcohol, with slight opalescence	

These properties demonstrate that the West Indian nutmeg oils differ markedly from the East Indian oils. The odor of the former is weaker, less characteristic and spicy than that of the East Indian oils. This difference results from the fact that West Indian oils contain a larger amount of terpenes, which lower their quality.

PHYSICOCHEMICAL PROPERTIES OF OIL OF MACE

Odor and flavor, as well as physicochemical properties of mace oil, closely resemble those of nutmeg oil (see above). On exposure to the air, the oil partly resinifies and takes on a turpentine-like, rather disagreeable odor.

---

[28] J. Assocn. Official Agr. Chem. **18** (1935), 613.
[29] Ibid., 615.
[30] Bull. Imp. Inst. **35** (1937), 289.

A. East Indian Oil.—Gildemeister and Hoffmann [31] reported the following properties for mace oils, apparently of East Indian origin:

Specific Gravity at 15°......	0.890 to 0.930
Optical Rotation..........	+10° 0′ to +22° 0′
Solubility................	Clearly soluble in 2 to 3 vol. of 90% alcohol

Distilling East Indian mace from Banda (I) and from Padang (II), Clevenger [32] obtained volatile oils with these values:

	I	II
Yield of Oil (cc. per 100 g. of spice)...	10.4 to 16.4	17.0 to 27.0
Specific Gravity at 20°/20°..........	0.923 to 0.947	0.917 to 0.936
Optical Rotation at 20°.............	+2° 42′ to +11° 48′	+7° 36′ to +11° 24′
Refractive Index at 20°.............	1.486 to 1.494	1.485 to 1.491
Acid Number.....................	2.0 to 3.9	1.4 to 3.0
Ester Number....................	1.2 to 7.3	3.5 to 8.5

Clevenger [33] noted that volatile oils distilled from mace exhibit a lower dextrorotation than oils derived from nutmegs of corresponding geographical origin. This may be the result of the loss, in mace, of the more volatile fractions of oil, which have a high dextrorotation.

B. West Indian Oil.—Clevenger [34] also distilled a number of oils from West Indian mace and reported the following properties:

Yield of Oil (cc. per 100 g. of spice)	8.5 to 15.0
Specific Gravity at 20°/20°.......	0.860 to 0.892
Optical Rotation at 20°..........	+21° 18′ to +41° 30′
Refractive Index at 20°..........	1.472 to 1.479
Acid Number..................	1.5 to 6.2
Ester Number.................	2.8 to 12.8

As in the case of the nutmeg oils, the West Indian mace oils exhibited lower specific gravities and refractive indices and higher optical rotations than the East Indian mace oils. Odor and flavor of the former type of oil are inferior to those of the East Indian oils.

CHEMICAL COMPOSITION OF NUTMEG AND MACE OIL

Oils of nutmeg and mace are so similar that their chemical composition can be discussed in one section. In fact, a number of researchers in the past did not clearly define the origin of the oil which they examined.

The earliest investigations of nutmeg (or mace) oil date back to the first

[31] "Die Ätherischen Öle," 3d Ed., Vol. II, 597.
[32] *J. Assocn. Official Agr. Chem.* **18** (1935), 614.
[33] *Ibid.*   [34] *Ibid.*

decades of the past century, but were so inconclusive that they require no discussion.[35] We owe our present knowledge of the chemical composition of this oil chiefly to the work of Wallach,[36] Semmler,[37] Thoms,[38] Power and Salway,[39] and Schimmel & Co.[40] These investigators have reported the presence of the following compounds (listed approximately according to their boiling points):

d- and *l*-α-Pinene. As far back as 1862 Schacht [41] observed in oil of mace a hydrocarbon which gave a solid hydrochloride; he named this hydrocarbon "Macene." Years later, Wallach identified "Macene" as α-pinene (nitrolbenzylamine m. 123°). It occurs in the lowest boiling fraction of the oil as an optically almost inactive mixture of *d*- and *l*-α-pinene.

Camphene. Identified by Power and Salway, who hydrated the terpene to isoborneol m. 207°–212° (phenylurethane m. 138°).

β-Pinene. Present only in small quantities. Characterized by oxidation to nopinic acid m. 126°–128° (Schimmel & Co.).

Dipentene. Identified by means of its tetrabromide m. 124°–125° (Wallach; Power and Salway; and Schimmel & Co.).

p-Cymene. Characterized by oxidation to *p*-hydroxyisopropylbenzoic acid m. 155°–156° (Schimmel & Co.).

d-Linaloöl. Oxidation to citral, the latter identified by preparation of the α-cityrl-β-naphthocinchoninic acid m. 200° (Power and Salway).

1-Terpinen-4-ol. Identified in the fraction b. 205°–215° (Schimmel & Co.).

Borneol. Oxidizing the fraction b. 205°–215°, Power and Salway, among other products, obtained camphor (semicarbazone m. 238°); it was probably formed by oxidation of borneol originally present in the oil.

dl-α-Terpineol. The same authors also noted that the oil contains *dl*-α-terpineol, which they characterized by preparation of dipentene dihydroiodide m. 80°, and by oxidation to the ketolactone $C_{10}H_{16}O_3$, m. 62°–63°.

About eighty years ago Wright [42] expressed the opinion that the oil contains an alcohol b. 212°–218°, to which he referred as "Myristicol." In reality this alcohol is a mixture of 1-terpinen-4-ol, borneol, and α-terpineol.

Geraniol. Identified by means of its diphenylurethane m. 81°–82° (Power and Salway.)

Safrole. Characterized by oxidation to piperonal m. 34°–35° (Power and Salway).

[35] For a listing of these publications, see Gildemeister and Hoffmann, "Die Ätherischen Öle," 3d Ed., Vol. II, 598, footnote 1.
[36] *Liebigs Ann.* **227** (1885), 288; **252** (1889), 105.
[37] *Ber.* **23** (1890), 1803; **24** (1891), 3818.
[38] *Ber.* **36** (1903), 3446.
[39] *J. Chem. Soc.* **91** (1907), 2037. (These two authors used a "normal" and a "heavy" [i.e., rectified] oil in their investigations.)
[40] *Ber. Schimmel & Co.,* April (1910), 75.
[41] *Arch. Pharm.* **162** (1862), 106.
[42] *J. Chem. Soc.* **26** (1873), 549.

An Aldehyde(?). The substance in question had an odor reminiscent of citral and yielded a β-naphthocinchoninic acid compound m. 248°. Its identity, however, was not established (Power and Salway).

Myristicin. This phenolic ether $C_{11}H_{12}O_3$ occurs in the highest boiling fractions of nutmeg and mace oils. It is one of the most important constituents of the oils. The chemical constitution of myristicin was elucidated by Thoms.[43] (For details, see Vol. II of the present work, p. 531.) Myristicin is toxic and acts as a narcotic. When taken in sufficient quantities, it is liable to cause fatty degeneration of the liver.

The oil also contains several phenols, of which Power and Salway identified:

Eugenol. Benzoate m. 69°, diphenylurethane m. 107°–108°.

Isoeugenol. Benzoate m. 105°.

In the saponification lyes of the oil Power and Salway noted the presence of these acids:

Formic Acid. Identified as barium salt.

Acetic Acid. Identified as barium salt.

Butyric Acid. Identified as barium salt.

n-Caprylic Acid. Identified as silver salt.

A Monocarboxylic Acid $C_{12}H_{17}OCOOH$. Nonvolatile, insoluble in water; m. 84°–85°.

Myristic Acid. M. 54°; present in the oil free and esterified. Depending upon the length of distillation and the steam pressure applied, smaller or larger amounts of this acid and its esters occur in the oil. On evaporation of the oil, the acid remains in the residue. If large quantities are present, the acid may separate from the oil in crystalline form. In the early literature this crystalline deposit was called "Myristicin." It must not be confused with true myristicin, i.e., the phenolic ether $C_{11}H_{12}O_3$ described above.

As a result of their work, Power and Salway [44] reported the following quantitative composition for the nutmeg oil which they investigated:

d-Pinene ⎫	
d-Camphene ⎭	About 80 per cent
Dipentene....................................	About 8 per cent
d-Linaloöl ⎫	
d-Borneol ⎟	
dl-Terpineol ⎬	About 6 per cent
Geraniol ⎭	
1-Terpinen-4-ol [45].............................	Small quantities
An Aldehyde with citral odor....................	Traces
Safrole......................................	About 0.6 per cent

[43] *Ber.* **36** (1903), 3446.

[44] *J. Chem. Soc.* **91** (1907), 2037.

[45] This compound identified by Schimmel & Co. (see above) appears to be identical with the alcohol found by Power and Salway, which, on oxidation, yielded a diketone $C_8H_{14}O_2$, dioxime m. 140°.

(*Top*) A cardamom plant in Southern India. (*Bottom Left*) A ginger planting on the Cochin oast (Southern India). A native holding a ginger plant, showing the rhizome. (*Bottom ight—Above*) A young pepper grove in Banka (Indonesia). The young shoots are sup- orted by poles; (*Below*) bunches of pepper on the left, and pepper berries on the right side. Lampong (Indonesia). *Photos Fritzsche Brothers, Inc., New York.*

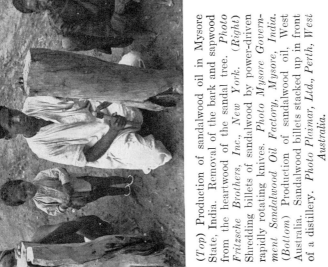

(Top) Production of sandalwood oil in Mysore State, India. Removal of the bark and sapwood from the heartwood of the sandal tree. Photo Fritzsche Brothers, Inc., New York. (Right) Shredding billets of sandalwood by power-driven rapidly rotating knives. Photo Mysore Government Sandalwood Oil Factory, Mysore, India. (Bottom) Production of sandalwood oil, West Australia. Sandalwood billets stacked up in front of a distillery. Photo Plaimar, Ltd., Perth, West Australia.

Myristicin... About 4 per cent

Eugenol)
Isoeugenol } .... About 0.2 per cent

Myristic Acid, free........................... About 0.3 per cent

Myristic Esters............................... Small quantities

Formic Esters ⎤
Acetic Esters ⎟
Butyric Esters ⎬ .... Small quantities
n-Caprylic Esters ⎟
Esters of Monocarboxylic Acid $C_{13}H_{18}O_3$ ⎦

Power and Salway emphasized that these proportions are not absolute; in fact, they vary greatly with the quality and origin of the spice from which an oil is derived. The oil investigated by Power and Salway (d_{15} 0.869, α_D +38° 4′) was distilled from Ceylon nutmegs. Its relatively low specific gravity and elevated optical rotation permit the conclusion that the oil was particularly rich in terpenes. In other nutmeg oils the content of oxygenated compounds is probably much higher than in the oil examined by Power and Salway.

USE OF NUTMEG AND MACE OILS

Oil of nutmeg and the almost identical oil of mace are used widely for the flavoring of numerous food products, particularly baked goods, cakes, cookies, custards, puddings, pickles, etc. The oils find application also in table sauces, tomato catsup, and all kinds of savory preparations and dishes. If well blended, they lend a pleasant smoothness to flavor combinations.

As Power and Salway [46] found, oils of nutmeg and mace are somewhat poisonous, the toxicity being caused by the presence of myristicin (see above). In pharmaceutical preparations—the oil has been recommended for treatment of inflammations of the bladder and urinary tract [47]—large doses must be avoided.

Oil of nutmeg is used also in certain types of perfumes, and for the flavoring of dentifrices.

OIL OF *Myristica Fragrans* HOUTT., FROM THE LEAVES

Meyer [48] steam-distilled dried leaves of the true nutmeg tree *Myristica fragrans* Houtt., and obtained 1.56 per cent of a colorless volatile oil with these properties:

[46] *Am. J. Pharm.* **80** (1908), 563.
[47] Fühner, *Med. Welt* **30** (1940), 779. *Merck's Jahresber.* **55** (1943), 157.
[48] *Ing. Nederland-Indië* **8** (1941), No. 1, VII, 7. *Chem. Abstracts* **35** (1941), 4549.

Specific Gravity at 27.5°/4°...... 0.8772
Optical Rotation at 27°.......... −3° 30′
Refractive Index at 26°.......... 1.4742

The oil contained 80 per cent of α-*pinene* and about 10 per cent of *myristicin*.

The oil is not produced on a commercial scale.

OIL OF *Myristica Succedanea* BL.

As was pointed out in the monograph on "Oil of Myristica," there are several species of *Myristica* other than *fragrans*, the kernels of which enter the spice trade as nutmegs. The botanical origin of these is not always easy to establish, but there are a few quite well-known species which will be described briefly here.

The most important species of these is probably *Myristica succedanea* Bl., locally called "Pala Maba," which originates chiefly from the island of Ternate in the Moluccas. According to Nijholt,[49] nutmegs and mace derived from *Myristica succedanea* Bl. are only slightly, if at all, inferior to first quality Banda nutmegs and mace. The kernels of the species *succedanea* are somewhat more elongated, and often smaller than those of the Banda type (which is true *Myristica fragrans* Houtt.). For this reason adulteration of the species *succedanea* with the inferior Papua nutmegs, which are also elongated and small, cannot easily be detected.

Van der Wielen and Hermans [50] distilled nutmegs of *Myristica succedanea* Bl. with superheated steam and reported these data:

	Kernel Oil
Yield of Oil	4.921%
Specific Gravity at 15°/4°	0.9227
Optical Rotation at 15°	+31° 14′
Refractive Index at 20°	1.4901

OIL OF *Myristica Argentea* WARB.

The kernels derived from *Myristica argentea* Warb., a native of New Guinea, have for a long time been known in the spice trade as "New Guinea," "Papua," or "Long Nutmegs." Occasionally (but wrongly) they are also called "Macassar Nutmegs" because the nuts (which originate from New Guinea) are often shipped first to Macassar, from there entering international trade. The term "Long Nutmegs" arose from the fact that

[49] Private communication from Dr. J. A. Nijholt, Director of the Laboratorium voor Scheikundig Onderzoek, Buitenzorg, Java.
[50] *Festschrift für Alexander Tschirch,* Leipzig (1926), 328.

they possess a longer and narrower shape than the genuine nutmegs derived from *Myristica fragrans* Houtt. The "New Guinea" or "Papua Nutmegs" measure about 3 to 3.7 cm. in length, and about 1.5 to 2 cm. in width. They can easily be recognized by their elliptical shape and weak, rather poor odor and flavor. There are numerous varieties of false nutmegs, but commercially the varieties can be reduced to two, the difference being one of size: the larger ones are usually designated "Papua Nutmegs," the smaller ones "Macassar Nutmegs." Although known in the Far East and Europe for a long time, they have been imported to the United States only since World War II. However, they are unsuitable for distillation purposes, the yield and quality of the volatile oil being subnormal. The same holds true of the "Papua" and "Macassar" mace, which is also of inferior quality. On commercial distillation, "Papua Mace" yields only about 4 per cent of volatile oil.

Van der Wielen and Hermans [51] distilled nutmegs and mace of *Myristica argentea* Warb. with superheated steam and reported these data:

	Kernel Oil	*Mace Oil*
Yield of Oil	1.624%	6.956%
Specific Gravity at 15°/4°	0.9126	0.9272
Optical Rotation at 15°	+14° 52'	+26° 2'
Refractive Index at 20°	1.4848	1.4936

OIL OF *Myristica Malabarica* LAM.

The dried seed of *Myristica malabarica* Lam., a native of India, is known in the trade as "Bombay Nutmegs," the mace as "Bombay," "False" or "Wild Mace." Like the "Papua Nutmegs," the "Bombay Nutmegs" are of elongated, elliptical shape (3 to 4 cm. long); hence the use of the latter as an adulterant of the former. The same is true of the corresponding mace, which has a dark red color, and narrow, long bands. Both "Bombay Nutmegs" and "Bombay Mace" are practically devoid of aroma, and so deficient in volatile oil that they cannot be used for distillation purposes.

[51] *Ibid.*

CHAPTER III

ESSENTIAL OILS OF THE PLANT FAMILY *ZINGIBERACEAE*

THE CARDAMOM OILS

Essence de Cardamome   *Aceite Esencial Cardamomo*
        Cardamomenöl   *Oleum Cardamomi*

OIL OF *Elettaria Cardamomum* MATON VAR. *Minuscula* BURKHILL

Introduction.—The cardamom of commerce consists of the dried ripe fruit of *Elettaria cardamomum* Maton (fam. *Zingiberaceae*). There are two varieties of this species:

1. Var. *minuscula* Burkhill, also called *α-minor*. This, the smaller and more valuable of the two varieties, is cultivated chiefly in southern India (Malabar Coast and adjacent regions), and lately also in Ceylon; hence the trade name "Malabar," "Malabar-Ceylon," and "Mysore-Ceylon" cardamoms for this variety. Large quantities are used as spice and for extraction of the essential oil by distillation.

2. Var. *β-major* Thwaites, also called *Elettaria major,* or "Long Wild Cardamoms." This, the larger and less valuable of the two varieties, grows wild in Ceylon and is cultivated there on a much smaller scale than var. *minuscula*. Until about 1900 only the var. *β-major* was used for distillation of the essential oil, but since the beginning of the century the picture has changed completely, and now the var. *α-minor* has largely replaced the other as regards distillation.

Cardamom fruit is official in most pharmacopoeias. However, the United States and the British Pharmacopoeias confine the definition of cardamom to the decorticated seed only; moreover, the British Pharmacopoeia requires that the seed must have been *recently* removed from the capsules. This latter point is important, because cardamom seed removed from the shell loses a substantial amount of volatile oil by evaporation (about 30 per cent in eight months), whereas husk-protected seed loses only a comparatively small quantity (Clevenger [1]).

The commercial term cardamom has been applied also to the aromatic capsules (fruit) of other plants belonging to the family *Zingiberaceae* (particularly to those of the genus *Amomum*) which are cultivated in Indonesia, Siam, Indo-China, and southern China. These, however, are little known

---

[1] *J. Assocn. Official Agr. Chem.* **17** (1934), 285.

in Europe and America, being used mostly in the Far East. Among them are the "Round Siam," and the "Round Chinese Cardamoms." The present monograph will deal exclusively with the official cardamom, *Elettaria cardamomum* Maton var. *minuscula* Burkhill; other cardamom types will be treated in separate monographs.

According to a report of the Colonial Institute,[2] Amsterdam, the export figure for cardamoms from 1930 to 1940 averaged 1,600 metric tons per year. Of these 60 per cent were "round" cardamoms (*Amomum*), and 40 per cent true cardamoms (*Elettaria*). The Netherlands East Indies exported about 100 tons, Siam 450 tons, and Indo-China 400 tons of "round" cardamoms per year, while British India shipped about 500 tons and Ceylon 150 tons of true cardamoms per year.

History.—The use of cardamom as a highly esteemed spice, a masticatory and an aphrodisiac by the wealthier classes of India goes back to early times. It was known to the Greeks and Romans. The first definite record of cardamom as a spice from the Malabar Coast, written by a European, appears to be that of the Portuguese navigator Barbose,[3] who in 1514 explored the west coast of southern India for spices. About thirty years later, Valerius Cordus submitted cardamoms to distillation, thus isolating the essential oil for the first time.

Description, Habitat, and Range.—*Elettaria cardamomum* Maton is a large perennial herb with a tuberous horizontal rhizome, which grows from 8 to 20 erect perennial leafy stems, 8 to 12 ft. in height. The rhizome also thrusts out flowering stems up to 3 ft. in length, which tend to spread horizontally and produce numerous pretty flowers arranged in panicles. The fruit is an ovoid, three-celled capsule, containing numerous seeds covered by an aril. The inside of the fruit is soft, almost cotton-like, protecting the seeds. The latter contain most of the essential oil, the hard pericarps very little (about 0.2 per cent). On drying, the fruit loses approximately 75 per cent of its weight.

To thrive, the cardamom plant requires—more than anything else—elevation, moisture, and shade. The herb grows abundantly, both wild and under cultivation, in the mountainous districts of southern India, Ceylon, and other tropical countries, at altitudes ranging from 2,500 to 5,000 ft. above sea level. Sections with an average temperature of 22° C., and an average yearly rainfall of 120 in. are best suited for the cultivation of cardamom. Equally important is shade (the plant thrives in moist mountain jungles).

The soil should be rich, with a high content of humus; even swampy ground is suitable. Loamy soil, such as that used for pepper and betel

---

[2] *Koloniaal Instituut, Amsterdam. Mededeeling* No. 58 (1942), 79.
[3] Gildemeister and Hoffmann, "Die Ätherischen Öle," 3d Ed., Vol. I, 122.

plantings may be satisfactory; in fact, cardamom is often planted as a secondary crop with pepper. In South India many growers raise cardamom along with coffee. If, for instance, a swampy stretch runs through a coffee plantation, it is sometimes planted with cardamom.

Planting and Cultivating.—In the mountain forests of tropical countries where the plant occurs naturally, it springs up spontaneously after removal of underbrush. In sections where conditions are still primitive, the natives simply locate a spot in a moist shady mountain forest where some cardamom plants are growing wild. Shortly before the rainy season starts, the natives root out any surrounding brush and weeds that may be harmful to the cardamom plants, at the same time making sure that a certain amount of light, necessary to proper growth of the cardamom, is admitted. The seeds already present in the ground germinate. When the young plants appear precautions are taken for their protection and they are then left to themselves for a year. The plants commence to flower at the end of the second year, and bear fruit a year later. A good crop can be expected in the fourth year. The plants remain productive for quite a few years. If growing in too great abundance, they may have to be thinned out; if the opposite condition holds, others may have to be planted. Each year the clearings should be weeded, the weeds being left on the ground to rot and to act as manure.

In the forests of Coorg and western Mysore cardamom plantings are started in February or March. For the purpose, a part of the forest is simply cut down, but rows of trees 20 to 30 yards wide are left standing between the cardamom plantings, to act as shade trees. The first harvest can be gathered in the fifth year; after this the planting may remain productive for eight years, when it starts to decline and has to be renewed.

These, in general, are the old and primitive methods of planting cardamom; they have been greatly improved in the course of the last fifty years, particularly in Ceylon, and along the Malabar Coast. Cardamom is now propagated either by division of the rhizomes ("bulbs"), or by seed grown in seed beds and nurseries. Seeds, however, germinate very slowly, and four months may elapse before the young shoots appear. In Ceylon the soil is thoroughly cleaned prior to planting, and holes are dug 12 to 15 in. deep, 1.5 to 2 ft. wide, and 7 ft. apart in both directions. The rhizomes should not be planted too deep, lest they start to rot. At present cardamom in Ceylon is propagated largely by seedlings, after they have been in a nursery bed for about a year. To thrive, they require rich and moist soil, with a humus content and ample shading. The two principal harvests take place in February/March and in August/September; smaller quantities can be harvested throughout the year, at two-month intervals. One acre produces from 150 to 300 lb. of cardamoms per year. After eight years a planting starts to decline; in the twelfth year it has to be renewed.

Harvest and Preparation of the Fruit.—In general, the capsules should be gathered from the fruit stems just before complete maturity, when still somewhat green but beginning to turn yellow. If left on the fruit stems to ripen, the capsules will split open and eject their seeds. In modern practice each fruit is carefully cut off with scissors, a portion of its pedicel being left attached to each capsular fruit. Formerly, it was customary to pull off whole racemes of fruit—a wasteful procedure, because not all fruits ripen at the same time.

As was mentioned above, the essential oil and the flavoring principles of cardamom reside almost exclusively in the seed borne within the capsules; nevertheless the spice trade attaches great (perhaps too great) value to the external appearance of the capsules themselves. These are prepared for the market by curing, which can be achieved in several ways. If possible the capsules are slowly dried on clean dry mats by exposure to the sun in dry weather. Great care has to be exercised, however, because over-rapid or too long drying may cause the capsules to swell, burst, and lose part of their seeds. Sun-drying also bleaches the capsules to some extent. In wet weather the capsules are placed on trays in racks, in special curing houses, and exposed to gentle heat. Artificially dried cardamoms retain their original green color. Occasionally the capsules are sprinkled with water, then bleached in the sun; this may improve the color of the pericarp, but usually increases the number of split fruits. When sufficiently dried, the pieces of attached stalks (pedicels) are clipped or rubbed off the capsules, and removed by winnowing or by cleaning in machines. The capsules are finally graded into different sizes by means of sieves. Split fruits, broken shells, and loose seeds must be removed. The final product may have to be air-dried once more, before it is packed into cases for shipment.

Since white or pale-colored cardamoms are preferred by some spice dealers, several methods have been devised for improving the color of the sun-bleached product. One of these methods consists in imparting a light coating of starch to the dried fruit; another practice is to bleach the cardamoms by fuming with burning sulfur.

Still another method of bleaching, frequently practiced in India, consists of a quite elaborate process: [4]

The cardamom pods are dropped into a large earthenware jar filled with clean water and some fruit of the soapberry tree (*Sapindus saponaria* L.). The mass is stirred vigorously by hand for a few minutes, allowed to rest,

---

[4] The author is greatly obliged to Dr. Karl Landes, New York, for the description of this process.

and stirred again. The fruits are then removed from the jar, washed clean in another tub of fresh water, taken out and placed into wicker baskets. After all the water has been drained off, the fruits are spread out and sprinkled with clean water, at half-hour intervals, throughout the night. In the morning the fruits are spread on drying mats, exposed to the sun for 4 to 5 hr., taken inside of a hut, sprinkled again, and finally exposed to the afternoon sun for 4 to 5 hr. This produces a good grade of bleached cardamoms.

As a supplement to our general description of the growing, harvesting and curing of cardamom in India and Ceylon, we shall add some observations by A. H. Khan [5] on the collection and drying of the spice as carried out in Madras Presidency:

Cardamoms occur naturally in the evergreen forests of the western Ghats, in Madras Presidency, flourishing best in rich loamy soil from 2,000 to 4,000 ft. above sea level. A typical area of natural cardamoms is the Silent Valley in Palghat Division, where the cardamoms are of the Malabar variety. Most private plantations, however, are of the Mysore variety, this being the more favored type. The plants begin to yield a small crop in the third year, and full crops from the fourth year on, after natural expansion of the plants has taken place, and the gaps have been filled in by additional hand planting.

The collection, which should begin when the seeds in the fruits start to turn dark, lasts from August through October. If the harvest is delayed too long, the over-ripe capsules may split on drying.

In regular plantations the cardamoms are collected very carefully, only the ripe capsules being picked throughout the harvesting period. This method, however, is too expensive for the collection of the wild cardamoms, which are spread over extensive forest areas. In the Silent Valley, therefore, harvesting consists of pulling out whole panicles of fruits; each panicle should bear at least 75 per cent ripe fruits. This may appear wasteful, but under the conditions given there is no alternative. The coolies hand over their collection to the forester after removing the fruits from the rachis, but since their wages are based on a volume measure, they are likely to remove as much of the main stalk as possible. The subordinate in charge of collection, therefore, should see that nothing of the pedicel (the direct stalk of the fruit) is included for measurement. Although the green cardamoms from the Silent Valley can stand two days of rough storing plus an additional day in transit without any appreciable deterioration in quality, it is best to send them down immediately after collection to the drying sheds in the plains, about 10 miles distant. Here they are given an intense

[5] *The Indian Forester* (April 1944). Cf. *Flavours* **7**, No. 6 (1944), 20.

cleaning, which consists of nipping the stalks from individual capsules, a job done very cheaply by women, girls, and boys.

The original sheds were thatched buildings with bamboo *thatti* walls 12 ft. high, made air-tight by plastering with mud. The portions of the walls abutting on the furnaces were of sheet zinc. The ceiling was constructed of unplastered bamboo mats to allow the vapor to escape. Each shed consisted of a drying room, two small side rooms for feeding the furnaces, and a veranda for the coolies engaged in cleaning the green cardamoms. The drying room was 15 × 30 ft., and consisted of two rooms—one above the other, and separated by a strong lattice-work partition 6 ft. above the floor. Two iron furnaces, 2 × 4 × 3 ft. each, were placed diagonally opposite each other in the corners of this room, 3 ft. away from the side walls. A smoke-tight zinc pipe, 9 in. in diameter, ran from each of these furnaces along each side wall, through the ends and vertically as an outside flue. Thus in the lower room, there were two parallel pipes 8 ft. apart and running the length of the room.

After removal of the stalks, the green cardamoms are spread evenly on the floor of the drying room and on mats on the middle partition, the capsules touching each other. About 1,250 lb. of green cardamoms can thus be spread at one time in a drying room of 15 × 30 ft. The furnaces are fed with thick green fuel (two cartloads being necessary to dry 1,250 lb. of cardamoms, yielding about 135 lb. of dried cardamoms), and the doors of the shed are closed. Care should be taken that there are no leaks in the pipes, as the smoke spoils the color of the spice. If the furnaces are fed constantly, it takes about 30 hr. for the green cardamoms to dry completely, those in the upper room drying about 4 to 6 hr. earlier than those below. The cardamoms near the furnaces on the floor, and those directly above them dry the fastest. It is essential that the cardamoms be removed from the drying room as soon as the drying is complete, as they lose their color if allowed to remain in the heat. Experience in the Silent Valley has proved that if the cardamoms are transported to the drying shed within two days of their collection, and are put in the drying shed the same day (provided all precautions are taken in the shed) the best quality of dried cardamoms —green in color, and smooth to the touch—is produced. The secret of getting the right color consists in placing the green cardamoms in the drying sheds as quickly as possible after collection. It is important to regulate the collection in the hills according to the capacity of the sheds.

Although sun-drying is cheaper and is practiced by small cultivators, the price obtained for this quality is considerably lower than that obtained for good kiln-dried quality, and the extra money spent on kiln-drying is covered by better prices.

As soon as the cardamoms are dried they are removed to the storeroom, where they are first rubbed on a wide-meshed sieve and then winnowed, to separate the grainless, shriveled fruits and other rubbish. They are then packed immediately in good double gunny sacks and stored in a dark room, after weighing. They should never be left heaped-up in the storeroom as, in wet weather, they soon swell up, absorbing moisture, and becoming mildewed if packed in bags in this condition. The dried cardamoms should be stored in a dark room, as light spoils their color.

Loss can occur in two stages: first, when the stalks are removed from the capsules, and second, in the course of drying. The loss after removal of the stalks from the capsules should be about 25–30 per cent. A higher loss indicates that the collection is not being done properly in the hills. In the case of the Silent Valley cardamoms (Malabar variety), the weight is reduced from 25 lb. green to 2¾ lb. dried. The loss in the case of the Mysore variety is less, as the Malabar cardamoms are more juicy and exhibit wider variance with the locality of production. It is worth while, therefore, to determine the standard loss to be expected for a new locality by careful records at the very first consignment, as this affords a good check for later control in the depot.

Trade Varieties.—The dried fruit of *Elettaria cardamomum* Maton var. *minuscula* Burkhill is a three-angled, ovoid or oblong, three-celled capsule, about 1 to 2 cm. long, slightly rounded at the base, and shortly beaked at the apex. The color of the green or unbleached capsules ranges from gray to green-brown; that of the bleached capsules from cream to buff. Each cell contains two rows of seeds, which are mostly agglutinated into groups of 2 to 7 by the adhering membranous aril. The length of the individual seeds ranges from 3 to 4 mm., their color from pale orange to dark brown. They have an aromatic, pungent and slightly bitter taste, and an odor reminiscent of eucalyptus. The pericarps are tough, and possess very little odor and flavor.

Formerly the trade terms "lesser," "middle" and "larger" or "shorts," "short-longs" and "long-longs," respectively, were applied to cardamoms of various origin, the terms "larger" or "long-longs" including the large wild-growing cardamoms (*Elettaria cardamomum* Maton var. *β-major* Thwaites) from Ceylon. These terms are no longer employed in their old sense, and others have come into use. The official cardamoms (from *Elettaria cardamomum* Maton var. *minuscula* Burkhill) are now often classified as "shorts" and "short-longs." The former are usually broad and plump, the latter finer-ribbed, and lighter than the "shorts." The Malabar cardamoms are of highest value; they consist of both types. The Mysore cardamoms are considered of next best grade; they consist mostly of "shorts," but are less pungent in flavor than the Malabar. Both the Malabar and the Mysore cardamoms are now grown also on the island of Ceylon; hence the terms

"Malabar," "Malabar-Ceylon," and "Mysore-Ceylon" cardamoms. Aside from these, there are the Mangalore and the Alleppi cardamoms, grown near the port of Mangalore, and in Travancore and Cochin, respectively.

Some spice exporters in the ports of southern India evaluate cardamoms according to their origin:

1. Travancore State. Producing regions lie along the Malabar coast. The spice is shipped from the ports of Cochin and Alleppi; hence the designation "Alleppi Greens." Often the spice is first sent to Bombay and Madras, the headquarters of many Indian spice dealers, and exported from there. "Alleppi Greens" are the cardamoms most suitable for distillation. The best time to purchase the spice is right after the harvest, i.e., in October/November, when it is still relatively fresh and freely available.

2. Mysore State. Cardamoms grown here are also of very good quality; they resemble those from the neighboring province of Coorg.

3. Province of Coorg. This quality often consists of the second or third picking of the cardamom-bearing plant. The pods are not green, but usually yellow, and often split, thus exposing the seed to the air. Coorg cardamoms, although slightly inferior to those from the Malabar coast as regards appearance, are said to have a fair content of essential oil. Principal shipping ports of the "Coorg Greens" are Mangalore and Cochin; often the spice is sent to Bombay or Madras for export.

As regards exports from the island of Ceylon, a well-known exporter in Colombo ships cardamoms under these designations:

(1) Green Cardamoms:
    Kandy Type—relatively large, of dark greenish color.
    Copernicus Type—slightly smaller than the "Kandy"; color generally green, but some capsules have an off-color.
    Green Faq. Type—small cardamoms with a gray-green color.
(2) Bleached Cardamoms:
    Malabar Half-Bleached—fair average quality of the season; rather small capsules.
    Curtius—fair-sized, rather long capsules.
    Cleophas—fair-sized, roundish capsules.
(3) Seeds:
    Crispus Type—freshly removed seeds, obtained by the husking of either green or bleached capsules.

In general, the trade distinguishes between decorticated cardamoms, green cardamoms, and bleached cardamoms.

Decorticated Cardamoms, viz., the seed removed by hand from the capsules, are supplied chiefly by Alleppi and Mangalore in India, and by Guatemala (C.A.). The seed is used mostly by the spice and bakery trade for the flavoring of sausages, pastry, and confectionery. Principal consumers are the Scandinavian countries. Some of the seed is employed for the flavoring of "Aquavit," a well-known Swedish liqueur.

Green Cardamoms are most suited for distillation. They are firm to the touch, and tightly closed, thereby protecting the aromatic, oil-containing seeds inside of the pods. Green cardamoms are shipped chiefly from Alleppi and Mangalore, and to a lesser extent from Ceylon.

Bleached Cardamoms are prepared by bleaching of the green fruits. They are classified according to size:

1. Bold
2. Medium Bold
3. Medium
4. Small

The bolder the cardamoms, the more expensive they are. In the United States bleached cardamoms are used mostly in pickling spices and in packaged goods.

Distillation.—As has been mentioned, perhaps too much importance is attached to the appearance and color of the capsules, particularly if they are to be used for distillation purposes. Since the odorous principles are contained almost exclusively in the seed, it would seem sufficient that the capsules enclosing the seed be firmly closed (not partially split), and that the spice itself be of recent harvest. Clevenger [6] proved that:

1. For all practical purposes the husks from cardamom fruits may be considered inert.[7]
2. Seed from green cardamoms yields on the average more volatile oil than seed from bleached cardamoms.
3. Cardamom seed, imported as such, yields on the average less volatile oil than that recently removed from the husks. This is undoubtedly due to the lack of protection by the husks.
4. The loss of volatile oil in husk-protected seed is comparatively small, even after eight months.
5. The loss of volatile oil in cardamom seed removed from the shells is considerable, amounting to about 30 per cent in eight months.

---

[6] *J. Assocn. Official Agr. Chem.* **17** (1934), 283.
[7] An experimental distillation of cardamom hulls by Fritzsche Brothers, Inc., New York, yielded 2.5 per cent of essential oil.

From the findings of Clevenger, and from the experience of commercial producers, it appears that the best material for distillation of cardamom oil is "green," "Malabar," or "Ceylon-Malabar" cardamoms of recent harvest. Fruit from Mysore or Coorg is also well suited for distillation. The capsules do not have to be bleached; in fact, bleached cardamoms are only a fancy grade in the spice trade. The most important factor is that the capsules be tightly closed, and not split, and that they contain as much seed as possible. External appearance of the spice plays no role in distillation. Prior to distillation, the fruits should be crushed and charged into the stills immediately afterward. They yield from 3.5 to 7.0 per cent of essential oil.

Note that no cardamom oil is produced in India and Ceylon; the spice is exported to Europe and America for use as such, or for distillation.

Cardamom Oil from Guatemala.—In 1945 Guatemala started to produce small quantities of cardamom oil from domestically grown seed. Since this new industry has grown substantially in the course of the last years, it may interest the reader to learn some details about its development, present status, and future possibilities.

According to Ippisch,[8] cardamom bulbs, for reproduction purposes, were first introduced into Guatemala in the early years of this century. They were imported from Ceylon or India, with the assistance of a New York broker, and planted in the northern part of Guatemala, in the area of Alta Verapaz (Cobán). The altitude here is about 2,000 ft., and rainfall exceeds 140 in. per annum.

Production of seed increased considerably after World War I, largely in the German estates located in the vicinity of Alta Verapaz. Decorticated seed material in tins was exported to Europe (Hamburg being chief port of entry). When the United States became engaged in World War II, the German estates were expropriated by the Guatemalan Government and all exports went to the United States. Production of the crop soon decreased, since the new managers, appointed by the government, lacked sufficient knowledge to grow it properly; however, tempted by the prevailing high prices, farmers in the neighborhood of the confiscated estates began to grow cardamoms. New plantations, in distant regions, were started also, particularly in Nebaj (Department of Quiché), and in Colomba (Department of Suchitepéquez). The seed was exported in wooden boxes lined with waxed paper; these containers replaced the tins which became unavailable during the war. The largest quantities of seed originating from the former German estates are sold at public auction; smaller lots are shipped

---

[8] Private communication, courtesy Mr. F. J. Ippisch, Manager of Oficina Controladora de Aceites Esenciales, Guatemala City, Guatemala, C.A.

through local export companies, chiefly to the United States, and to a smaller extent to Belgium and Sweden.

The plantations are located at altitudes ranging from 2,000 to 4,000 ft., in areas where the soil is fairly moist, and retentive of water. Valleys sheltered from the wind are good locations for planting. Old coffee plantations, protected by the shade of leguminous trees, are frequently used. Common practice is to plant the bulbs at intervals of 8 by 8 ft., or even less (if possible, 6 by 6 ft.). The holes are about 15 in. deep and 2 ft. square. Several pounds of cattle manure are placed into each hole. The latter is then filled with good surface soil, the soil and manure being well mixed. The most common method of propagation is by division of the whole stool of a mature plant. At least two tubers are planted in each hole. Propagation by seed is never practiced. The soil around the plant must be cleaned two or three times a year, until the plant is sufficiently large to resist excessive weed growth. Stems which turn yellow or brown, or which have racemes growing from their bases, should be cut out. Any leaves, either from the cardamom itself or from the protecting shade trees, are removed from the hollow inside the stool, and earth is scraped up around the base of the stems to be placed over the tubers.

The best sign of a flourishing crop is the presence of a large number of healthy shoots, since each of these shoots produces a flowering raceme from its base. In fertile soil stools of mature plants may attain 5 ft. in diameter. The first crop is obtained in the second year after planting. A really fair crop, however, can be expected only in the third or fourth year. The harvest increases, from this time, until the eighth or ninth year, when it begins to decline.

The flowers on the racemes remain open for about a week; but three months elapse before the fruit matures. In Guatemala cardamoms ripen in December/January, and occasionally as late as February. In some years it is possible to harvest a secondary crop in March/April, before the beginning of the rainy season.

The green pods (capsules) containing the seeds are collected and carried by Indians to the farmsteads, where they are dried in airy concrete patios, similarly to coffee. When harvested at the proper time, the seeds are black; each pod contains up to 24 seeds. The most primitive way of separating the seed from the pods is by grinding them by hand on stones—a very ancient method used by the Mayans in treating corn. Nearly all Guatemalan seed is produced in this way, mechanical means (a specially adapted hammer mill and a special seed cleaner) being limited to two farms.

Production of small quantities of oil began in 1945, the amount representing an unimportant percentage of the crop. Two types of raw material are employed: (a) good seed, properly ground. The resulting fine powder is distilled in thin layers. (b) Broken seed, powder and pods—i.e., waste material obtained in the separation of seed from the pods. The yield of oil from type (a) averages 6 per cent; that from type (b) ranges from 2.5 to 3 per cent.

The oil derived from the two types of material just described has not yet found full approval in the United States, perhaps because local methods of distillation are backward; or perhaps because green capsules chiefly are used for distillation in the United States. However, the Guatemalan oil has been accepted in Europe. Only limited quantities are exported.

Exports of decorticated seeds from Guatemala approximate 80,000 lb. annually, representing about one-third of the consumption in the United States. Compared with those from Ceylon and India, the exports from Guatemala are gradually increasing. This fact encourages efforts to enlarge production, both of the seed and the oil, and to improve material and equipment.

Physicochemical Properties of Cardamom Oil.—Oil of cardamom derived from the fruit or seed of *Elettaria cardamomum* Maton var. *minuscula* Burkhill is a colorless, or almost colorless liquid, with a spicy odor, reminiscent of eucalyptus.

Gildemeister and Hoffmann [9] reported these properties for the oil:

Specific Gravity at 15°...... 0.923 to 0.941
Optical Rotation........... +24° 0' to +41° 0'
Refractive Index at 20°..... 1.462 to 1.467
Acid Number.............. Up to 4.0
Ester Number............. 92 to 150
Solubility................ Soluble in 2 to 5 vol. and more
           of 70% alcohol

Oils distilled under the author's supervision in Seillans (Var), France, from fruit imported from South India exhibited the following values:

Specific Gravity at 25°/25°...... 0.922 to 0.927
Optical Rotation.............. +22° 0' to +25° 55'
Refractive Index at 20°........ 1.4640 to 1.4672
Acid Number................. 1.4 to 5.6
Ester Number................ 85.4 to 98.0
Solubility.................... Soluble in 2.5 to 3.5 vol. of 70%
           alcohol and more

Oils produced by Fritzsche Brothers, Inc., New York, from imported fruit, chiefly "Alleppi Greens," had properties varying within these limits:

Specific Gravity at 25°/25°...... 0.924 to 0.931
Optical Rotation.............. +22° 0' to +27° 23'

[9] "Die Ätherischen Öle," 3d Ed., Vol. II, 445.

Refractive Index at 20°......... 1.4630 to 1.4672
Acid Number.................. 1.4 to 5.6
Ester Number................. 88.2 to 121.8
Solubility..................... Soluble in 2.5 to 3.5 vol. and
more of 70% alcohol

Two samples of cardamom oil produced in Guatemala, and examined in the laboratories of Fritzsche Brothers, Inc., New York, exhibited the following characteristics:

	I	II
Specific Gravity at 25°.....	0.931	0.924
Optical Rotation...........	+39° 14'	+30° 10'
Refractive Index at 20°.....	1.4640	1.4642
Acid Number..............	0.7	1.3
Ester Number.............	151.6	125.2
Saponification Number.....	152.3	126.5
Cineole Content...........	...	About 20%
Solubility................	Soluble in 2 to 2.5 vol. and more of 70% alcohol	Soluble in 3 vol. of 70% alcohol; slightly opalescent in 10 vol.

The two oils had normal properties; their odor was very good. Other samples of Guatemala-distilled cardamom oil examined by the same firm had somewhat abnormal properties, the result, probably, of improper or careless distillation.

Distilling cardamoms (obtained from various crude drug firms) in a Clevenger apparatus, Fischer, Tornow and Proper [10] obtained oils with these properties:

Specific Gravity at 25°/25°....... 0.9213 to 0.9318
Optical Rotation at 25°.......... +21° 49' to +33° 27'
Refractive Index at 20°.......... 1.4624 to 1.4660
Acid Number.................. 1.26 to 3.59
Ester Number................. 113.08 to 176.29

Clevenger [11] experimentally distilled cardamom seed separated from *bleached* fruit (I), cardamom seed separated from *green* fruit (II), cardamom seed imported from India (III), and cardamom seed imported from Guatemala (IV). The oils thus obtained had properties varying within the following limits:

[10] *Bull. Natl. Form. Comm.* **13,** No. 1–2 (1945), 7.
[11] *J. Assocn. Official Agr. Chem.* **17** (1934), 283.

	I	II	III	IV
Yield of Oil, in cc. per 100 g. of Seed	5.2 to 11.2	6.6 to 11.2	3.4 to 8.0	3.5 to 8.6
Specific Gravity at 20°/20°........	0.922 to 0.938	0.923 to 0.930	0.923 to 0.932	0.923 to 0.932
Optical Rotation at 20°.............	$+27°$ 0' to $+36°$ 42'	$+20°$ 6' to $+32°$ 42'	$+20°$ 42' to $+36°$ 42'	$+27°$ 42' to $+36°$ 6'
Refractive Index at 20°.............	1.461 to 1.467	1.461 to 1.464	1.461 to 1.467	1.463 to 1.467
Acid Number......	1.8 to 4.8	1.1 to 3.4	1.0 to 4.5	0.8 to 5.6

Other data reported by Clevenger show that the physicochemical characteristics of the volatile oils obtained from the capsules (fruit) including the seed, and from the seed alone, are practically identical.

In his work on cardamom oil, Clevenger [12] also found that the time required for complete saponification of the oil approximates 3 hr. The reason for this prolonged period of saponification is the presence of substantial quantities of terpinyl acetate in the oil.

Chemical Composition.—Although oil of cardamom derived from *Elettaria cardamomum* Maton var. *minuscula* Burkhill has been a very important spice oil for a long time, relatively little is known about its chemical composition. More than a century ago, Dumas and Péligot [13] observed the presence of crystals of terpin hydrate in an old Malabar cardamom oil; this compound probably originated from terpineol, a substance identified in the oil more than fifty years later. The oil has been investigated by Schimmel & Co., [14] Parry, [15] Wallach [16] and Moudgill, [17] who reported the presence of the following compounds in the oil:

Limonene. Noted by Parry.

Sabinene(?). Presence possible (Moudgill).

Cineole. Identified by Schimmel & Co., who prepared the iodol compound m. 112°–113°.

d-α-Terpineol. Isolated from the fraction b_{14} 150°–164° of the saponified oil by Schimmel & Co., who identified the terpineol, m. 35°–37°, α_D $+81°$ 37' (in the superfused state), by means of the phenylurethane m. 112°–113°, $[\alpha]_D^{20}$ $+33°$ 58' (in 10 per cent alcoholic solution), and by preparation of the nitrosochloride. The latter was converted to a nitrolpiperidine m. 151°–152°, which melting point was 8° lower than that of the optically inactive terpineol nitrolpiperidine.

Terpinyl Acetate. Schimmel & Co., and Wallach found that the *d*-α-terpineol occurs in the oil in large part as acetate.

Borneol. Reported as a constituent of the oil by Moudgill; m. 198°.

[12] *Ibid.*
[13] *Ann. chim. phys.* [2], **57** (1834), 335.
[14] *Ber. Schimmel & Co.*, October (1897), 9.
[15] *Pharm. J.* **63** (1899), 105.
[16] *Liebigs Ann.* **360** (1908), 90.
[17] *J. Soc. Chem. Ind.* **43** (1924), 137T.

An Acid(?). The same author also noted that the oil contains an acid with an odor reminiscent of cuminaldehyde. Its molecular weight was determined as 182(?).

Use.—Like the spice itself, the essential oil is employed widely for the flavoring of various food products, such as cakes, confectionery, gingerbread, sausages, and pickles. It finds use also in spicy table sauces, curry preparations, and in certain bitters and liqueurs.

Medicinally, the oil is frequently employed as an adjuvant or corrective of tonic, carminative, and purgative preparations.

SUGGESTED ADDITIONAL LITERATURE

Arno Viehoever and Le Kya Sung, "Common and Oriental Cardamoms," *J. Am. Pharm. Assocn.* **26** (1937), 872.

OIL OF *Elettaria Cardamomum* MATON VAR. *β-Major* THWAITES

The so-called "Long Wild Cardamoms," also known as *Cardamomum majus* or *Cardamomum longum,* are derived from *Elettaria cardamomum* Maton var. *β-major* Thwaites, and originate on the island of Ceylon; hence the popular name "Ceylon Cardamoms," or "Long Ceylon Cardamoms." The plant grows wild in the forests in the interior of Ceylon. It is also cultivated on a small scale near Kandy, the capital of the island. The dried fruit consists of a lanceolate-oblong, triangular capsule, up to 40 mm. long and 6 to 8 mm. broad, grayish-brown or dark brown in color. The fruit offered on the market frequently retains the long, cylindrical, three-lobed calyx at one end, and at the other end the fruiting stalk. The seeds are angular, yellowish-red, and possess the characteristic odor and flavor of cardamom.

Until about 1900 almost all oil of cardamom was distilled from this variety; since then, however, the var. *minuscula* Burkhill has been used for this purpose (see the preceding monograph). Nevertheless, a brief description of the oil derived from the var. *β-major* Thwaites is included here.

According to Gildemeister and Hoffmann,[18] distillation of the dried crushed fruits yields from 4 to 6 per cent of volatile oil.

Physicochemical Properties.—The oil distilled from long Ceylon cardamoms is a light yellow, slightly viscous liquid with an odor and flavor characteristic of cardamom. Gildemeister and Hoffmann [19] reported these properties for the oil:

Specific Gravity at 15°....... 0.895 to 0.906
Optical Rotation............ +12° 0′ to +15° 0′

[18] "Die Ätherischen Öle," 3d Ed., Vol. II, 446.   [19] *Ibid.*

Saponification Number....... 25 to 70
Solubility.................. Turbid in 70% alcohol. Soluble
in 1 to 2 vol. and more of 80%
alcohol

An oil derived from *Elettaria cardamomum* var. *β-major* and analyzed by Sage [20] exhibited the following values:

Specific Gravity at 15.5°..... 0.909
Optical Rotation at 20°...... +16° 30'
Refractive Index at 25°...... 1.474
Acid Number.............. 1.1
Ester Number.............. 12
Solubility.................. Soluble in 1 to 2.5 vol. of 70%
alcohol

Compared with cardamom oil from the var. *minuscula* Burkhill, the oil derived from the var. *β-major* Thwaites has a considerably lower specific gravity and optical rotation. It is particularly by the difference in their optical rotations that the two oils can be distinguished.

Chemical Composition.—The oil has been investigated by Weber,[21] and by Wallach,[22] who reported the presence of these compounds:

Sabinene. In the fraction b. 165°–167°; identified by oxidation to sabinenic acid m. 56°–57° (Wallach).

Terpinene. Passing gaseous hydrogen chloride through the fraction b. 170°–178°, Weber obtained terpinene dihydrochloride m. 52°. The fraction b. 178°–182° yielded terpinene nitrosite m. 155°. (Terpinene as a new terpene was first reported by Weber.)

1-Terpinen-4-ol. Investigating the fraction b. 205°–220°, Weber believed he had identified α-terpineol as a constituent of the oil. Later, however, Wallach proved that the terpene alcohol described by Weber was not α-terpineol, but the isomeric 1-terpinen-4-ol (cf. Vol. II of the present work, p. 198). Wallach identified this compound by means of its dihydrobromide m. 59°, and by oxidation to *p*-menthane-1,2,4-triol, m. 128°–129°.

1-Terpinen-4-yl Formate and Acetate. The oil contains formates and acetates of 1-terpinen-4-ol. Calculated on the basis of the ester number, the ester content of the oil ranges from 8 to 24 per cent.

A Solid Compound(?). Separated from the distillation residue. It melted at 60°–61°, after recrystallization from alcohol.

Use.—The small quantities of oil distilled from long Ceylon cardamoms have the same use as the oil derived from the var. *minuscula* Burkhill (see the preceding monograph).

[20] *Perfumery Essential Oil Record* **15** (1924), 150.
[21] *Liebigs Ann.* **238** (1887), 98.
[22] *Ibid.* **350** (1906), 168. *Nachr. Ges. Wiss. Göttingen* (1907), Sitzung July 20th.

Oil of *Amomum Cardamomum* L.

As was mentioned in the introduction to the monograph on "Oil of Carda-mom," there are other members belonging to the fam. *Zingiberaceae* (and particularly the genus *Amomum*), the fruits of which resemble those of *Elettaria cardamomum* Maton, the true cardamom. One of them is *Amomum cardamomum* L., the so-called "Siam," "Round" or "Cluster Carda-mom," which grows in Siam, Java, Sumatra, and other parts of the East Indian Archipelago. The fruits are now seldom encountered in Europe and America. At one time the fruit was official in the French Codex.

Round or Siam cardamoms occur in small compact bunches; they are smaller than a cherry, roundish, somewhat ovate, and possess a strong camphoraceous and aromatic flavor, resembling that of true cardamoms (*Elettaria cardamomum* Maton).

Distilling the seed of *Amomum cardamomum* L., Schimmel & Co.[23] ob-tained 2.4 per cent of a semisolid liquid m. 42°, with an odor reminiscent of camphor and borneol. The oil had these properties:

Specific Gravity at 42°.................. 0.905
Optical Rotation at 42°................ +38° 4'
Saponification Number................. 18.8
Saponification Number after Acetylation. 77.2
Alcohol Content, Calculated as Borneol.. 22.5%
Solubility............................ Soluble in 1.2 vol. of 80%
                                    alcohol

Another oil of *Amomum cardamomum* L., investigated in Buitenzorg [24] (Java), had the following properties:

Specific Gravity at 26°.............. 0.909
Optical Rotation.................... −0° 20'
Acid Number....................... 0.8
Saponification Number.............. 14
Cineole Content (Resorcinol Method).. 12%

In the oil which they had distilled from the seed of *Amomum cardamomum* L., Schimmel & Co.[25] identified *d-borneol* and *d-camphor*. The crystalline mass separating from the oil consisted of approximately equal parts of *d*-borneol and *d*-camphor.

Oil of *Amomum cardamomum* L. is not produced on a commercial scale.

[23] *Ber. Schimmel & Co.*, October (1897), 9.
[24] *Jaarb. Dep. Landb. in. Ned.-Indië*, Batavia (1911), 48.
[25] *Ber. Schimmel & Co.*, October (1897), 9.

OIL OF *Amomum Globosum* LOUR.

Another of the unofficial cardamoms is the "Large Round Chinese Carda-mom," derived from *Amomum globosum* Lour., an evergreen said to re-semble members of the genus *Myristica* in appearance. The plant is indi-genous to southern China and now grows wild and under cultivation in the province of Kwangtung. The fruits are round or globular, pale yellow, with an average diameter and thickness of 15 mm., longitudinally streaked, tapering at both ends, these latter covered with numerous long nonglandular hairs. Each fruit contains about 24 seeds, possessing a pleasant aromatic odor and flavor, similar to that of true cardamom (*Elettaria cardamomum* Maton), but more camphoraceous. Viehoever and Sung [26] suggested that the fruit of *Amomum globosum* Lour. might be used as a substitute for the fruit of the official Malabar cardamom.

According to the same authors,[27] the seed of the large round Chinese cardamoms yields from 4 to 6 per cent of a volatile oil with these properties:

Specific Gravity............ 0.965 to 0.975
Refractive Index........... 1.462
Acid Number.............. 1.5 to 3
Saponification Number....... 25 to 35
Solubility.................. Soluble in 4 to 5 vol. of 80%
                            alcohol
Congealing Point........... Does not congeal at $-5°$
Odor...................... Distinctly aromatic
Taste..................... Camphoraceous and cooling

Years earlier, Schimmel & Co.[28] had distilled "wild" cardamoms from Indo-China, reported to have been derived from *Amomum globosum* Lour. The oil (yield 4 per cent) had the following properties:

Specific Gravity at 15°....... 0.9455
Optical Rotation............ $+43° 54'$
Refractive Index at 20°...... 1.47141
Acid Number.............. 0.8
Ester Number.............. 128.4
Solubility.................. Not soluble in 70% alcohol;
                            soluble in 1 and more vol.
                            of 80% alcohol

The values resembled those of true cardamom oil (from *Elettaria carda-momum* Maton), but the odor was quite different, recalling camphor oil, rather than true cardamom oil.

Oil of *Amomum globosum* Lour. is not produced on a commercial scale.

[26] *J. Am. Pharm. Assocn.* **26** (1937), 872.   [28] *Ber. Schimmel & Co.*, April (1913), 108.
[27] *Ibid.*, 881, 882, 884.

OIL OF *Amomum Aromaticum* ROXB.

The fruits of *Amomum aromaticum* Roxb., also known as "Bengal Cardamoms" or "Winged Bengal Cardamoms" contain an essential oil with a strongly camphoraceous and cineole-like odor and taste.

Steam-distilling the fruit, Schimmel & Co.[29] years ago obtained 1.12 per cent of a volatile oil with these properties:

Specific Gravity at 15°.......	0.920
Optical Rotation............	−12° 41'
Solubility..................	Clearly soluble in 1 and more vol. of 80% alcohol

The oil contained *cineole*, identified by the preparation of several derivatives.

The odor of the oil differed markedly from that of true cardamom oil (from *Elettaria cardamomum* Maton). The oil offers no practical interest.

OILS OF *Aframomum Angustifolium* (SONN.) K. SCHUM. AND *Amomum Korarima* PEREIRA

There is much confusion in literature [30] regarding the taxonomy and nomenclature of the plant or plants from which the so-called "Madagascar" and "Cameroon Cardamoms" are derived. It is probably the fruit of *Aframomum angustifolium* (Sonn.) K. Schum., syn. *Amomum angustifolium* Sonnerat. (Formerly the fruit was known as *Cardamomum majus* Geiger.) Other species may also be involved. The plants occur in West Africa, East Africa, on the Seychelles Islands, and in Madagascar. The seeds possess an aromatic odor and flavor, slightly resembling that of true cardamom seed, but much more cineole- and cajuput-like.

Several oils have been described in literature:

I. An oil distilled by Schimmel & Co.[31] from East African seed (yield 4.5 per cent).

II. An oil distilled by the same firm [32] from Cameroon cardamoms (yield 2.33 per cent).

III. An oil distilled by Haensel [33] from the fruits of *Amomum korarima* Pereira (the *Cardamomum majus* fruit of Geiger), yield 1.72 per cent.

[29] *Ber. Schimmel & Co.,* April (1897), 48.
[30] Cf. Gildemeister and Hoffmann, "Die Ätherischen Öle," 3d Ed., Vol. II, 451.
[31] *Ber. Schimmel & Co.,* April (1912), 132.
[32] *Ibid.,* October (1897), 10.
[33] *Pharm. Ztg.* **50** (1905), 929. *Chem. Zentr.* (1905), II, 1792.

IV. An oil distilled from the same type of material as III (yield 1.2 per cent), and described by Holmes.[34]

These oils had the following properties:

	I	II	III	IV
Specific Gravity at 15°	0.9017	0.907	0.903	0.9038
Optical Rotation	−16° 50′	−20° 34′	−6° 49′	−3° 0′
Refractive Index at 20°	1.46911	...	...	...
Acid Number	0.4	...	...	3.6
Ester Number	4.2	...	...	22.1
Saponification Number	...	...	50	...
Saponification Number after Acetylation	...	...	107	...

Oils of this type are not produced on a commercial scale, chiefly because their odor is not sufficiently interesting and distinctive.

OIL OF *Aframomum Melegueta* (ROSCOE) K. SCHUM.

Aframomum melegueta (Roscoe) K. Schum. (syn. *Amomum melegueta* Roscoe) is a herbaceous perennial, native to Africa; it was introduced also to the West Indies, probably during the days of the slave trade. The plant occurs on the west coast of Africa from the Congo to the Sierra Leone; hence the name "Pepper Coast" or "Melegueta Coast" for this section of Africa. The shrub furnishes the "Grains of Paradise," "Grana Paradisi," in the trade also referred to as "Guinea Grains," "Melegueta" or "Mallaguetta Pepper." The name probably originates from the old empire of Melle in the upper Niger region, where countless slaves were captured and brought down the river to be sold on the coast.

"Grains of Paradise" in many aspects resemble the seeds of official cardamom. When rubbed between the fingers they give off a faint aromatic odor; their taste is hot and peppery. African natives have always esteemed these grains as a most wholesome spice; in fact, newly captured Negroes were so dependent upon the spice that slaving ships had to carry ample supplies on board. In Europe and America "Grains of Paradise" are now seldom used, except in veterinary preparations, and for the flavoring of certain types of liqueurs and vinegars. Formerly they were employed quite widely as a condiment or spice.

On steam distillation, the grains yield from 0.3 to 0.75 per cent of a yellowish or slightly brown volatile oil, with a spicy, but not too characteristic, odor. An oil distilled by Schimmel & Co.[35] had these properties:

[34] *Perfumery Essential Oil Record* **5** (1914), 302.
[35] *Ber. Schimmel & Co.*, April (1915), 38.

Specific Gravity at 15°........... 0.8970
Optical Rotation................ −3° 10′
Refractive Index at 20°.......... 1.49116
Acid Number................... 2.7
Ester Number.................. 41.2
Ester Number after Acetylation... 63.9
Solubility....................... Not soluble in 90% alcohol. Soluble in 1 vol. of 95% alcohol, slightly turbid with more

When treated with a 3 per cent solution of sodium hydroxide, about 16 per cent of the oil went into solution.

Nothing is known about the chemical composition of the oil. The oil is not produced on a commercial scale.

OIL OF *Zingiber Nigrum* GAERTN.

The so-called "Bitter-Seeded Cardamoms" are believed to be derived from *Zingiber nigrum* Gaertn., a plant growing in moist soil in the shady forests of the Far East, particularly in the province of Kwangtung (southeastern China). It is perhaps identical with *Alpinia allughas* Roscoe, a Cochin-Chinese species, about which there is still some controversy.[36] This evergreen herbaceous perennial is said to resemble *Amomum melegueta* (Roscoe) K. Schum. ("Grains of Paradise"), in having a long, slender, branched rhizome and an erect stem. The fruits are oval or ovate-oblong, dusky brown, pointed at both ends, and contain 25 to 30 seeds 3.5 mm. in length and 2 mm. in diameter.

Steam-distilling bitter-seeded cardamoms, Viehoever and Sung [37] obtained from 2 to 3 per cent of a volatile oil with these properties:

Specific Gravity at 40°...... 0.985
Refractive Index at 30°..... 1.471
Acid Number.............. 7.50
Ester Number............. 79.53
Solubility................. Soluble in 5 vol. of 80% alcohol

Investigating the oil derived from the fruit of *Zingiber nigrum* Gaertn., Karyione and Matsushima [38] isolated a terpene $C_{10}H_{16}$, b. 173°–176°, d_4^{23} 0.8564, $[\alpha]_D^{22} + 1°\ 16′$, and a sesquiterpene $C_{15}H_{24}$, b_{13} 139°–142°, d_4^{24} 0.9233,

---

[36] Cf. Viehoever and Sung, "Common and Oriental Cardamoms," *J. Am. Pharm. Assocn.* **26** (1937), 875, 877, 884.
[37] *Ibid.*
[38] *J. Pharm. Soc. Japan* (1927), 96. *Chem. Zentr.* (1927), II, 2405.

$[\alpha]_D^{30} + 16° 21'$. The oil apparently contained also a sesquiterpene alcohol which, however, was not examined.

The oil derived from bitter-seeded cardamoms has a camphoraceous and bitter taste. It is not produced on a commercial scale.

OIL OF GINGER

Essence de Gingembre   *Aceite Esencial Jengibre*   *Ingweröl*
Oleum Zingiberis

Description, Origin, and History.—Ginger, one of the most important and oldest of spices, consists of the prepared and sun-dried rhizomes of *Zingiber officinale* Roscoe (fam. *Zingiberaceae*). The rhizomes, known in the trade as "hands" or "races," reach the spice trade either with the outer cortical layers intact ("coated," or "unscraped ginger"), or with the outer coating partially or completely removed ("uncoated" or "scraped ginger"). To improve their appearance, some grades of ginger are bleached by various means, e.g., by liming.

Ginger possesses a warm pungent taste and a pleasant odor, hence its wide use as a flavorant in numerous food preparations and beverages, savory dishes, curries, baked goods, confectionery, gingerbread, soups, pickles, and many popular soft drinks. Like most pungent spices, ginger is consumed all over the world, particularly in tropical or warm countries. It dilates the superficial vesicles of the skin, resulting first in a feeling of warmth, then in increased activity of the sweat glands and perspiration, and finally in a marked cooling effect on the skin.

The odor of the rhizomes is caused by the presence of a volatile oil (1 to 3 per cent), which can be isolated by steam distillation of the comminuted spice. The pungent principles, on the other hand, are nonvolatile and must be extracted by percolation with suitable solvents, which procedure yields the so-called oleoresin of ginger (see below). Since the essential oil is contained chiefly in the epidermal tissue, great care should be exercised in the peeling of the rhizomes, and excessive scraping must be avoided; indeed unpeeled ginger constitutes a much more suitable raw material for distillation purposes than peeled ginger.

According to the historical researches of Hoffmann,[1] ginger was certainly

[1] Gildemeister and Hoffmann, "Die Ätherischen Öle," 3d Ed., Vol. I, 119.

known to, and highly esteemed by, the ancient Greeks and Romans who obtained the spice from Arabian traders via the Red Sea. It was introduced to Germany and France in the ninth century, and to England in the tenth century. The Spaniards brought ginger to the West Indies and to Mexico soon after the conquest, and as early as 1547 the spice was exported from Jamaica to Spain. Since the rhizomes can easily be transported in a living state for considerable distances, the plant has been introduced to many tropical and subtropical countries and is now cultivated in several parts of the world, the most important producing regions being Jamaica (B.W.I.), Cochin and Calicut (Malabar Coast, South India), Sierra Leone and Nigeria (West Africa), southern China, and Japan. Of these, Jamaica produces what most connoisseurs consider the finest grade, possessing the most delicate aroma and flavor. The Cochin quality ranks perhaps second. It exhibits a characteristic lemon-like by-note, for which reason some experts prefer the Cochin ginger even to that from Jamaica. As a matter of fact, Cochin ginger often brings a somewhat higher price on the world market than the Jamaica quality. West African ginger is usually considered to rank third. Lately, however, the appearance of the West African spice seems to have improved. Nevertheless, compared with Jamaica and Cochin ginger, the West African exhibits a slightly camphoraceous, somewhat coarser odor and flavor, and darker color. Of all ginger grades, it possesses the greatest pungency and gives the highest yield of essential oil—hence its present wide use for the extraction of oleoresin and for the distillation of oil. Moreover, the African ginger is usually lower priced than the other two grades.

Chinese ginger, produced in the northwestern part of Kwangtung (near Shiuking and Szechwan) and in the central provinces, is usually not exported as dried spice, but preserved in sugar syrup; hence it cannot be used for distillation or extraction.

Japanese ginger possesses a certain pungency, but lacks the characteristic ginger aroma, in fact, the plant is not true ginger (*Zingiber officinale* Roscoe), but *Zingiber mioga* Roscoe. The dried rhizomes yield an essential oil different from that of true ginger.

There are other countries producing ginger—Malaya and Indonesia among them—but compared with the countries mentioned previously they are not important and may be omitted in this discussion. Attempts to grow ginger have been made recently in Cuba and on several West Indian islands; the principal obstacle has been the difficulty of drying the rhizomes in the humid climate of the producing regions. In spite of this drawback, however, small quantities have been produced on some of the islands.

Botany, Planting, Cultivating, and Harvesting.—*Zingiber officinale* Roscoe is a herbaceous plant with a perennial horizontal tuberous and creeping rhizome, and an annual stem which rises 2 to 3 ft. in height. The plant does not exist in many forms or varieties, because it is usually propagated from cuttings (i.e., by asexual reproduction), and only very seldom, if at all, by seed.

According to Ridley,[2] ginger requires a tropical or subtropical climate, where the temperature is high for at least part of the year. Bright sunshine, as well as heavy rains, are necessary. In Jamaica, e.g., the annual rainfall averages 88 in.

The most suitable soil consists of a light, free, sandy loam. Stiff clays or coarse sands are not conducive to the cultivation of ginger. Sandy soils are apt to pack after a heavy rain and to become too dense for the rhizomes to develop. Wet swampy ground does not suit the plant at all, and areas liable to floods should be avoided. If the ground becomes too dry during the dry season, a system of irrigation will be needed. Swampy terrain may be utilized by systematic and careful drainage. The ideal ground for ginger is good garden soil, rich in humus, light and well worked, friable and fairly dry.

As regards elevation, in India the plant is grown successfully both in the low country and up to an altitude of 4,000 to 5,000 ft.

Prior to planting, the soil must be broken up finely with a hoe or plow, and if possible harrowed afterward. Ginger is always grown from cuttings of the rhizome. The points of the rhizome each contain a bud ("eye"), from which the plant grows. Frequently a portion of the crop is retained for planting stock. The cuttings should be from 1 to 2 in. long. Robles, Cernuda, and Loustalot[3] found that storage of planting stock of ginger under coarse river sand reduces losses from rot.

According to Ridley,[4] planting in India and the West Indies generally takes place in March and April—though occasionally until June—depending upon the time of onset of the wet season. In Jamaica the planting process consists of burying the cuttings in trenches or holes a few inches below the surface and about 1 ft. apart. The small grower simply digs a hole in a convenient spot. The thrifty planter first burns over his plot to destroy weeds and insects, plows it over, and lays it out in beds and trenches. Much of the ginger in Jamaica is cultivated as a garden plant, along with bananas, chillies, etc., in small lots. Best results are obtained by careful selection of the planting stock and by pulverizing and thoroughly breaking up the soil. Ginger being a soil-exhausting stock, a system of crop rotation should be pursued. Alemar and Pennock[5] found that the use of straw

[2] "Spices," London, Macmillan & Company, Ltd. (1912), 393.
[3] *U. S. Dept. Agr., Fed. Expt. Sta. Puerto Rico, Rept.* 1947 (1948), 73
[4] "Spices," London, Macmillan & Company, Ltd. (1912), 396.
[5] *U. S. Dept. Agr., Fed. Expt. Sta. Puerto Rico, Rept.* 1941 (1942), 11.

mulch effectively checked soil erosion. Alemar [6] also tested well decomposed sugar cane filter press cake, or cachaza, and well decomposed leaves of various tree species as mulches for the growing of ginger on clay loam soil containing a considerable amount of sand. He reported a doubling in the yield of rhizomes. Robles [7] found that the use of manure favored the yield of green ginger. Robles, Cernuda and Loustalot [8] noted that the application of composted manures resulted in significant increases in yields of fresh ginger. Moreover, higher yields of fresh and dry ginger were produced under shade. The results of these experiments indicate that ginger should be grown under partial shade to obtain the best yields of both fresh and dry ginger. The yield of ginger seven months after planting was double that harvested four months after planting.

Under favorable conditions the ginger plant appears aboveground about ten to fifteen days after planting, but as much as two months may be required before it begins to show. In Jamaica, planted ginger can be dug in December or January, or until March, whereas ratoon ginger is lifted between March and December. (The planters in Jamaica distinguish between "plantation ginger" and "ratoon ginger." Planted ginger gives the best results, and indeed represents the best method of cultivation. When the stock is left in the ground to throw up fresh stems, and produce fresh rhizomes, it is known as "ratoon ginger"—Ridley.)

The rhizomes can be lifted from the ground (harvested) when the green leafy stems turn yellow and wither; this usually happens after the flowering period. (However, the plant does not always produce flowers; indeed, in some regions flowers are very rarely seen.) A belated harvest results in the development of tough and fibrous rhizomes. In general, the rhizomes mature nine to ten months after planting. They are lifted from the soil by a single thrust of the fork, care being taken not to damage them. After removal from the ground, the rhizomes are piled into heaps, the roots broken off, and soil and other adhering matter removed. This must be done quickly; otherwise the rhizomes will dry with the roots and earth still adhering, and they will not become white. In Jamaica, the rhizomes are placed in water immediately after cleaning and are then ready for peeling.

Preparation for the Market.—There are two general types of ginger, viz., the fresh green ginger used for the preparation of candied ginger [9] (in sugar syrup), and the dried or cured ginger employed in the spice trade, for the preparation of extracts and oleoresins, and for the distillation of its volatile oil. The following will deal only with dried or cured ginger.

[6] *Ibid.* 1943 (1944), 31.
[7] *Ibid.* 1946 (1947), 49.
[8] *Ibid.* 1947 (1948), 73.
[9] For details see Cernuda, *ibid.* 1948 (1949), 24. Cf. Ridley, "Spices," London, Macmillan & Company, Ltd. (1912), 417.

The dried rhizomes contain a brownish cork beneath the epidermis, and a resinous, almost horny cortical layer. The parenchyma inside consists of a whitish mass, the cells being filled with starch. Minute sacs or glands containing essential oil and resin are scattered throughout the rhizomes, but are particularly numerous in the epidermal tissue. The dried rhizomes should be firm and full, free from any trace of mildew. They are rather brittle and crack easily; the presence of broken pieces lowers the value of a lot.

Commercial grades are known as "scraped" ("decorticated" or "uncoated"), and "coated ginger." Great care has to be exercised in the peeling operation because the essential oil and resin-bearing cells are located chiefly in the epidermal tissue. Excessive scraping depreciates the quality of the spice substantially. Scraped ginger is a grade from which the cortex has been removed partly or entirely. This is always the case with Jamaica ginger, and often with Cochin ginger. In "coated ginger," on the other hand, a good portion or all of the outer layer remains attached to the dried rhizome. African ginger is usually "coated," Calicut and Cochin ginger occasionally. In addition, there are bleached and unbleached gingers, the bleaching being accomplished by covering the rhizomes with a coat of lime or chalk. Liming has the effect of improving color and appearance, and of protecting the spice from mildew and attacks of weevils and other pests. In some producing regions the freshly dug rhizomes are parboiled (scalded) in water for 10 to 15 min. in order to destroy their vitality. The cleaned rhizomes ("hands" or "races") are then dried in the sun, without peeling. This procedure results in "black ginger," an unscraped, "coated" type, possessing a dark, ash-colored, wrinkled epidermis. In other producing regions sun-drying is supplemented by drying on trays, within huts, above a smouldering fire.

In Jamaica, which island produces the finest grade of peeled or "uncoated" ginger, the freshly dug, carefully washed and cleaned rhizomes are scraped by means of a special knife, with a narrow-edged blade riveted to the handle. This process requires care on the part of the worker. More skillful workers peel the coat between the toes of the rhizomes, while children or less experienced persons peel the less intricate parts. (Cernuda [10] tried blanching in a 10 per cent lye solution at 100° C. for 3 min. After removal from the lye solution the rhizomes bleached to a pale, whitish color; they had little ginger flavor left, but were still pungent. Peeling of ginger can be easily accomplished by immersion in lye, but whether the alkali can be completely removed is questionable.) After peeling, the "races" or "hands" are thrown into water and washed. The purer the

---

[10] *U. S. Dept. Agr., Fed. Expt. Sta. Puerto Rico, Rept.* 1948 (1949), 24.

water, and the more used, the whiter the ginger becomes (Ridley). The carefully washed "hands" are dried in the sun; this is usually done on barbecues, such as serve for the drying of coffee. Small producers use a framework of sticks, with boards, or palm or banana leaves, laid upon it; or a few large banana or palm leaves may be placed upon the ground and the ginger spread on them. The rhizomes are put out early in the morning, turned over at noon, and taken indoors in the evening. Unless proper care is exercised, they are apt to turn moldy in rainy or cloudy weather. Thorough drying requires six to eight days. In general, drying of ginger presents a delicate and difficult problem; in a very moist climate or during long spells of rain planters may lose their crops. Efforts have therefore been made to supplement sun-drying with artificial drying, and to dry the rhizomes without previous removal of the coat, which would save a considerable amount of labor. However, these experiments have resulted in rhizomes of a dark color and a flavor inferior to those of the sun-dried spice. Since Jamaica ginger is known and highly esteemed for its light color and fine flavor, these experiments must be regarded as unsuccessful. Nevertheless, further work along these lines, and with modern fruit-drying equipment, appears advisable.

On drying, fully developed rhizomes lose almost 70 per cent of their weight (about 40 per cent of this loss consists of surface moisture). The moisture content of dried ginger should not exceed 10 per cent, but is occasionally as high as 25 per cent in poor grades.

Production of Ginger in Jamaica.—In Jamaica (B.W.I.) ginger is grown at altitudes ranging from 2,500 to 3,500 ft., centers of production lying within a radius of 25 miles of Christiana, in the hills of Manchester Parish, north of Mandeville. A small, secondary crop of poorer quality originates from Hanover and Westmoreland. Harvest extends from January to May, with the peak in February and March. The roots are collected, peeled by hand, and then dried in the sun. In the mountain area frequent showers often render completion of the drying process difficult. Locally, ginger is classified into three grades, according to "boldness" and fullness of the "hands." The rhizomes of Jamaica ginger are laterally compressed, irregularly branched, from 4 to 16 cm. long and from 4 to 20 mm. thick, with the cork entirely removed. The external color ranges from weak orange to yellowish-orange. In Jamaica, ginger is not bleached artificially. "Plantation ginger" is of better quality than "ratoon ginger," the latter being smaller, more fibrous, and of darker color than the third grade of "plantation ginger." Some ratoon ginger is harvested after a dry season. It can be used for extraction or grinding.

The total production of ginger in Jamaica per year averages 3.5 million pounds. Because of the danger of worm infestation, growers and dealers do not store the spice longer than three to four months. On the other hand, high temperatures in summer threaten loss of weight by excessive evaporation. Hence exporters try to prevent overstocking and usually export the entire yearly production.

Production of Ginger on the Malabar Coast.—There are two types of ginger produced on India's Malabar Coast, viz., the Calicut and the Cochin ginger. Both are characterized by their lemon-like odor and flavor, the Calicut spice being even more "lemony" than the Cochin product. Some experts consider ginger from the Malabar Coast the finest in the world. Manufacturers of ginger ale, particularly in Great Britain, use large quantities of this type, because it imparts not only pungency, but also an agreeable lemon-like flavor to beverages.

Calicut ginger is very bold, less fibrous, and more starchy than No. 1 bold Jamaica ginger, and usually well dried, and of creamy color when scraped.

Cochin ginger reaches the market bleached or unbleached; when bleached and scraped, it resembles the Jamaica spice, except for its lemon-like top note. The smaller pieces are simply sun-dried, without peeling.

According to Sennhauser,[11] on the Malabar Coast the freshly unearthed rhizomes are scraped by means of short, sharp knives, and dried in the sun for about four days. (Actual peeling has never been practiced in India.) The rhizomes so prepared enter the market as unbleached ginger. Exporters dry the spice for another two days and subsequently assort it; this process, called "garbling," consists of removing pieces that are too light, or the skin of which has not been removed properly. All ginger is then washed in tanks and properly dried in the sun on barbecues for two or more days, depending upon the strength of the sunlight. This yields the unassorted quality.

Bleached ginger is prepared as follows: ungarbled, rough rhizomes are washed and kept in air-tight rooms for a night. The following day the ginger is dipped in an aqueous solution of lime and placed on an iron frame above a kiln. Beneath the iron frame are iron pans containing sulfur. On heating of the sulfur, the fumes pass through perforated baskets holding the limed ginger. The process goes on for three days and then the rhizomes are dried. Bleached and limed ginger is intended chiefly for local consumption in India. This method of preparation to some extent helps to preserve the spice during the wet monsoon season.

On the Malabar Coast the terms "peeled," "coated" or "uncoated" are not used in connection with ginger, the most popular classifications now being:

[11] Private communication from Mr. E. H. Sennhauser, Volkart Brothers, Inc., New York.

Cochin Ginger, rough, washed and unassorted.
Cochin Ginger, bleached, unassorted.
Calicut Ginger, rough, washed, and unassorted.
Calicut Ginger, rough, bleached and unassorted.

Prior to World War II ginger was exported from India, chiefly to Germany and Scandinavian countries as grades A, B, C, D, and T. Grades A to C inclusive, were so-called "scraped" ginger, the grading being done according to size. Grade C thus consisted of very small pieces. Some of the thin parts of the rhizomes were cut off entirely during the scraping process; they were called "nibs" and sold as "Calicut Ginger D." The scrapings themselves entered the trade as "Calicut Ginger T." Calicut ginger A, B and C, after being bleached with sulfur fumes, were washed and bleached with lime; grades D and T were not limed. Since scraping was never practiced on the lower Malabar Coast, the Cochin D ginger did not consist of actual cuttings from larger pieces, but simply of small pieces of unscraped, unassorted and washed Cochin ginger. An average lot of ginger yielded from 50 to 60 per cent of A, B and C grades, about 43 per cent of D, and 7 per cent of T grade.

The following quantities of ginger, in tons, have been exported by steamer and rail (1934 to 1948 inclusive):

	Cochin	*Calicut*
1934	3,900	1,250
1935	1,400	600
1936	1,750	450
1937	3,200	275
1938	4,200	900
1939	5,500	1,000
1940	3,200	1,000
1941	3,900	1,600
1942	4,900	1,600
1943	...	...
1944	...	...
1945	...	...
1946	3,443	...
1947	4,122	873
1948	3,434	...

Production of Ginger in West Africa.—There are two principal qualities of West African ginger, viz., the better grade which originates from Sierra Leone, and the lower quality coming from Nigeria.

Annual production in Sierra Leone now averages 4.5 million pounds; at one time (1938) it reached more than 6 million pounds. Agricultural experts estimate that an additional million pounds (or a total of 5.5 million pounds) can and will eventually be grown in Sierra Leone, but this is probably the largest quantity the export market can absorb. A large part, up to 75 per cent, of the present production goes to the United States. Over 3,000 acres are now devoted to the production of ginger, areas of cultivation being centered in the southern provinces along the railway line, to facilitate transportation.

As regards Nigerian ginger, the crop is small, only about 250,000 lb. having been exported in 1947 and in 1948. In Nigeria ginger is grown on small, scattered holdings. The crop is marketed from January to February in only two market places in the entire country; both are located in Zaria Province, the towns being Katchia and Zonkwa. The markets are opened only twice, usually for a period of from seven to ten days.[12]

African ginger has a very pungent flavor; its odor is slightly camphoraceous, somewhat coarser than that of the Jamaica or Cochin spice. The rhizomes usually contain the skin (coated quality); or the cork may be partly removed on the flattened sides, leaving light brownish areas. The portions containing cork are longitudinally or reticulately wrinkled, and grayish brown. Internally, the color ranges from light yellow to brown.

According to Landes,[13] ginger from the Sierra Leone can easily be recognized by its dark color and unbleached condition. It is smaller than the third grade of Jamaica ginger, and usually much darker than even the Jamaica "ratoon ginger." Prices of the Sierra Leone ginger have for a long time been about 10 or 15 per cent lower than those of the Jamaica product, but lately this situation has changed somewhat because of the large demand for the Sierra Leone spice. The latter is popular for distillation purposes, for preparation of oleoresins, and for flavoring of cattle feed.

Nigerian ginger, although of fairly good quality, is somewhat too fibrous for grinding. The spice trade, and distillers in the United States, use only small quantities. Because of its low grade the Nigerian product is shipped largely to Europe.

Distillation and Yield of Ginger Oil.—As has been pointed out on several occasions, the most economical raw material for isolation of the essential oil by distillation is dried African ginger. It is not only lower priced than the other qualities (Jamaica and Cochin ginger), but also gives a higher yield of oil. In general the yield varies from 1.5 to 3 per cent of volatile oil, averaging 2 per cent. Prior to distillation the dry material should be comminuted, then charged into the stills immediately. Care must be taken

[12] Information through the courtesy of Mr. T. W. Delahanty, Chief, Consumers Merchandise Branch, Office of International Trade, United States Department of Commerce, Washington, D. C.

[13] Private communication from Dr. Karl H. Landes, New York.

to distribute the mass evenly on several trays within the still. Distillation is carried out with direct steam; depending upon the charge in the still and the steam pressure applied, distillation of one charge may require up to 20 hr. The distillation waters may require cohobation.

Peelings and shavings of the rhizomes constitute an excellent material for distillation, giving a good yield of oil, provided distillation is carried out immediately after peeling.

Physicochemical Properties of the Essential Oil.—The volatile oil derived from dried ginger is a mobile (viscous on aging!), greenish to yellowish liquid, possessing the characteristic aromatic odor, but not the pungent flavor ("bite"), of the spice. The odor of the oil is quite lasting.

Gildemeister and Hoffmann [14] reported these properties for oil of ginger:

Specific Gravity at 15°.......... 0.877 to 0.886; oils of lower and higher specific gravity have been observed, however

Optical Rotation............... $-26°\ 0'$ to $-50°\ 0'$
Lower rotations have been observed, however. For example, an oil distilled from old roots that had been stored for a long time, exhibited d_{15} 0.8924 and α_D $-16°\ 58'$

Refractive Index at 20°.......... 1.489 to 1.494
Acid Number.................. Up to 2
Ester Number................. Up to 15
Ester Number after Acetylation... 24 to 50
Solubility..................... Only sparingly soluble in alcohol. Up to 7 vol. of 95% alcohol are required for solution, which is not always clear. In 90% alcohol the oils are generally, but not always, completely soluble

Genuine oils of ginger distilled by Fritzsche Brothers, Inc., New York, and under the author's supervision in Southern France (Seillans, Var) exhibited properties varying within the following limits:

Specific Gravity at 15°/15°....... 0.876 to 0.884, occasionally as high as 0.896
Optical Rotation............... $-30°\ 10'$ to $-44°\ 20'$, occasionally as low as $-23°\ 58'$
Refractive Index at 20°.......... 1.4876 to 1.4917
Acid Number.................. 1.1 to 3.7
Saponification Number.......... 0.9 to 11.2, occasionally as high as 20.6

Clevenger, Kenworthy and Lubell [15] reported these properties for ginger oils distilled experimentally from coarsely ground African and Jamaica ginger:

[14] "Die Ätherischen Öle," 3d Ed., Vol. II, 440.
[15] *J. Assocn. Official Agr. Chem.* **20** (1937), 410.

	Clevenger		Kenworthy *		Lubell	
	1	*2*	*1*	*2*	*1*	*2*
Yield of Oil (cc. per 100 g. of spice)..	3.25	1.39	2.46	1.22	3.30	1.37
Specific Gravity at 25°/25°........	0.879	0.884	0.888	0.884	0.879	0.884
Optical Rotation at 25°............	−42° 12'	−32° 12'	−21° 48'	−16° 36'	−41° 6'	−35° 30'
Refractive Index at 20°............	1.491	1.493	1.494	1.495	1.492	1.494
Acid Number.....	2.93	4.29	2.6	4.8	3.5	3.46
Ester Number.....	11.02	18.7	35.3	25.3	16.03	17.7

* Kenworthy carried out his assays ten months after the ginger had been ground and stored in sealed glass jars. According to Clevenger, the results obtained by all three of the workers mentioned can be considered satisfactory. The variations in data reported are probably accounted for by the differences in time elapsed between grinding of the ginger and the assay of the material.

The Imperial Institute [16] in London steam-distilled dried ginger *peelings* from Sierra Leone and obtained 4.0 per cent of essential oil (I); Varier [17] submitted air-dried scrapings of northern Travancore ginger to water distillation and obtained 0.8 per cent of oil (II). The two oils had the following properties:

	I	II
Specific Gravity.....................	d_{15}^{15} 0.881	d_{30} 0.8905
Optical Rotation.....................	−43° 45'	−5° 12'
Refractive Index.....................	n_D^{20} 1.492	n_D^{30} 1.4859
Acid Number........................	1.5	0.9
Ester Number........................	2.9	6.1
Ester Number after Acetylation........	33.1	72.2
Solubility in 95% Alcohol at 15°........	Soluble in 4 vol. and more	...

In various fractions of the oil which he investigated, Varier [18] noted the presence of *camphene, β-phellandrene,* and *zingiberene.*

Chemical Composition of Ginger Oil.—The early investigations of ginger oil by Papousek,[19] and Thresh [20] require no discussion since they yielded no practical results. More recently the volatile oil derived by steam distillation of dried ginger rhizomes, and its chief sesquiterpenic constituent (zingiberene), have been examined quite thoroughly by a number of researchers, among them Bertrand and Walbaum,[21] von Soden and Rojahn,[22]

[16] *Bull. Imp. Inst.* **24** (1926), 651.
[17] *Current Science* **14** (1945), 322. *Chem. Abstracts* **40** (1946), 5206.
[18] *Ibid.*
[19] *Sitzb. Akad. Wiss. Wien* **9** (1852), 315. *Liebigs Ann.* **84** (1852), 352.
[20] *Pharm. J.* [3], **12** (1881), 243.
[21] *J. prakt. Chem.* [2], **49** (1894), 18.
[22] *Pharm. Ztg.* **45** (1900), 414.

Schimmel & Co.,[23] Dodge,[24] Brooks,[25] Ruzicka and van Veen,[26] Soffer and his collaborators,[27] and still more recently by Eschenmoser and Schinz.[28] The following compounds have been reported as constituents of the oil:

n-Decylaldehyde. In the forerun of the oil, Dodge noted the presence of an optically inactive aliphatic aldehyde d_{15} 0.828, which he believed to be *n*-decanal; it gave an oxime m. 63°, and a semicarbazone m. 98°. The odor of the aldehyde was reminiscent of orange peel oil. Since the melting points of the derivatives obtained by Dodge were not identical with those of pure *n*-decanal, it is possible that Dodge had isolated a mixture of *n*-decanal and *n*-nonanal.

n-Nonylaldehyde. That *n*-nonanal actually occurs in the oil was shortly afterward reported by Brooks.

d-Camphene. The terpene fraction b. 155°–165° of the oil exhibits dextrorotation (α_D +63° 13′), whereas the oil itself is laevorotatory (see above). In the terpene fraction Bertram and Walbaum identified *d*-camphene by hydration to isoborneol m. 212°, and by preparation of the bromal compound m. 71°.

d-β-Phellandrene. In the fraction b. 170° of the oil Bertram and Walbaum identified β-phellandrene by means of its nitrite m. 102°. More recently West [29] reported the presence of *d*-β-phellandrene in African ginger oil. West characterized this terpene by preparation of the nitrosochloride and the nitrosite; oxidation yielded 4-isopropyl-2-cyclohexen-1-one.

Methyl Heptenone. Reported as a constituent of the oil by Brooks.

Cineole. Identified by Schimmel & Co.; iodol compound m. 112°.

d-Borneol. Also identified by Schimmel & Co. (m. 204°, acid bornyl phthalate m. 164°), and later confirmed by Brooks.

Geraniol(?). Schimmel & Co. noted that the oil contains another alcohol, perhaps geraniol.

Linaloöl. Reported as a constituent of the oil by Brooks.

Acetates and Caprylates. According to Brooks, the alcohols occur in the oil partly free, partly as acetates and caprylates.

Citral. Identified by Schimmel & Co., who prepared the β-naphthocinchoninic acid compound m. 197°. Presence later confirmed by Brooks. Oils distilled from Cochin ginger appear to contain more citral than oils derived from African or Jamaica gingers.

Chavicol(?). According to Brooks, the oil also contains a phenol, perhaps chavicol.

Zingiberene. This sesquiterpene $C_{15}H_{24}$, the principal constituent of ginger oil, has been the subject of several investigations. For details the reader is referred to Vol. II of the present work, p. 87. Note, however, that the structural formula given in

[23] *Ber. Schimmel & Co.*, October (1905), 34.
[24] *8th Intern. Congr. Appl. Chem.* **6** (1912), 77 (Washington and New York).
[25] *J. Am. Chem. Soc.* **38** (1916), 430.
[26] *Liebigs Ann.* **468** (1929), 143.
[27] *J. Am. Chem. Soc.* **66** (1944), 1520.
[28] *Helv. Chim. Acta* **33**, I (1950), 171.
[29] *J. Soc. Chem. Ind.* **58** (1939), 123T.

Vol. II (and suggested by Soffer et al.[30]) has recently been declared erroneous by Eschenmoser and Schinz,[31] who have postulated the following configuration for zingiberene:

Zingiberol. According to Brooks, this sesquiterpene alcohol $C_{15}H_{26}O$ imparts to the oil its mild characteristic odor. From 150 g. of terpeneless and sesquiterpeneless ginger oil, Brooks isolated 24 g. of zingiberol $b_{14.5}$ 154°–157°. On heating, zingiberol lost water with simultaneous formation of a sesquiterpene $C_{15}H_{24}$, b. 255°–257°, which was either zingiberene or isozingiberene (cf. Vol. II of the present work, pp. 87, 91 and 261).

Use.—Oil of ginger is employed widely in the flavoring of all kinds of food products, particularly baked goods, confectionery, and spicy table sauces. It finds limited use also in perfumery, where it imparts an individual note to compositions of the oriental type.

Oleoresin Ginger

As with black pepper, the volatile (essential) oil derived by steam distillation of ginger represents only the aromatic, odorous constituents of the spice; it does not contain the nonvolatile, pungent principles for which ginger is so highly esteemed. To obtain all of these constituents in concentrated form, ground ginger is percolated with volatile solvents—acetone, alcohol, or ether. Concentration of the solution and careful removal of the solvent *in vacuo* yield the so-called oleoresin of ginger, in which the full pungency of the spice is preserved. The quantitative composition of the oleoresin depends upon the solvent used. Oleoresin of ginger (commercially known also as "Gingerin") generally contains these compounds:

1. *"Gingerol"* and *"Zingerone."* About seventy years ago, Thresh [32] isolated the pungent principles from ginger and named the mixture of substances *"Gingerol."* In his work, he encountered considerable difficulty in extracting gingerol in pure form, because the compound is easily affected by various reagents. (On *heating* ginger with alkaline hydroxides, the pungency of the spice, unlike that of capsicum, is destroyed.) Years later,

---

[30] *J. Am. Chem. Soc.* **66** (1944), 1520.   [32] *Pharm. J.* [3], **12** (1882), 721.
[31] *Helv. Chim. Acta* **33**, I (1950), 171.

Nomura [33] extracted ginger with ether and, on subsequent treatment with alkali, obtained a substance with a very pungent flavor, which appeared to be a ketone and which he named *"Zingerone."* Almost simultaneously Lapworth, Pearson and Royle,[34] and Nelson [35] found that zingerone occurs in ginger not as free ketone, but in the form of compounds in which the ketone is a product of condensation (in molecular proportions) with saturated aliphatic aldehydes, principally enanthaldehyde (*n*-heptanal). Gingerol, a very pungent yellow oil, appears to be a mixture of homologues of this type:

$$\begin{array}{c} HO \\ \diagdown \\ \diagup \\ H_3CO \end{array} C_6H_3 \cdot CH_2 \cdot CH_2 \cdot CO \cdot CH_2 \cdot \overset{\overset{\displaystyle OH}{|}}{CH} \cdot (CH_2)_n \cdot CH_3$$

In the chief homologue n seems to be 5, in other homologues n is 3 or 4.

On treatment of gingerol with alkali (as in the method of extraction employed by Nomura—see above) decomposition takes place, and free ketone (zingerone) is formed.

According to Nomura,[36] zingerone $C_{11}H_{14}O_3$, m. 40°–41°, possesses the following structural formula:

$$CH_2 \cdot CH_2 \cdot CO \cdot CH_3$$

1-[4-Hydroxy-3-methoxyphenyl]-3-butanone

Zingerone has been synthesized by Mannich and Merz.[37]

2. *"Shogaol."* Nomura and his collaborators [38] isolated another pungent substance from ginger, which they named *"Shogaol"* (from the Japanese term "Shoga" for ginger). Shogaol $C_{17}H_{24}O_3$, a homologue of zingerone, has this structural formula:

[33] *J. Chem. Soc.* **111** (1917), 769.
[34] *Ibid.*, 777.
[35] *J. Am. Chem. Soc.* **39** (1917), 1466.
[36] *Science Repts. Tôhoku Imp. Univ.* [1], **6** (1917), 41.
[37] *Arch. Pharm.* **265** (1927), 15. Cf. Cotton, U. S. Patent 2,381,210, August 7, 1945. *Chem. Abstracts* **40** (1946), 362.
[38] *Science Repts. Tôhoku Imp. Univ.* [1], **7** (1918), 67; [1], **18** (1929), 661. Cf. *Proc. Imp. Acad. Tokyo* **3** (1927), 159.

$$\begin{array}{c} HO \\ \diagdown \\ C_6H_3\cdot CH_2\cdot CH_2\cdot CO\cdot CH{=}CH\cdot (CH_2)_4\cdot CH_3 \\ \diagup \\ H_3CO \end{array}$$

[4-Hydroxy-3-methoxyphenyl]-ethyl-*n*-α,β-heptenyl Ketone

Shogaol can be synthesized from zingerone and hexaldehyde.

3. Aside from the pungent principles mentioned above, oleoresin of ginger also contains the volatile oil of ginger.

4. Unidentified resins are also contained in the oleoresin.

SUGGESTED ADDITIONAL LITERATURE

R. Tzucker and C. B. Jordan, "A Study of the Assay of Ginger," *J. Am. Pharm. Assocn., Sci. Ed.* **29**, No. 6 (1940), 265.

Morris B. Jacobs, "Pungent Compounds Used in Flavoring," *Am. Perfumer* **48** (July 1946), 60.

THE CURCUMA OILS

OIL OF *Curcuma Longa* L.

There are several species of the genus *Curcuma* (fam. *Zingiberaceae*), native to southeastern Asia, the tuberous roots of which contain a valuable yellow coloring principle (curcumine). One of the most important of these species is the common curcuma, *Curcuma longa* L., the dried tubers of which are used also for the flavoring of spicy native dishes. The plant is cultivated in India (Madras, Bombay, Bengal),[1] southern China, Formosa, Java, and the Philippines. On steam distillation the dried tubers yield from 1.3 to 5.5 per cent of an essential oil. (Rupe and collaborators[2] claimed that steam distillation of the dried tubers gives at the most 2 per cent of oil; Dieterle and Kaiser[3] reported a yield of 3.5 per cent from the rhizomes of *Curcuma domestica* Valeton.) The essential oil referred to in literature and in the trade as "Oil of Curcuma" is produced from *Curcuma longa* L.

[1] Re cultivation in India, see *Chemist Druggist* **88** (1916), 953.
[2] *Helv. Chim. Acta* **17** (1934), 372.
[3] *Arch. Pharm.* **270** (1932), 413.

Physicochemical Properties.—The volatile oil derived from the tubers of *Curcuma longa* L. is an orange-yellow, occasionally slightly fluorescent liquid, with an odor reminiscent of the tubers.

According to Gildemeister and Hoffmann,[4] the oil has these properties:

Specific Gravity at 15°............	0.938 to 0.967
Optical Rotation.................	−13° 0′ to −25° 0′, or dextrorotatory up to +28°[5]
Refractive Index at 20°...........	1.512 to 1.517
Acid Number....................	0.6 to 3.1
Ester Number...................	6.5 to 16
Ester Number after Acetylation....	28 to 53
Solubility.......................	Soluble in 4 to 5 vol. of 80% alcohol; soluble in 0.5 to 1 vol. of 90% alcohol

Chemical Composition.—Oil of *Curcuma longa* has been the subject of numerous chemical investigations. The early works of Bolley, Suida and Dembe,[6] Kachler,[7] and Flückiger [8] require no discussion here because they gave no tangible results. Jackson and collaborators [9] isolated as chief constituent a substance of the empirical molecular formula $C_{13}H_{18}O$ or $C_{14}H_{20}O$, $b_{11\text{-}12}$ 158°–163°, $[\alpha]_D$ + 24° 35′, which they believed to be an alcohol and named turmerol. Later Rupe and collaborators [10] investigated turmerol more thoroughly but could not isolate it in pure form. Whenever they treated this compound with metallic sodium, hydrogen chloride or phosphorus trichloride, they obtained curcumone, a ketone $C_{12}H_{16}O$, which Rupe and Wiederkehr [11] succeeded in synthesizing.

(Curcumone is a slightly viscous liquid with a sharp odor reminiscent of ginger. It has these properties: $b_{8\text{-}11}$ 119°–122°, d_{20} 0.9566, $[\alpha]_D^{20}$ +80° 33′; phenylhydrazone m. 92°, *p*-bromophenylhydrazone m. 71°, benzylidenecurcumone m. 106°, piperonalcurcumone m. 130°. Regarding the structural formula of curcumone, see Vol. II of the present work, p. 450).

It should be emphasized that curcumone is not a natural constituent of oil of curcuma, but an artifact obtained by treating the main fraction b_{11} 155°–158° of the oil with alkali or acids.

Real progress in the elucidation of the chemical composition of curcuma

---

[4] "Die Ätherischen Öle," 3d Ed., Vol. II, 425.

[5] Bacon [*Philippine J. Sci.* **5** (1910), A, 262] noted that on aging of the oils the optical rotation of dextrorotatory oils increases, whereas that of laevorotatory oils decreases.

[6] *J. prakt. Chem.* **103** (1868), 474.

[7] *Ber.* **3** (1870), 713.

[8] *Ber.* **9** (1876), 470.

[9] *Am. Chem. J.* **4** (1882), 368; **18** (1896), 111. *Pharm. J.* [3], **13**, (1883), 839.

[10] *Ber.* **40** (1907), 4909; **42** (1909), 2515; **44** (1911), 584, 1218. Cf. Luksch, "Über Curcumaöl," Inaugural Dissertation, Basel (1906).

[11] *Helv. Chim. Acta* **7** (1924), 654.

oil was made in 1934, when Rupe, Clar, Pfau and Plattner,[12] and Kelkar and Rao [13] found that more than 50 per cent of the oil consists of turmerone (see below). These two groups of researchers reported the presence of the following compounds in the oil derived from the tubers of *Curcuma longa* L.:

d-α-Phellandrene. In the lowest boiling fraction; first identified by Schimmel & Co.,[14] later confirmed by Luksch.[15] Nitrite m. 108°. Kelkar and Rao reported that the oil contains about 1 per cent of *d-α*-phellandrene.

d-Sabinene. About 0.6 per cent (Kelkar and Rao).

Cineole. About 1 per cent (Kelkar and Rao).

Borneol. About 0.5 per cent (Kelkar and Rao).

Zingiberene. About 25 per cent (Kelkar and Rao).

Sesquiterpene Alcohols(?). Rupe et al.[16] found that 5.9 per cent of the oil consists of sesquiterpene alcohols. Kelkar and Rao,[17] on their part, reported that the oil contains approximately 9 per cent of tertiary alcohols $C_{15}H_{24}O$ or $C_{15}H_{20}O$ which, on boiling with alkali, yield curcumone (see above).

Turmerone and *ar*-Turmerone. According to the two last-named authors, about 58 per cent of the oil is composed of a mixture of sesquiterpene ketones $C_{15}H_{22}O$ or $C_{15}H_{20}O$, the so-called turmerones.

Examining the turmerones more thoroughly, Rupe and collaborators found that the chief fraction (about 59 per cent) of the oil consists of 50 per cent of turmerone (an alicyclic sesquiterpene ketone $C_{15}H_{22}O$) and of 40 per cent of *ar*-turmerone (an aromatic ketone $C_{15}H_{20}O$). Cf. Vol. II of the present work, p. 450.

ar-Turmerone has been prepared synthetically by Colonge and Chambion,[18] and by Mukherjee.[19]

α- and *γ*-Atlantone. Rupe and collaborators reported the presence of small quantities of *α*- and *γ*-atlantone in the oil. (Cf. Vol. II of this work, p. 448.)

Use.—Oil of curcuma is used in the flavoring of spicy food products and, to a smaller extent, in perfumes of heavy oriental character. However, the oil appears to have lost its former importance, and today only small quantities are produced.

Oil of *Curcuma Aromatica* Salisb.

The yellow rhizomes of *Curcuma aromatica* Salisb. (fam. *Zingiberaceae*), known in India as "wild yellow root" are cultivated in Mysore, Travancore, and Cochin for medicinal purposes.

---

[12] *Ibid.* **17** (1934), 372.
[13] *J. Indian Inst. Sci.* **17A** (1934), 7.
[14] *Ber. Schimmel & Co.,* October (1890), 17.
[15] "Über Curcumaöl," Inaugural Dissertation, Basel (1906).
[16] *Helv. Chim. Acta* **17** (1934), 372.
[17] *J. Indian Inst. Sci.* **17A** (1934), 7.
[18] *Compt. rend.* **222** (1946), 557.
[19] *J. Indian Chem. Soc.* **24** (1947), 341. *Chem. Abstracts* **42** (1948), 5439.

Physicochemical Properties.—Steam-distilling the triturated rhizomes, Rao, Shintre and Simonsen [20] obtained 6.1 per cent of a greenish-brown oil with a camphoraceous odor, exhibiting these properties:

Specific Gravity at 30°/30°	0.9139
Specific Optical Rotation at 30°	−12° 30′
Refractive Index at 30°	1.5001
Acid Number	0.9
Ester Number	2.03
Ester Number after Acetylation	58.66

Chemical Composition.—Rao, Shintre and Simonsen [21] identified the following compounds in the oil:

d-Camphene (0.8%). Characterized by means of its hydrochloride m. 149°–150°.

d-Camphor (2.5%). Identified by its melting point 175°, and by means of its oxime m. 118°.

Sesquiterpenes (65.5%, chiefly *l*-Curcumene). A few years after the first investigation, Rao and Simonsen [22] studied *l*-curcumene more thoroughly and found that it consists of two isomeric monocyclic sesquiterpenes $C_{15}H_{22}$, viz., *l*-α- and *l*-β-curcumene (cf. Vol. II of this work, p. 89).

Sesquiterpene Alcohols (22%). These alcohols b_7 142°–144°, and b_7 152°–154° are probably of tertiary nature.

Acids (0.7%, Caprylic Acid and *p*-Methoxycinnamic Acid). The caprylic acid was identified by means of its silver salt, and the *p*-methoxycinnamic acid through its m. p. 170°.

Unidentified Compounds (8.5%).

Use.—Oil of *Curcuma aromatica* is not produced on a commercial scale.

OILS OF *Curcuma Domestica* VALETON AND *Curcuma Xanthorrhiza* ROXB.

Curcuma domestica Valeton (fam. *Zingiberaceae*), a native of Indonesia, is probably closely related to *Curcuma xanthorrhiza* Roxb.; the dried tuberous roots of both plants are locally called "Temoe Lawak." Because of its cholagogic effect, the drug has been used for centuries in native medicines against liver and gall bladder diseases. The active principle of the rhizome is *p*-tolylmethylcarbinol.[23]

[20] *J. Indian Inst. Sci.* **9A** (1926), 140.
[21] *Ibid.*
[22] *J. Chem. Soc.* (1928), 2496.
[23] Grabe, *Arch. expt. Path. Pharmakol.* **176** (1934), 673. Czetsch-Lindenwald, *Süddeut. Apoth. Ztg.* **81** (1941), 555. *Pharm. Ind.* **9** (1942), 262. Siebert, *Deut. med. Wochschr.* **67** (1941), 679.

Steam-distilling dried rhizomes of *Curcuma domestica* Valeton imported from Indonesia, Dieterle and Kaiser [24] obtained 3.5 per cent of a yellowish oil (d 0.941, α_D^{20} −19° 18′, n_D^{20} 1.5025) which had a peculiar odor and a burning taste.

Meijer and Koolhaas [25] found that the rhizomes of *Curcuma xanthorrhiza* Roxb. contained a maximum of volatile oil (and curcumine) at the beginning of their development. In the course of a year the oil content varied between 7.3 and 29.5 per cent, calculated upon the dried rhizomes. The highest yield of oil was obtained in October.

Physicochemical Properties.—Meijer and Koolhaas [26] reported these properties for the volatile oils of *Curcuma xanthorrhiza* Roxb., which they had obtained in the course of their investigation:

Specific Gravity at 27.5°/4°...... 0.9099 to 0.9250
Optical Rotation at 27°.......... −9° 0′ to −14° 40′
Refractive Index at 26°.......... 1.5024 to 1.5079

Chemical Composition.—Dieterle and Kaiser [27] identified the following compounds in the volatile oil derived from the dried tubers of *Curcuma domestica* Valeton:

d-Camphor (1%). Characterized by means of its monobromide m. 76°.

Cyclo-isoprenemyrcene (85%). The chief constituent, identified by means of the hydrochloride m. 83°.

p-Tolylmethylcarbinol (5%). The medicinally active principle. Characterized by preparation of the phenylurethane m. 97.5°.

Meijer and Koolhaas [28] reported that the volatile oil from *Curcuma xanthorrhiza* Roxb. contained from 1 to 2 per cent of *camphor*.

Use.—The oil is not produced on a commercial scale.

OIL OF *Curcuma Amada* ROXB.

Curcuma amada Roxb. is a medicinal herb with a mango-like odor, growing in India. Steam-distilling the dried rhizomes, Dutt and Tayal [29] obtained 1.1 per cent of a volatile oil which contained:

[24] *Arch. Pharm.* **270** (1932), 413.
[25] *Ibid.* **277** (1939), 91.
[26] *Ibid.*
[27] *Ibid.* **270** (1932), 413; **271** (1933), 337.
[28] *Ibid.* **277** (1939), 91.
[29] *Indian Soap J.* **7** (1941), 200. *Chem. Abstracts* **35** (1941), 6393.

	Per Cent
d-α-Pinene	18
Ocimene	47.2
Linaloöl	11.2
Linalyl Acetate	9.1
Safrole	9.3
Unidentified Substances	3.5
Loss during Distillation	1.7

The oil is not produced on a commercial scale.

OIL OF *Curcuma Caesia* ROXB.

Steam-distilling the air-dried and powdered rhizomes of *Curcuma caesia* Roxb. (fam. *Zingiberaceae*), Dutt [30] obtained about 1.5 per cent of a volatile oil (which in India is called "Kachri oil"). On standing, *d*-camphor crystallized from the oil. The rectified oil was dextrorotatory and had a pronounced camphoraceous odor. The following compounds were identified in the oil:

	Per Cent
d-Camphor	76.6
Camphene ⎱ Bornylene ⎰	8.2
Sesquiterpenes(?)	10.5
Unidentified Substances	4.7

The oil is not produced on a commercial scale.

OIL OF *Curcuma Zedoaria* ROSCOE
(Oil of Zedoary)

Curcuma zedoaria Roscoe (*Curcuma zerumbet* Roxb.), fam. *Zingiberaceae*, is cultivated in Ceylon where the leaves are relished as a vegetable. On steam distillation the dry rhizomes yield from 1 to 1.5 per cent of volatile oil, which in the trade is known as "Oil of Zedoary" ("Essence de Zédoaire" in French, and "Zitwerwurzelöl" in German).

Physicochemical Properties.—Oil of zedoary is a somewhat viscous greenish-red liquid with an odor reminiscent of ginger, and to a small extent of cineole and camphor.

Gildemeister and Hoffmann [31] reported the following properties:

Specific Gravity at 15°	0.982 to 1.01
Optical Rotation	+8° 0′ to +17° 0′

[30] *Indian Soap J.* **6** (1940), 248. *Chem. Abstracts* **34** (1940), 6015.
[31] "Die Ätherischen Öle," 3d Ed., Vol. II, 429.

Refractive Index at 20°........ 1.50233 to 1.50882
Acid Number................. 0.3 to 2.4
Ester Number................ 16 to 22.4
Ester Number after Acetylation. 56 to 73.4
Solubility.................... Soluble in 1.5 to 2 vol. of
                               80% alcohol

Steam-distilling the dried and powdered rhizomes of *Curcuma zedoaria* Roscoe, Rao, Shintre and Simonsen [32] obtained 0.94 per cent of a viscous dark green oil with these properties:

Specific Gravity at 30°/30°........ 0.9724
Refractive Index at 30°.......... 1.5002
Acid Number.................... 1.3
Saponification Number........... 3.0
Saponification Number after Acety-
      lation...................... 66.3

Chemical Composition.—In this oil Rao, Shintre and Simonsen noted the following compounds:

d-α-Pinene (1.5%). Identified by means of its nitrosochloride m. 104°–105°, and nitroylbenzylamine m. 122°–123°.

d-Camphene (3.5%). Characterized by hydration to *d*-isoborneol m. 208°.

Cineole (9.6%). Characterized by means of the phosphoric acid compound and the hydrobromide.

d-Camphor (4.2%). Identified through the semicarbazone m. 235°–237°.

d-Borneol (1.5%). Phenylurethane m. 138°.

Alcohols(?) (traces). Not identified.

Sesquiterpenes(?) (10%). Among them probably zingerberene (nitrosate m. 82°–83°).

Sesquiterpene Alcohols(?) (48%). Not identified.

Residue (21%). Consisting probably also of sesquiterpene alcohols.
      Dehydrogenation of the sesquiterpene fraction gave cadalene.

Use.—According to the author's knowledge oil of zedoary is at present not produced on a commercial scale.

[32] *J. Soc. Chem. Ind.* **47** (1928), 171T.

OIL OF GALANGAL

Essence de Galanga   *Aceite Esencial Galangal*   *Galgantöl*
Oleum Galangae

Botany and Origin.—Galangal, the so-called Chinese ginger or China root, is the dried rhizome of *Alpinia officinarum* Hance. A member of the family *Zingiberaceae*, galangal is botanically (and in its pharmacological effects) related to true ginger (*Zingiber officinale* Roscoe). Among the various *Alpinia* species described in literature, *Alpinia officinarum* Hance, also known as *Radix galangae minoris* or small galangal, is now the most important one.

Alpinia officinarum, the true or official galangal, is grown in southeastern China, particularly on the island of Hainan, and near Pak-hoi, on the adjacent coast of China proper. The reed-like plant attains a height of about 3 ft. The rhizomes, irregularly branched pieces, externally of reddish or rusty dark brown, internally of light orange-brown color, are dug up in the fall, washed, cut into pieces, and dried. In the trade the spice occurs in the form of irregularly shaped branching pieces. Its odor is aromatic and appealing, the flavor warm, spicy and pungent, resembling that of ginger and pepper.

The growing, handling, and marketing of galangal are entirely in Chinese hands. The dried rhizomes pass through many intermediaries before they are shipped on junks to Hong Kong, where European export houses buy the merchandise from Chinese dealers. From Hong Kong the root is exported in three qualities, assorted according to thickness, the thickest pieces fetching the best price. Prior to World War II, Marseilles was a heavy buyer of galangal root, especially of the lower grades, which represent the most economical distillation material.

The rhizomes contain from 0.5 to 1 per cent of volatile oil which can be extracted by steam distillation.

Physicochemical Properties.—Oil of galangal has a greenish-yellow color and a camphoraceous spicy odor, resembling that of cardamom and myrtle. The taste is faintly bitter, with a cooling after-effect.

Gildemeister and Hoffmann [1] reported the following properties for oil of galangal:

Specific Gravity.............. 0.915 to 0.924
Optical Rotation.............. −1° 30′ to −7° 55′
Refractive Index at 20°........ 1.476 to 1.482

[1] "Die Ätherischen Öle," 3d Ed., Vol. II, 434.

Acid Number	Up to 4
Ester Number	12 to 17
Ester Number after Acetylation	40 to 64
Eugenol Content	3 to 4%
Solubility	Miscible with 0.2 to 0.5 vol. and more of 90% alcohol; soluble in 10 to 25 vol. of 80% alcohol

Oils of galangal distilled by Fritzsche Brothers, Inc., New York, from imported root material had properties varying within these limits:

Specific Gravity at 15°/15°	0.913 to 0.923
Optical Rotation	$-3°\ 5'$ to $-6°\ 50'$
Refractive Index at 20°	1.4770 to 1.4810
Saponification Number	12.6 to 14.0
Solubility at 20°	Soluble in 0.5 vol. and more of 90% alcohol

Steam-distilling rhizomes of *Alpinia officinarum* Hance in East India, Rao and collaborators [2] obtained 0.3 per cent of an oil with the following properties:

Specific Gravity at 30°/30°	0.9061
Specific Optical Rotation at 30°	$+3°\ 24'$
Refractive Index at 30°	1.4801
Acid Number	2.6
Ester Number	19.1
Ester Number after Acetylation	78.0%

The odor of this oil was camphoraceous.

Chemical Composition.—The chemical composition of oil of galangal was investigated by Schimmel & Co.,[3] Schindelmeiser,[4] Fromm and Fluck,[5] and more recently by Ruzicka, Capato and Huyser.[6] The following compounds have been reported as constituents of the oil:

d-α-Pinene. Identified by means of the nitrosochloride and the nitrolpiperidine (Schindelmeiser).

Cineole. Characterized by means of the hydrogen bromide addition compound (Schimmel & Co.).

Eugenol. Gildemeister and Hoffmann [7] reported a eugenol content of 3 to 4 per cent. Fromm and Fluck could not find any eugenol in their oil. Horst [8] claimed a eugenol content of 25 per cent, which permits the conclusion that the oil he investigated was not pure.

[2] *Perfumery Essential Oil Record* **28** (1937), 413.
[3] *Ber. Schimmel & Co.*, April (1890), 21. Horst, *Pharm. Zeitschr. f. Russland* **39** (1900), 378.
[4] *Chem. Ztg.* **26** (1902), 308.
[5] *Liebigs Ann.* **405** (1914), 181.
[6] *Rec. trav. chim.* **47** (1928), 379.
[7] "Die Ätherischen Öle," 3d Ed., Vol. II, 434.
[8] *Pharm. Zeitschr. f. Russland* **39** (1900), 378.

A Compound $C_{10}H_{16}O(?)$. Fromm and Fluck isolated 1 to 2 per cent of a substance $C_{10}H_{16}O$, b. 208°–210° which, however, was not identified.

A Sesquiterpene(?). From the fraction b. 230°–240°, Schindelmeiser obtained a viscous dihydrochloride $C_{15}H_{24}\cdot 2HCl$, m. 51°, b_{10} 145°–150°.

A Sesquiterpene(?). Fromm and Fluck isolated a sesquiterpene b_{12-15} 138°–140°, which contained two double bonds, and added four atoms of bromine. Since this sesquiterpene does not react with gaseous hydrogen chloride, it was not identical with that described by Schindelmeiser (see above).

A Sesquiterpene(?). Fromm and Fluck isolated another sesquiterpene which yielded a dihydrochloride $C_{15}H_{24}\cdot 2HCl$, melting at the same temperature as cadinene dihydrochloride. Determination of a mixed melting point, however, showed a depression of 10°.

Cadinene Isomers. Investigating the sesquiterpenes occurring in the higher boiling fractions b_{20} up to 145° of galangal oil, Ruzicka, Capato and Huyser found that they consisted of a mixture of bicyclic sesquiterpenes which did not give a solid hydrochloride. Dehydrogenation with sulfur yielded considerable quantities of cadalene, which indicated that the sesquiterpenes in question were isomers of cadinene.

Hexahydrocadalene Hydrates(?). From the viscous greenish sesquiterpene alcohol fraction b_7 140°–150°, the same authors isolated needles m. 167° which were apparently identical with those observed years ago by Fromm and Fluck. The liquid chief portion of the fraction had the empirical molecular formula $C_{15}H_{26}O$. Dehydration with formic acid yielded a sesquiterpene b_7 124°–125°, d_4^{15} 0.9250, n_D^{15} 1.5102, which, upon dehydrogenation with sulfur, gave cadalene.

Thus the sesquiterpene alcohols present in oil of galangal consist of a mixture of hexahydrocadalene hydrates. Dehydration yields isomers of cadinene.

Use.—Very small quantities of oil of galangal are used in certain types of flavor and perfume compositions to which it imparts warm, unique, and somewhat spicy notes.

OIL OF *Alpinia Galanga* WILLD.

The dried rhizomes of *Alpinia galanga* Willd. (*Galanga officinalis* Salisb.), fam. *Zingiberaceae,* known also as *Radix galangae majoris* or large galanga, are no longer encountered in the drug trade. In Malaya, particularly in Java, they serve for the flavoring of curries.

Physicochemical Properties.—On steam distillation, the *fresh* roots yield 0.04 per cent of a volatile oil with a peculiar strong and spicy odor. Schimmel & Co.[9] examined an oil from Salatiga in Java and found these properties:

> Specific Gravity at 15°...... 0.9847
> Optical Rotation........... +4° 20′
> Refractive Index at 20°..... 1.51638

[9] *Ber. Schimmel & Co.,* October (1910), 138.

Acid Number.............. 1.8
Ester Number............. 145.6
Solubility................. Soluble in 1 vol. of 80% alcohol;
  opalescent on addition of 3
  vol. of 80% alcohol

Chemical Composition.—Another oil from Salatiga was investigated by Ultée,[10] who reported the presence of the following compounds:

d-Pinene. Presence probable.

Cineole (20–30%). Characterized by means of its iodol compound m. 112°.

Camphor. M. 170°–175°.

Methyl Cinnamate (48%). M. 34°.

Use.—Oil of *Alpinia galanga* is not produced on a commercial scale.

OIL OF *KAEMPFERIA GALANGA* L.

Kaempferia galanga L. (fam. *Zingiberaceae*) is cultivated in Malaya and India. The dried rhizomes, known in Java as "Kentjoer" or "Tjekoer," are used in Malaya for medicinal purposes and for the flavoring of curries; in India they are employed chiefly in perfumery.

Physicochemical Properties.—Steam-distilling dried and triturated rhizomes in India, Panicker, Rao and Simonsen [1] obtained from 2.4 to 3.88 per cent of volatile oil which, after removal of the crystalline ethyl ester of *p*-methoxycinnamic acid, had these properties:

Specific Gravity at 30°/30°......... 0.8792 to 0.8914
Specific Optical Rotation at 30°..... −2° 36′ to −4° 30′
Refractive Index at 30°............ 1.4773 to 1.4855
Acid Number.................... 0.5 to 1.3
Saponification Number............ 99.7 to 109.0
Saponification Number after Acety-
  lation....................... 110.1 to 116.3

[10] *Mededeelingen van het Algemeen-Proefstation op Java te Salatiga* II. Serie Nr. 45 (reprinted from *Cultuurgids* [1910], II. Lfr. 8). *Ber. Schimmel & Co., April* (1911), 19.

[1] *J. Indian Inst. Sci.* **9A** (1926), 133.

Chemical Composition.—The chemical composition of oil of *Kaempferia galanga* L. was investigated by van Romburgh,[2] and by Panicker, Rao and Simonsen [3] who reported the presence of the following compounds:

n-Pentadecane $C_{15}H_{32}$. M. 10°, b_6 125°–127°; identified first by van Romburgh and later by Panicker, Rao and Simonsen.

Ethyl *p*-Methoxycinnamate. Presence reported by the same authors. Ethyl *p*-methoxycinnamate, the chief constituent of the oil, is the ethyl ester of *p*-methoxycinnamic acid, m. 50° (cf. Vol. II of the present work, p. 604). It separates from the oil in crystalline form. Identified by van Romburgh, and by Panicker, Rao and Simonsen.

l-Δ^3-Carene. Characterized by its nitrosate m. 147.5° (with decomposition); conversion into *l*-sylvestrene dihydrochloride and into carvestrene dihydrochloride m. 52° (Panicker, Rao and Simonsen).

Camphene. Small quantities only; characterized by conversion into isoborneol m. 209°–210° (Panicker, Rao and Simonsen).

Borneol. Small quantities only; identified by means of its phenylurethane m. 138°–139° (Panicker, Rao and Simonsen).

p-Methoxystyrene. Identified by preparation of the pseudonitrosite, m. 117° with decomposition, and of the nitroxime m. 112°–113°. In the opinion of Panicker, Rao and Simonsen, it is possible that the *p*-methoxystyrene, which they observed in the oil, was not an actual constituent but may have been formed (via *p*-methoxycinnamic acid) by distillation of the ethyl *p*-methoxycinnamate (see above).

Use.—Oil of *Kaempferia galanga* is not produced on a commercial scale.

CONCRETE AND ABSOLUTE OF LONGOZE

The longoze, *Hedychium flavum* Roxb. (fam. *Zingiberaceae*), is a native of India, but has been introduced to Madagascar and adjacent islands. The flowers, which emit a heavy, exotic perfume, are processed in Nossi-Bé and on Réunion Island for the extraction of the natural flower oil. The blooming period takes place in February and March; the blossoms have to be picked early in the morning.

[2] "On the crystallized constituent of the essential oil of *Kaempferia galanga* L." *Koninkl. Akad. Wetenschappen Amsterdam.* Reprinted from: Proceedings of the Meeting of Saturday, May 26, 1900, 38. "On some constituents of the essential oil of *Kaempferia galanga* L.," *Koninkl. Akad. Wetenschappen Amsterdam*, April 19, 1902, 618.

[3] *J. Indian Inst. Sci.* **9A** (1926), 133.

According to Muller,[1] the yield of concrete is very low, extraction with petroleum ether giving from 0.05 to 0.07 per cent of concrete. The latter melts at about 37° and has the consistency of an orange flower concrete. The color is dark brown. On treatment with alcohol in the usual way the concrete yields 50 per cent of an alcohol-soluble absolute, a viscous liquid of brownish color. Steam-distilling a concrete (*I*), and an absolute (*II*) of longoze, Trabaud and Sabetay[2] obtained 22 and 29.6 per cent, respectively, of volatile oils with these properties:

	I	II
Specific Gravity	d_{15} 0.9153	d_{18} 0.9666
Optical Rotation	$-7°\,20'$	$-10°\,40'$
Refractive Index at 20°	1.4872	1.5002
Acid Number	5.1	11.2
Ester Number	54.6	94.1
Ester Number after Cold Formylation	188	...
Phenol Content	20%	...
Methoxy Content (Zeisel Method)	3.05%	3.5%
Aldehyde or Ketone Content, Calculated as $C_{10}H_{16}$	2.73%	...

Nothing definite is known about the chemical composition of longoze flower oil. Trabaud and Sabetay expressed the opinion that it contains indole and methyl anthranilate. The volatile oils obtained by these authors exhibited an odor reminiscent of vanilla, pepper, orange flowers, tuberose, and jasmine.

Only about 20 kg. of concrete of longoze are produced in Nossi-Bé per year; it is shipped to France for use in high-grade perfumes, to which this flower oil imparts heavy and exotic tonalities.

[1] Private communication from Mr. Charles Muller, Ambanja, Madagascar.
[2] *Perfumery Essential Oil Record* **29** (1938), 142.

CHAPTER IV

ESSENTIAL OILS OF THE PLANT FAMILY *PIPERACEAE*

OIL OF PEPPER

Essence de Poivre   *Aceite Esencial Pimienta Negra*   *Pfefferöl*
Oleum Piperis

OIL OF PEPPER BERRIES
(*Piper nigrum* L.)

Introduction and History.—The trade distinguishes between two principal types of pepper, viz., the black and the white, both derived from the same plant, *Piper nigrum* L. (fam. *Piperaceae*), a climbing or trailing vine-like shrub native to southern India. The flowers are whitish, perfect, and arranged on elongated spikes; the fruit is a globular drupe of red color when ripe. Black pepper is the dried, whole, unripe fruit of this plant; white pepper consists of the ripe fruit from which the dark hull (outer and inner coating) has been removed.

Pepper is one of the most important and oldest spices. It was known to the Greeks as far back as the fourth century B.C. The Romans valued it highly and imported large quantities from the East. Trade in the spice continued throughout the Middle Ages until the overland caravan routes were cut off by advancing Turkish tribes, who leveled exorbitant duties upon merchandise passing through their territories. It was chiefly the disruption of the old overland routes that caused European spice traders to look for new ways—by sea—to India. Toward the end of the fifteenth century Vasco da Gama discovered an all-sea passage to the Malabar Coast, one of the principal spice-producing territories of that time. With this important discovery the modern spice trade came into being. At first a monopoly of the Portuguese, the trade was soon contested by the Hollanders, French, and English, and out of this struggle developed the long naval wars between these great sea powers. Ships loaded with precious spices were captured, spice plantations were raided, and planting material was imported to other tropical coasts in order to start new plantations. Cultivation of pepper expanded from the Malabar to the Malacca Coast, and to the East Indian Archipelago (where the bulk of pepper is produced today).

Botany, Planting and Harvesting.—*Piper nigrum* L. is a climbing, vine-like plant attaining a length of as much as 30 ft. The stems, which on aging become woody and measure about ½ in. in thickness, develop numer-

135

ous short roots; by means of these latter the vine clings to its support, which may be either a pole of hard, rot-resistant wood with a rough surface, or a live tree (*Erythrina corallodendron* L., fam. *Leguminosae*, e.g.).

The plant requires a warm, humid, tropical climate, with heavy rainfall and intermittent spells of dry weather, and some shade. In Madagascar the trees ("Bois Noir")[1] supporting the pepper vines are used to shade coffee plants. Pepper, therefore, represents only an auxiliary crop in Madagascar. In Malaya pepper is often cultivated in conjunction with *Ourouparia* (= *Uncaria*) *gambir* (Hunter) Roxb. (fam. *Rubiaceae*), a catechu-producing shrub, the leaves of which shade the ground and, dropping, fertilize it.

The most suitable soil is a well-drained vegetable loam. The land should be flat; hillsides, if used for the growing of pepper, should be terraced.

Plants may be raised from seed selected from vigorous vines. For this purpose, well-matured fruits are chosen and soaked in water for a few days until the pericarp can be removed. The seed (each fruit contains one seed) is sown in a seedbed and kept there until the young plant can be set out in the field.

A better method of propagation, however, is by means of cuttings taken from the tops of vigorous vines. The flowers of wild-growing pepper plants are usually unisexual, but for cultivation hermaphroditic forms are preferred, as they will develop into more prolific vines. The cuttings may be planted in nursery beds and left there until the root system has developed sufficiently. Or the cuttings may be set out directly in the field, at the beginning of the rainy season, three or four cuttings being planted around one support tree or pole. As the young plants grow up they are attached to the support tree or pole with soft twine. Some producers nip off the ends of the vines when these have attained a height of about 2 ft.; this induces the vine to grow lateral shoots. The plant grows quite rapidly. When it has reached the top of the support it is pulled down and wound around the base, or it is coiled up and buried in the ground to make it more vigorous and prolific. During all this time the plant must be well manured.

About six months after the cuttings have been planted, the first flowers develop. They are very small and borne in catkins (close, bracted spikes), but the petals do not last long and soon after pollination of the flowers drop off. The first fruits develop only two to three years after planting of the cuttings; in some sections the vines do not bear before the fifth year. In

---

[1] *Acacia Lebbek* Willd. (=*Albizzia Lebbek*) has lately been attacked by certain diseases and is now gradually being replaced by *Acacia stipulata* De Candolle (=*Albizzia stipulata* Boivin) and *Cassia fistula* L.

general, six to seven years are required before a really good crop can be harvested. Full productivity lasts from about the fifth to the twentieth year. In other words, a vine bears fruit for approximately fifteen years, after which it must be renewed.

The fruit is borne on spikes 4 to 5 in. long, each spike carrying from 50 to 60 berries. Each fruit contains a single seed, enclosed in a pulpy layer within the pericarp. At first the color of the berries is dull green, but on ripening it changes to red, fully matured berries being of bright red color. Since the berries on one spike do not ripen at the same time, great discretion must be exercised as to the proper moment for harvesting. In most cases harvesting commences when the first red berries appear. The spikes on which one or two berries have turned red are then gathered by natives, who use ladders and drop the spikes into baskets hanging from their shoulders. If the spikes are left on the vines too long, all berries will ripen; many of them will be overripe and drop to the ground at a touch. This may mean a substantial loss in the yield. Harvesting, therefore, represents a particular problem in countries where labor is scarce. In Madagascar, for example, picking must start as early as possible; otherwise a high percentage of the berries will have dropped to the ground by the time the crew of harvesters has worked through the plantation. Moreover, Madagascar swarms with a certain type of blackbird which relishes mature pepper corns; in fact the bird devours so many berries that, when cooked, its meat has a distinct peppery flavor.

The only reason for allowing the berries to remain on the vines beyond their usual maturation period is to prepare white pepper (see below). White pepper can be produced only from well-ripened berries. Harvesting of the berries for this purpose requires great care, and is always connected with some loss.

The time of the harvest depends upon the climatic conditions prevailing in the various producing regions. The principal harvest in Indonesia and Madagascar takes place between the end of August and the beginning of November but, in any given territory, lasts no more than a few weeks. In addition, there is a small harvest in March/April. (According to Landes,[2] the harvest in India commences in late December and continues until April.)

A fully grown vine covers the supporting stake like a curtain, and carries numerous spikes bearing an abundance of berries. The quantity of berries harvested from the vines on one support depends greatly upon the age of the vines and the height of the support; on the average it amounts to 5 to 6 kg.

[2] *Coffee and Tea Industries* **73** (April 1950), 61.

Drying and Preparation of the Spice.—The drying of the berries for the preparation of black pepper is a relatively simple matter. Native producers spread the freshly picked spikes (with attached berries) on reed mats, exposing them to the sun for about a week; large producers place the material on concrete floors, such as are used for the drying of coffee or cloves. To remove the berries from the stalks the heaped-up material is beaten with sticks, or natives tread upon it barefooted, the latter procedure entailing a minimum of waste. On drying in the sun, the color of the berries changes from green (or red) to dark brown or almost black. At the same time the skin becomes tough and wrinkled. During the drying process the material must be turned over frequently to prevent formation of mildew. After completion of the drying, women and children separate the berries from the stalks, leaflets and other impurities. For this purpose each worker, sitting on the floor next to a small heap, goes through the material by hand, picking out coarse particles, brushing any adhering berries off the stalks, and removing stalks, leaflets and other impurities by winnowing in a flat basket, until only dark, well-dried berries remain. On some of the large estates machines have been introduced for separation of the berries. Here the freshly picked spikes are often dried artificially, in specially constructed smokehouses by means of an open fire maintained below perforated platforms. In Java, damp wood is employed for the fire; this develops a heavy smoke and may impart a peculiar smoky odor to the berries. In all cases drying is carried out as rapidly as possible, to prevent formation of mold in the final product. Since not all the berries on the spikes are of the same ripeness (some being quite immature and green), it is customary in some countries to immerse the freshly harvested spikes in boiling water for a short time, prior to drying. This is supposed to ripen the green berries and speed up the drying process in general.

The dried berries are packed in sacks for shipment abroad.

As regards *white* pepper, this can be prepared only from well-matured (red) berries, unripe berries being unsuitable. Hence production of white pepper is feasible only where cultivation is carried out with great care, as, for example, on the island of Banka, which supplies about 80 per cent of all the white pepper produced in the Indonesian Archipelago. Harvesting of the spikes has to be delayed until all berries are fully developed and well matured. On the large estates white pepper is now often prepared by decorticating fully matured *and dried* black pepper in specially constructed machines. Different degrees of decortication can thus be obtained. The older and more frequently practiced procedure is the following: freshly picked spikes bearing ripe (red) berries are piled on mats or on a concrete floor and beaten with sticks, or treaded upon, until the berries drop off the stalks. The berries are then placed into concrete tanks filled with slowly running water warmed by the sun. After eight to ten days of soaking, the skins decay and come off the berries. The removal is hastened by trampling upon the thick mass in the water. After cleaning with running water, the white berries are spread in the sun to dry, frequently turned over, and finally winnowed in the usual way.

Producing Regions and Qualities of Pepper.—On the basis of geographical origin and quality, the trade recognizes a number of grades of black pepper. By far the most important grade, as regards quantity, is the so-called "Lampong Black Pepper" which includes not only the pepper produced in the Lampong district of southern Sumatra, but also the black pepper originating from other parts of Indonesia (western Borneo, Riouw Archipelago, Banka, Billiton, northern Sumatra, and western Java). Prior to World War II, the harbor of Oosthaven, near Teluk Betung, on the south coast of Sumatra, handled about 90 per cent of the total pepper exports from the Netherlands East Indies.

Lampong black pepper is the most popular of all black pepper types, the principal source for grinders, particularly in the United States. The corns are of small size, of dark, somewhat grayish color, and wrinkled appearance. The shells are firmly attached to the berries. The flavor is more pungent than aromatic.

Singapore and Penang pepper resemble Lampong pepper in general. In fact, much of the pepper handled through these ports originates from adjacent Sumatra (Lampong district and Achin).

Some grades of pepper are of large size and excellent appearance and aroma, hence very expensive. Among these are the Tellicherry and Alleppi black pepper from the Malabar Coast in southern India. These are too valuable to be used for distillation purposes.

Siam, Saigon (Indo-China), and Madagascar black pepper are also of very good quality, and highly aromatic.

Space does not permit a discussion of all details pertaining to the various grades of pepper; they can be found in books on spices. Besides, for the extraction of the essential oil by distillation, only the lower priced qualities can be used, appearance being unimportant. As a matter of fact, the best and most economical material for distillation consists of broken hulls; but these should be of as recent harvest as possible, because they lose a great deal of volatile oil by evaporation when exposed to the air for a prolonged time (see below).

As regards *white* pepper, by far the most important type is the so-called "Muntock White Pepper" and, to a much lesser degree, the "Sarawak White Pepper." The former originates from the island of Banka (off Sumatra), the latter from the British part of Borneo.

White pepper cannot be used for distillation purposes, first because of its high price, and second because the hulls (which contain most of the essential oil) have been removed. As a matter of fact, the hulls obtained as waste product in the preparation of white pepper would make excellent distillation material, but are usually thrown away as worthless in the producing regions.

To convey some idea of the importance of the various types of pepper exported from the different countries of production, the following figures [3] (in metric tons) are cited:

Countries of Origin	1937	1938
Africa:		
Madagascar......................	113	181
Nigeria...........................	2	22
Sierra Leone.....................	21	92
Asia:		
British India.....................	1,215	703
Ceylon...........................	74	3
Malacca and other British possessions.	865	...
North Borneo.....................	4	3
Sarawak..........................	2,209	3,061
Indo-China.......................	3,851	5,705
Siam.............................	2	?
Total of all countries, except the Netherlands East Indies...............	8,356	9,770
Netherlands East Indies..............	31,060	54,502
Grand Total......................	39,416	64,272

These statistics illustrate that prior to World War II pepper from the Netherlands East Indies (Lampong black pepper and Muntock white pepper) was by far the most important type. The occupation of Indonesia by Japanese forces during the war, however, resulted in a great deal of damage to the pepper industry of the archipelago. Exact figures are not yet available, but according to latest reports [4] production has decreased considerably. In Banka, for example, there were about 15,500,000 pepper vines prior to the war; of these only 0.5 per cent survived the occupation. In 1950 the number was brought up to some 1,500,000 vines, viz., about 10 per cent of the prewar figure. Economic and political conditions permitting, a further increase can be expected.

Yield and Physicochemical Properties of the Oil.—On steam distillation, crushed black pepper yields from 1 to 2.6 per cent of volatile oil (Gilde-

[3] *Mededeel.* No. 175 *Centraal Kantoor Statistiek,* Batavia, Java.
[4] Private communication from Dr. J. A. Nijholt, Buitenzorg, Java.

meister and Hoffmann [5]). The yield depends greatly upon the age of the dried berries submitted to distillation. On storage, or during transport abroad, the spice loses a considerable amount of volatile oil by evaporation.[6] Such loss is particularly noticeable in the hulls (skins) of the berries, and in damaged or broken berries, the most economical raw material for distillation. It appears that oil of pepper should actually be distilled in the growing areas, where hulls and broken berries are freely available at very low cost; however, this has never been accomplished nor even seriously attempted. Practically all of the pepper oil offered on the market is produced in Europe or America from imported berries.

Prior to distillation the berries should be crushed, then distilled immediately. In the course of distillation development of ammonia takes place.

The volatile oil derived by steam distillation of pepper is an almost colorless to slightly greenish liquid with the characteristic odor of pepper, reminiscent also of phellandrene, one of the principal constituents of the oil. The taste of the oil is mild, *not at all pungent*.

Gildemeister and Hoffmann [7] reported these properties for pepper oil:

Specific Gravity at 15°........	0.873 to 0.916
Optical Rotation..............	−10° 0′ to +3° 0′
Refractive Index at 20°........	1.480 to 1.499
Acid Number.................	Up to 1.1
Ester Number................	0.5 to 6.5
Ester Number after Acetylation.	12 to 22.4 (4 determinations)
Solubility...................	Not readily soluble in alcohol. Usually soluble in 10 to 15 vol. of 90% alcohol; soluble in 3 to 10 vol. of 95% alcohol
Phellandrene Test............	Usually strongly positive, even without previous fractionation of the oil

Oils distilled by Fritzsche Brothers, Inc., New York, from imported Lampong black pepper had properties varying within the following limits:

Specific Gravity at 15°/15°......	0.874 to 0.881
Optical Rotation..............	−13° 30′ to −16° 5′
Refractive Index at 20°.........	1.4807 to 1.4840
Saponification Number.........	Up to 2.8
Phellandrene Test.............	Strongly positive

Note the relatively high laevorotation of these oils.

Oils distilled under the author's supervision in Seillans (Var), France from imported Saigon black pepper, exhibited these values:

Specific Gravity at 15°..........	0.880 to 0.884
Optical Rotation..............	−2° 40′ to −3° 36′

[5] "Die Ätherischen Öle," 3d Ed., Vol. II, 457.
[6] Cf. Griebel and Hess, *Z. Untersuch. Lebensm.* **79** (1940), 184.
[7] "Die Ätherischen Öle," 3d Ed., Vol. II, 458.

Refractive Index at 20°......... 1.4849 to 1.4877
Saponification Number.......... Up to 1.9
Phellandrene Test.............. Positive

The oils had a very good odor and flavor, characteristic of pepper.

Oils distilled experimentally in Nossi-Bé (Madagascar) during the author's visit to that island, from freshly dried black pepper of local production, had the following properties:

Specific Gravity at 15°.......... 0.874 to 0.886
Optical Rotation............... +0° 56' to +1° 55'
Refractive Index at 20°......... 1.4818 to 1.4870
Saponification Number.......... 1.8 to 3.7
Phellandrene Test.............. Only slightly positive

Note the (slight) dextrorotation of these oils. Apparently they contained less *l*-phellandrene than other types of pepper oil. Odor and flavor of these Madagascar-produced oils were excellent.

Some distillers use siftings and dust (refuse obtained in the drying of pepper) for the production of pepper oil. Such refuse distilled experimentally under the author's supervision in Seillans (Var), France gave oils with these properties:

	From Siftings	From Dust
Yield of Oil......................	1.14%	0.85%
Specific Gravity at 15°.............	0.911	0.911
Optical Rotation..................	−1° 20'	−2° 0'
Refractive Index at 20°............	1.4961	1.4980
Saponification Number.............	7.5	2.8
Phellandrene Test................	Negative	Negative

Note the low yield of oil, the relatively high specific gravity and refractive index of these two oils, and the negative phellandrene reaction in both cases. These features can probably be explained as a result of evaporation of low boiling terpenes on the large surface of the broken raw material. Odor and flavor of the two oils were harsher and coarser than those of the oils distilled from sound pepper berries.

An oil distilled from *white* pepper by Fritzsche Brothers, Inc., New York, exhibited the following properties:

Specific Gravity at 15°/15°...... 0.875
Optical Rotation............... −7° 45'
Refractive Index at 20°......... 1.4818
Saponification Number.......... 0.5
Phellandrene Test.............. Positive

Adulteration.—Oil of pepper is occasionally adulterated with low priced and readily accessible terpenes and sesquiterpenes, such as phellandrene, dipentene, and caryophyllene. Because these compounds are natural components of the oil, it is most difficult to prove that they have been added to an oil. The expert must rely upon careful odor and flavor tests and upon a comparison with standard samples of guaranteed purity.

Chemical Composition.—The first investigations of volatile pepper oil carried out many years ago by Dumas,[8] Subeiran and Capitaine,[9] and Eberhardt[10] brought no tangible results except, perhaps, the fact that the oil is almost free of oxygenated compounds. Treating the fraction b. 176° with alcohol and acid, Eberhardt obtained terpin hydrate (cf. Vol. II of the present work, p. 200). It is not known, however, from which terpene the terpin hydrate originated.

Despite its importance as a flavoring oil, the volatile oil derived from the dried fruit of *Piper nigrum* L. was not thoroughly examined until very recently. Work carried out more than fifty years ago by Schimmel & Co.[11] and by Schreiner and Kremers[12] revealed the presence of *l*-phellandrene, caryophyllene, and perhaps dipentene. It was only in 1951 that Hewitt, Ritter, Konigsbacher and Hasselstrom[13] established the presence of the following compounds in the volatile pepper oil derived by steam distillation of ground Malabar pepper (yield 3.2 per cent):

α-Pinene. Identified by preparation of the nitrosochloride and of the nitrolpiperidine m. 117°. Oxidation with selenium dioxide gave myrtenal, the semicarbazone of which melted at 206°.

β-Pinene. Characterized by oxidation to nopinic acid m. 126°–127°.

l-α-Phellandrene. Identified by means of the maleic anhydride adduct m. 125°.

dl-Limonene. Tetrabromide m. 124°–125°.

Piperonal (Heliotropin). Identified by preparation of the semicarbazone m. 218°–219°, and of the 2,4-dinitrophenylhydrazone m. 260°.

Dihydrocarveol. 3,5-Dinitrobenzoate m. 120.5°–121.5°.

A Compound $C_{18}H_{34}O_4(?)$. In minute amounts only. M. 161°.

---

[8] *Liebigs Ann.* **15** (1835), 159.
[9] *Ibid.* **34** (1840), 326.
[10] *Arch. Pharm.* **225** (1887), 515.
[11] *Ber. Schimmel & Co.*, October (1890), 39.
[12] *Pharm. Arch.* **4** (1901), 61.
[13] Paper presented at the XIIth International Congress for Pure and Applied Chemistry, New York, September 1951. Publication forthcoming. Investigation carried out at the Evans Research and Development Corporation in New York, in cooperation with the Quartermaster Pioneering Research Laboratories, Philadelphia, under the sponsorship of the Quartermaster Food and Container Institute for the Armed Forces, Chicago, Illinois.

β-Caryophyllene. Hydration yielded β-caryophyllene alcohol m. 94°–95°, the phenylurethane of which melted at 135°–136°.

A Piperidine Complex (?). On vacuum distillation of the benzene extract of ground pepper, a solid collected toward the end of the distillation; after drying on a porous plate it melted at 60°–64°. Treatment with phenylisothiocyanate yielded a solid phenylthiourea m. 99°. A mixed melting point with authentic N-phenyl-N', N'-pentamethylenethiourea showed no depression. According to the above-named authors, this piperidine complex requires further study (see also below).

In their work on the chemical composition of the volatile pepper oil, Hewitt, Ritter, Konigsbacher and Hasselstrom also obtained some evidence of other constituents, a tertiary alcohol, a monoterpene aldehyde, and heavier unidentified odor components among them.

Use.—Oil of pepper is a valuable adjunct in the flavoring of sausages, canned meats, soups, table sauces, and certain beverages and liqueurs. The oil is used also in perfumery, particularly in bouquets of the oriental type, to which it imparts spicy notes difficult to identify.

Oleoresin of Pepper

The volatile (essential) oil yielded by steam distillation of dried black pepper represents only the aromatic, odorous constituents of the spice; it does not contain the pungent, nonvolatile principles for which pepper is so highly esteemed as a condiment. To obtain all these constituents in concentrated form, ground pepper berries must be extracted repeatedly (percolated) with volatile solvents—alcohol, acetone, or ether. Concentration of the solutions and removal of the solvent *in vacuo* yields the so-called oleoresin of pepper. Quantitative composition of the oleoresin depends upon the solvent used. Oleoresin of pepper generally contains these compounds:

1. *Piperine,* an alkaloid of the empirical molecular formula $C_{17}H_{19}NO_3$. It is a crystalline compound m. 129°–130°, very sparingly soluble in water, neutral to litmus, and one of the principal substances responsible for the sharp taste of pepper. Piperine occurs in the berries in amounts ranging from 4 to 10 per cent. On hydrolysis by acids or alkalies, piperine decomposes into the strongly basic piperidine (see below), and piperic acid $C_{12}H_{10}O_4$, m. 215°, slightly soluble in water, soluble in alcohol or ether. Piperine is thus a piperidide of piperic acid.

2. *Chavicine,* also of the empirical molecular formula $C_{17}H_{19}NO_3$, and an isomer of piperine (see above). Chavicine, too, is a piperidide of an

unsaturated acid, viz., chavicinic acid,[14] the latter being a geometrical isomer of piperic acid. Chavicine is said to be the most pungent component of black pepper.

3. *Other piperidides.* Ott and Eichler's [15] work recognized the possibility of the existence of other pungent principles in oleoresin pepper; they differ from piperine only in the structure of the acid component and contain isopiperic and isochavicinic acids. These workers confirmed the presence only of the latter acid, and suggested that even this may have been formed by stereochemical rearrangement.

These combined principles are likely to be present in the amount of 5 to 10 per cent in the spice as such. In the oleoresin the amount ranges from 25 to 50 per cent, depending upon the method of preparation.

Piperettine. This homologue of piperine has recently been isolated and identified in *Piper nigrum* by Spring and Stark.[16] This piperidide is nearly insoluble in acids and alkalies. It has the molecular structural formula $C_{19}H_{21}O_3N$ and melts at $146°$. The following configuration has been established for piperettine:

4. *The volatile oil,* possessing the characteristic aroma, but not the pungency of pepper. For details see the monograph above.

5. *A volatile alkaloid,* which Pictet and Pictet [17] declared to be an optically active modification of β-methylpyrroline. Present in small quantities only.

6. *Resins,* not identified.

As regards the hydrolysis of piperine, it takes place according to the following scheme:

[14] Ott and Lüdemann, *Ber.* **57B** (1924), 214.
[15] *Ber.* **55B** (1922), 2653.
[16] *J. Chem. Soc.* (1950), 1177.
[17] *Helv. Chim. Acta* **10** (1927), 593.

Piperine

$$\downarrow H_2O$$

Piperidine $\quad$ + $HOOC \cdot CH{=}CH \cdot CH{=}CH$

Piperic Acid

Owing to the position of the two double bonds, there are four *cis-trans* isomerides of the acid resulting from the hydrolysis of piperine; the configuration of each isomeride has been established:

Piperic acid....................	*α-trans*	*γ-trans*	m. 217°
Isopiperic acid................	*α-cis*	*γ-trans*	m. 145°
Chavicinic acid................	*α-cis*	*γ-cis*	liquid
Isochavicinic acid.............	*α-trans*	*γ-cis*	m. 200°–202°

The most unstable of these isomerides is isochavicinic acid (cf. Karrer [18]).

OIL OF PEPPER LEAVES
(*Piper nigrum* L.)

Steam-distilling the leaves of wild-growing East Indian pepper vines (*Piper nigrum* L.), B. S. Rao and his collaborators obtained a volatile oil with these properties:

Specific Gravity at 30°/30°......	0.9035
Refractive Index at 30°.........	1.4969
Acid Number.................	1.2
Ester Number................	12.4
Ester Number after Acetylation..	37.9

[18] "Organic Chemistry," New York (1946), 825.

The oil had an odor reminiscent of lime oil. Nothing is known about its chemical composition. According to the author's knowledge, the oil has never been produced on a commercial scale.

OIL OF PEPPER BERRIES FROM "LONG PEPPER"

There are two distinct types of the so-called "Long Pepper," derived from two plants belonging to the family *Piperaceae* and closely allied to *Piper nigrum* L., the regular pepper plant:

1. *Piper officinarum* DC. (*Chavica officinarum* Miq.), a climbing shrub, indigenous to the Malayan Archipelago and cultivated chiefly in Java—whence the term "Javanese Long Pepper."
2. *Piper longum* L. (*Chavica roxburghii* Miq.), also a climbing shrub and resembling *P. officinarum* DC., except for the broader leaves, which are heart-shaped. It is cultivated in Bengal, Ceylon, Timor, on the Philippine Islands, and in southern India—whence the term "Indian Long Pepper."

The fruits of the two species are similar, but quite different from those of *Piper nigrum* L. The numerous tiny berries are fused together, forming a spike-like cylindrical cone.

Long pepper is very aromatic and pungent in taste, hence its use in native curries. In classical times and during the Middle Ages long pepper was highly esteemed in Europe, but today it is rarely encountered in Europe or America, where it has become an almost forgotten spice.

On steam distillation dried long pepper berries yield about 1 per cent of a somewhat viscous, light green volatile oil with an odor reminiscent of that of *Piper nigrum* L. and ginger oil.[19] Nothing is known about its chemical composition. According to the author's knowledge, the oil is not produced commercially.

SUGGESTED ADDITIONAL LITERATURE

Morris B. Jacobs, "Pungent Compounds Used in Flavoring," *Am. Perfumer* **48** (July 1946), 60.

H. Hadorn and R. Jungkunz, "Pepper and Cubeb," *Pharm. Acta Helv.* **26** (1951), 25.

"Abstracts of Some Articles Pertaining to the Cultivation of Black Pepper," Technical Collaboration Branch, Office of Foreign Agricultural Relations, United States Department of Agriculture, Washington 25, D. C., January 1951.

[19] Cf. *Ber. Schimmel & Co.,* April (1890), 48.

OIL OF CUBEB

Essence de Cubèbe   *Aceite Esencial Cubeba*   *Cubebenöl*
Oleum Cubebarum

Cubeb is the dried, nearly full-grown, unripe fruit of *Piper cubeba* L. f. (fam. *Piperaceae*), a climbing perennial native to the Malayan Archipelago. The fruits are borne on spikes, each of which carries about 50 or more. The latter are globose drupes, with a stem-like portion attached to the base of each drupe. When dried, cubeb berries closely resemble black pepper berries, except for the fact that the base of each cubeb is prolonged into a firmly attached stalk. This has given rise to the term "Tailed Pepper" for cubebs.

Cubebs are grown chiefly in coffee and cocoa plantations, where the shade trees serve as support for the climbing vines; in other respects planting, cultivating and harvesting are identical with those practiced with pepper (*Piper nigrum* L.). The principal harvest takes place in April and May, and a smaller one at the end of August. The fruiting spikes are collected while still green, though fully grown; the fruits are stripped from the rachis, dried in the sun, and cleaned (winnowed) like pepper berries.

Prior to World War II, cubebs were produced chiefly in the central part of Java and in the northwestern part of Sumatra (Achin). Java berries were shipped from Cheribon and Semarang, Sumatra berries from Meulaboh. Cubebs from Java went mostly to Singapore, those from Sumatra to Penang.

According to Nijholt,[1] exports of cubebs from Java and Sumatra before the occupation of the Netherlands East Indies by Japan were (in metric tons):

1936...... 138
1937...... 206
1938...... 118
1939...... 144
1940...... 111

Average exports from 1926 to 1936 amounted to 135 metric tons per year. Most of this quantity went to Singapore and Penang in British Malaya for transshipment; small quantities went directly to Holland, Germany, and France.[2] During and for sometime after the war, cubebs became quite

---

[1] Private communication, courtesy Dr. J. A. Nijholt, Laboratorium voor Scheikundig Onderzoek, Buitenzorg, Java, January 1950.

[2] *Koloniaal Instituut, Amsterdam. Mededeeling* No. 58 (1942), 27.

scarce on the American and European markets. Perhaps cultivation of cubebs by the native and Chinese planters in the interior of Java and Sumatra was neglected. Moreover, much of the drug has always been used in native medicines. It appears quite probable, however, that with the establishment of more normal conditions in the interior of Java and Sumatra production of cubebs may be increased again.

As was mentioned above, Singapore and Penang have always been very important transshipping ports for cubebs. There are a number of Chinese intermediaries in these ports, through whose hands large quantities of the drug pass, before it is shipped abroad by the well-known export houses. These Chinese experts are amazingly skilled in their trade, being able to separate true cubebs from admixed spurious berries by mere appearance. Based upon their long experience, they can guarantee that a shipment will consist of 95 per cent of true cubebs, with only 5 per cent of spurious berries and foreign matter. This guarantee is very important to the export houses, because the cubebs arriving from Java and Sumatra are often badly adulterated with other tailed pepper berries of similar appearance.[3]

The adulteration of cubebs has always been a most annoying problem and several methods for the detection of spurious berries have been suggested. The best known is the following:

> Crush a berry in a porcelain mortar and touch the crushed mass with a small quantity of sulfuric acid (80 per cent). If the berry is a true cubeb, a bright red color will develop, due to the presence of small quantities of cubebic acid. A spurious berry will develop a yellow-brown color.

Unfortunately, this test is not absolutely reliable, because certain spurious berries, *Piper ribesioides* Wall., e.g., give the same bright red color as do true cubebs.[4] According to Rosenthaler,[5] *Piper clusii* and *Piper guineense*, two species of the group loosely called "False Cubebs" in the trade, give a violet and a brownish color, respectively, when treated with sulfuric acid.

A more reliable, but more complicated, test consists of submitting a sample of berries to distillation according to the Clevenger method (cf. Vol. I of the present work, p. 317), and of determining the yield and physicochemical properties of the volatile oil. In the case of true cubebs, Clevenger[6] obtained yields (v/w) ranging from 12.5 to 20.0 per cent. Adultera-

[3] For details regarding spurious berries see Holmes, *Perfumery Essential Oil Record* **3** (1912), 125.

[4] *Ber. van de Afd. Handelsmuseum van het Koloniaal Instituut* No. 11. Rep. from *Indische Mercuur*, January 12, 1923.

[5] *Pharm. Acta Helv.* **2** (1927), 29.

[6] *J. Assocn. Official Agr. Chem.* **20** (1937), 140.

tion of the berries is indicated by the abnormal properties of the volatile oil (see below, section "Physicochemical Properties").

An expert botanist, examining them under the microscope, will have no difficulty in distinguishing between true cubebs and spurious berries.

As regards the external appearance of true cubebs, the upper portion of the fruit is almost globular, from 3 to 6 mm. in diameter, abruptly contracted into a slender stem-like portion, which seldom exceeds 7 mm. in length. The pericarp is dusky red to slightly brown, rarely grayish, in color. The berries possess a spicy, aromatic odor, and a somewhat bitter and acrid taste. According to trade standards, a given lot should contain not more than 10 per cent of shriveled and immature fruits, not more than 5 per cent of stems, not more than 2 per cent of foreign organic matter, and not more than 2 per cent of acid-insoluble ash. Each 100 g. of the drug should yield not less than 13.0 cc. of volatile oil of cubeb.

In commercial distillation, crushed cubebs yield from 10 to 18 per cent of essential oil (Gildemeister and Hoffmann [7]). In the course of distillation, development of ammonia takes place, as in the case of pepper, pimenta, ginger, and other spices; the reason for this has not yet been determined.

Physicochemical Properties.—The volatile oil derived from true cubebs is a somewhat viscous liquid with a spicy odor characteristic of the berries, and a warm, camphoraceous, slightly acrid taste. The color ranges from light green to blue-green. To obtain a colorless oil, the highest boiling fractions, which are blue, must be kept apart from the main oil, in the course of distillation of the berries.

According to Gildemeister and Hoffmann,[8] oil of cubeb has these properties:

Specific Gravity at 15°..........	0.915 to 0.930
Optical Rotation...............	−25° 0′ to −43° 0′
Refractive Index at 20°.........	1.4938 to 1.4981
Acid Number..................	Up to 0.8
Ester Number.................	1.9 to 5.6
Ester Number after Acetylation...	25 to 30
Solubility.....................	The solubility in 90% alcohol varies widely. Some oils (probably those distilled from old cubebs) are soluble in an equal or smaller volume of 90% alcohol; others require up to 10 vol. of 90% alcohol for clear solution; still others are not clearly soluble at all

[7] "Die Ätherischen Öle," 3d Ed., Vol. II, 460.
[8] *Ibid.*

Boiling Range.................. The bulk of the oil distills between 250° and 280°
at atm. pr. Distillation of an oil in a simple
glass still at 745 mm. gave these fractions:

	Per Cent
205°–225°......	4
225°–250°......	14
250°–260°......	16
260°–265°......	22
265°–275°......	24
275°–280°......	5
Residue........	15

Shipments of pure cubeb oil imported by Fritzsche Brothers, Inc., New York, had properties varying within the following limits:

Specific Gravity at 25°/25°.......	0.906 to 0.918
Optical Rotation at 25°..........	−20° 38′ to −29° 58′
Refractive Index at 20°..........	1.4948 to 1.4979
Saponification Number..........	0.9 to 3.2

The properties of genuine oils distilled by the same firm from imported cubeb berries exhibited these values:

Specific Gravity at 25°/25°.......	0.902 to 0.914
Optical Rotation at 25°..........	−12° 27′ to −30° 34′
	(usually about −23°)
Refractive Index at 20°..........	1.4919 to 1.4955
Saponification Number..........	0.9 to 2.8

In order to determine the boiling range of genuine cubeb oil, the laboratories of Fritzsche Brothers, Inc. distilled 50 g. each of two oils of own production, using a 125 cc. distilling flask with an air condenser 33 cm. long. The following fractions were obtained:

	Per Cent	
	b_{765}	b_{763}
Up to 200°..............	8	9
200° to 225°.............	6	6
225° to 250°.............	12	7
250° to 270°.............	60	64
270° to 280°.............	6	9
Residue................	8	5

Some decomposition was observed in both distillations.

Clevenger [9] examined numerous samples of imported cubebs, determining the yields of resins and volatile oil, and the properties of the latter. He found these limits for authentic cubebs:

[9] *J. Assocn. Official Agr. Chem.* **20** (1937), 140. Cf. *ibid.* **21** (1938), 566.

Resin........................... 6.44 to 8.47%
Yield of Volatile Oil (cc. per 100 g.
    of berries)..................... 12.5 to 20.0
Specific Gravity at 25°/25°........ 0.911 to 0.919
Optical Rotation at 25°........... −19° 42′ to −46° 0′
Refractive Index at 20°........... 1.492 to 1.498
Acid Number.................... 0.35 to 1.0
Ester Number................... 1.0 to 10.7

Clevenger found that about 10 per cent of the imported lots of cubebs contained spurious berries, in some cases as much as 50 per cent. Examination of adulterated lots gave the following data:

	50% Adul- *terated*	*12% Adul-* *terated*	*50% Adul-* *terated*
Resin...........................	7.71%	6.58%	6.6%
Yield of Volatile Oil (cc. per 100 g. of berries).....................	17.0	14.2	10.3
Specific Gravity at 25°/25°........	0.879	0.910	0.905
Optical Rotation at 25°...........	+22° 30′	−17° 18′	+2° 24′
Refractive Index at 20°...........	1.476	1.492	1.491
Acid Number....................	0.5	1.8	0.5
Ester Number...................	5.4	4.5	3.6

From these figures it appears that adulteration of cubebs can best be detected by examination of the physical properties of the volatile oil obtained.

Fischer, Tornow and Proper,[10] who also used the Clevenger procedure, observed these minimum and maximum values for the volatile oil distilled from imported cubeb berries:

Yield of Volatile Oil (cc. per 100 g.
    of berries)..................... 10 to 18
Specific Gravity at 25°/25°........ 0.9155 to 0.9206
Optical Rotation at 25°........... −32° 25′ to −34° 12′
Refractive Index at 20°........... 1.4946 to 1.4980

In general, the oils derived from true cubebs are characterized by a pronounced laevorotation and a relatively high specific gravity; in this respect they usually differ markedly from most oils distilled from spurious berries or false cubebs. The latter often exhibit dextrorotation and a low specific gravity.

True cubeb oils derived from old berries possess a higher specific gravity than oils distilled from berries of more recent harvest. This is probably due to the presence of "cubeb camphor" in the oils from old berries. This hydrate is gradually formed on aging of the berries and particularly on

[10] *Natl. Form. Comm. Bull.* **13**, No. 1–2 (1945), 9.

exposure to moist air. Oils which contain cubeb camphor lose water on distillation and partly decompose.

Chemical Composition.—Although large quantities of cubeb oil have been used for many years, and although the oil was once official in the United States and British Pharmacopoeias, little was known about its chemical composition until about 1928 (see the work of Rao, Shintre and Simonsen below).

The first investigations of the oil date back to the early part of the last century, when the so-called *"Cubeb Camphor"* was noted as a constituent of the oil. It has been described by Müller,[11] Blanchet and Sell,[12] Winckler,[13] Schmidt,[14] Schaer and Wyss,[15] and more recently by Elze.[16] Cubeb camphor was originally thought to be a sesquiterpene hydrate $C_{15}H_{24} \cdot H_2O$, the melting point of which was variously given as 65°, 67°, and 70°. It is odorless, laevorotatory, and crystallizes from old oils in the form of rhombic crystals. (It does not appear to occur in oils from recently harvested berries!) When kept above concentrated sulfuric acid, cubeb camphor loses water and forms a sesquiterpene. When heated for a prolonged time at 200° to 250° it is completely decomposed into water and a sesquiterpene. Elze[17] expressed the opinion that cubeb camphor is a saturated tertiary tricyclic sesquiterpene alcohol $C_{15}H_{25}OH$. He found that the repeatedly recrystallized product melts at 105°–106°, and that it boils at 275°–280°, with a small loss of water. The oil investigated by Elze contained about 1 per cent of cubeb camphor.

Aside from cubeb camphor, early workers noted the presence, in the oil, of a laevorotatory terpene[18] (presumably pinene or camphene), and *dipentene*[19] which yielded a dihydrochloride m. 48°–49°. Wallach[20] also found that the principal fraction (b. 250°–280°) of the oil consisted of two laevorotatory sesquiterpenes, one of which was identified as *l-cadinene*. The other (b. 262°–263°) had a lower laevorotation than the *l*-cadinene, and yielded no hydrochloride; it was not identified.

In 1928, Rao, Shintre and Simonsen[21] examined an oil distilled from fruits of *Piper cubeba* L. grown in Mysore (southern India). Strangely,

[11] *Liebigs Ann.* **2** (1832), 90.
[12] *Ibid.* **6** (1833), 294.
[13] *Ibid.* **8** (1833), 203.
[14] *Arch. Pharm.* **191** (1870), 23. *Ber.* **10** (1877), 188.
[15] *Arch. Pharm.* **206** (1875), 316.
[16] *Riechstoff Ind.* **3** (1928), 193.
[17] *Ibid.*
[18] Oglialoro, *Gazz. chim. ital.* **5** (1875), 467. *Ber.* **8** (1875), 1357.
[19] Wallach, *Liebigs Ann.* **238** (1887), 78.
[20] *Ibid.*
[21] *J. Soc. Chem. Ind.* **47** (1928), 92T. Cf. Rao, Sudborough and Watson, *J. Indian Inst. Sci.* **8A** (1925), 159.

this oil exhibited *dextrorotation* ($[\alpha]_D^{30}$ +25° 36′), and the yield of oil was only 7.5 per cent. The dextrorotation of this oil is so unusual as to lead to a suspicion that the oil was distilled from spurious, rather than genuine, cubebs. Nevertheless, the authors claim that the fruits used in their investigation were of undoubted authenticity. In their opinion, it is possible that they were slightly immature, however. (Another batch of oil, distilled from the same source a few years earlier, with a yield of 11.9 per cent, exhibited laevorotation $[\alpha]_D^{30}$ −29° 54′.) Investigating the chemical composition of the dextrorotatory oil, Rao, Shintre and Simonsen reported the presence of the following compounds:

d-Sabinene. Characterized by conversion to terpinene dihydrochloride m. 52°, and by oxidation to *d*-sabinenic acid m. 55°–57°.

d-Δ⁴-Carene. Oxidation to *cis*-caronic acid m. 172°–173°, and to *d*-1,1-dimethyl-2-γ-ketobutylcyclopropane-3-carboxylic acid, the semicarbazone of which melted at 178°–180°.

1,4-Cineole. Characterized by conversion to terpinene dihydrochloride m. 52°, and by oxidation with potassium permanganate to an acid m. 157°. (Cf. Wallach.[22])

d-Terpinen-4-ol. Conversion to terpinene dihydrochloride, and oxidation to the glycerol, *d*-1-methyl-4-isopropyl-1,2,4-trihydroxycyclohexane m. 114°–116°, anhydrous m. 126°–128°.

Other Terpene Alcohols(?). Not identified.

l-Cadinene. The sesquiterpene fraction consisted chiefly of *l*-cadinene, which yielded a dihydrochloride m. 117°–118°.

Other Sesquiterpenes(?). Not identified.

Sesquiterpene Alcohols(?). None of the sesquiterpene alcohol fractions gave a crystalline hydrochloride; the presence of cadinol is therefore improbable. The alcohols did not react with phthalic anhydride at 130°.

Other Compounds(?). Not identified.

The oil had the following quantitative composition:

	Per Cent
d-Sabinene	33
d-Δ⁴-Carene and 1,4-Cineole	12
d-Terpinen-4-ol and Other Terpene Alcohols	11
l-Cadinene and Other Sesquiterpenes	14
Sesquiterpene Alcohols	17
Unidentified Compounds	13

Rao, Shintre and Simonsen explained the laevorotation of the above-mentioned *second* cubeb oil distilled from Mysore fruit by the fact that it contained *dl*-sabinene (oxidation to *dl*-sabinenic acid m. 83°–84°)

[22] *Liebigs Ann.* **392** (1912), 62.

and a sesquiterpene fraction with a higher laevorotation ($[\alpha]_D^{30}$ $-37°$ 30′) than that of the oil which they investigated thoroughly.

Use.—Cubeb berries, as well as the essential oil, were once used as stimulants to the genita-urinary mucous membranes, especially in the latter stages of gonorrheal urethritis, and as stimulants to the bronchial mucous membranes in the treatment of bronchitis. At present, the oil is rarely employed medicinally, except as a local remedy in the form of lozenges for the relief of various throat conditions.

Although officially declared a drug, the oil is also used for the flavoring of bitters, and spicy table sauces.

Oils of False Cubebs

In addition to the true cubeb, *Piper cubeba* L. (fam. *Piperaceae*), there are numerous varieties of this species, and species of *Piper* other than *cubeba*, the berries of which resemble cubebs, and which enter the trade under the general name "False Cubebs." Natives often use them instead of true cubebs. Many lots of cubebs offered on the market contain admixed false cubebs.

Most important of the false cubebs appear to be *Piper crassipes* Korthals, which has a cajuput-like odor (it is perhaps identical with *Piper ribesioides* Wallich), and *Piper molissimum* Blume (also called "Keboe Cubeb" or "Karbauw Berries"). Aside from these false cubebs, native to Malaya, there are those from Africa, among them *Piper guineense* Schum. et Thonn. and *Piper clusii* C. DC. These are known as "African Cubebs," "Congo Cubebs," "African Black Pepper," "Guinea Pepper" and "Ashanti Pepper." For further information on false cubebs, the reader may consult the work of Holmes,[23] Dekker,[24] Rosenthaler,[25] and Goester and Steenhauer.[26]

Distilling false cubebs (which had a decided mace odor) along with the berries of another unknown *Piper* species, Umney and Potter [27] obtained 4 per cent of an oil with these properties:

Specific Gravity	0.894
Optical Rotation	$+16°$ 0′
Saponification Number	0
Ester Number after Acetylation	56.1

A comparative distillation test with true cubebs yielded twice as much oil as distillation of the false cubebs. The oil from the true cubebs had a

[23] *Perfumery Essential Oil Record* **3** (1912), 125.
[24] *Ibid.* **4** (1913), 89.
[25] *Pharm. Acta Helv.* **2** (1927), 29.
[26] *Pharm. Weekblad* **64** (1927), 870. Cf. U. S. Dispensatory 24th Ed. (1947), 342.
[27] *Perfumery Essential Oil Record* **3** (1912), 64.

much higher specific gravity (0.917) than that of the oil from the false cubebs; moreover, it exhibited pronounced laevorotation ($-42°$). As regards the boiling range, the false cubeb oil started to boil below 160°; about 50 per cent distilled from 160° to 200°; about 30 per cent from 200° to 270°. In the case of the true cubeb oil, only 5 per cent distilled below 200°, and 85 per cent between 200° and 257°.

Henderson and Robertson [28] examined the high boiling fractions of an oil derived from false cubebs, and reported the presence of the following compounds:

l-Cadinene. (Cf. Vol. II of the present work, p. 91).

A Sesquiterpene(?). B_{10} 120°–124°, b_{12} 124°–128°.

l-Cadinol(?). The sesquiterpene alcohol $C_{15}H_{26}O$, b_{10} 153°–155°, d_4^{16} 0.9727, $[\alpha]_{5461}^{15}$ $-54°$ 0', n_D^{16} 1.508, isolated by Henderson and Robertson, was probably identical with *l*-cadinol. It could be converted to cadinene dihydrochloride.

Cubebol. A sesquiterpene alcohol $C_{15}H_{26}O$, m. 61°–62° (cf. Vol. II of the present work, p. 287).

Another species of *Piper* usually included among the group of the so-called false cubebs is *Piper lowong* Blume. Steam-distilling berries of this species, Peinemann [29] obtained 12.4 per cent of an oil with a specific gravity of 0.865. Forty per cent of the oil boiled from 165° to 175°; this fraction had an optical rotation of $+22°$. Thirty-four per cent of the oil boiled from 230° to 255°; this fraction was optically inactive. The fraction distilling at 270° was of blue-green color.

So far as the above-mentioned African *Piper* species are concerned, Schimmel & Co.[30] distilled two samples of *Piper guineense* Schum. et Thonn. ("Ashanti Pepper") and obtained 11.5 and 10.96 per cent of two oils, which had these properties, respectively:

	I	II
Specific Gravity at 15°...............	0.8733	0.8788
Optical Rotation.....................	$-3°$ 43'	$-5°$ 34'
Refractive Index at 20°..............	1.48905	1.48847
Acid Number.........................	0.6	0.9
Ester Number........................	5.5	4.2
Ester Number after Acetylation........	...	22.1

Both oils gave a strongly positive phellandrene reaction.

False cubebs are not distilled commercially. Occasionally, however, they

[28] *J. Chem. Soc.* (1926), 2811.   [30] *Ber. Schimmel & Co.*, April (1914), 100.
[29] *Arch. Pharm.* **234** (1896), 238.

may be by inadvertence (or otherwise) intermixed with a lot of true cubebs. The reputable distiller must watch out for this.

SUGGESTED ADDITIONAL LITERATURE

H. Hadorn and R. Jungkunz, "Pepper and Cubeb," *Pharm. Acta Helv.* **26** (1951), 25.

THE MATICO OILS

Essence de Matico   Aceite Esencial Mático   Maticoöl
Oleum Foliorum Matico

The oil distilled from the leaves of the tree *Piper angustifolium* Ruiz et Pavon (*Artanthe elongata* Miq.), fam. *Piperaceae*, a native of South America, no longer has any commercial importance; therefore it will be discussed here only briefly.

There has been a great deal of confusion regarding the taxonomy of the trees which produce the matico leaves offered in the drug trade. Shipments of leaves that could hardly be distinguished from true matico leaves have been received, but on distillation the various lots have yielded oils with different physicochemical properties and of different chemical composition. Even as far back as 1905 it looked as if *true* matico leaves (from *Piper angustifolium* Ruiz et Pavon) were no longer available. Induced by Thoms,[1] Gilg and de Candolle then undertook a botanical study of the material offered in the trade as matico leaves. These prominent botanists arrived at the conclusion that the leaf material originated from the following species:

P. *angustifolium* Ruiz et Pavon
P. *camphoriferum* C. DC.
P. *lineatum* Ruiz et Pavon
P. *angustifolium* var. *ossanum* C. DC.
P. *acutifolium* Ruiz et Pavon var. *subverbascifolium*
P. *mollicomum* Kunth.
P. *asperifolium* Ruiz et Pavon

[1] *Arch. Pharm.* **247** (1909), 591.

On steam distillation the leaves from these species yielded from 0.3 to 6 per cent of volatile oil.

Physicochemical Properties and Chemical Composition.—According to Gildemeister and Hoffmann,[2] the *old* type of matico oil produced years ago had a specific gravity varying between 0.93 and 0.99, and was slightly dextrorotatory. It was a viscous liquid with an odor reminiscent of cubeb and mint. The chief constituent of the *old* type of matico oil was the so-called *matico camphor* $C_{15}H_{26}O$, m. 94° (cf. Vol. II of the present work, p. 762). This type of matico oil has not been produced for many years.

The *new* type of matico oil is distilled from the leaves of various species of *Piper* (see above); therefore the physicochemical properties of the different lots of oil vary to such an extent that they are practically meaningless. Gildemeister and Hoffmann [3] reported these properties for the *new* type of commercial matico oils:

Specific Gravity at 15°............ 0.940 to 1.135
Optical Rotation................. −27° 28′ to +5° 34′
Refractive Index at 20°.......... 1.496 to 1.529
Acid Number................... Up to 4
Ester Number.................. 2.5 to 5.1
Ester Number after Acetylation.... 26 to 47
Solubility...................... Solubility varies greatly; some oils are soluble in 3 vol. of 80% alcohol, others require 25 vol. and more for solution. Some oils soluble in 0.3 vol. of 90% alcohol; others require 6 vol. and even then exhibit turbidity

An oil of matico (yield 1.4 per cent) distilled under the author's supervision in Southern France (Seillans, Var) exhibited the following properties:

Specific Gravity at 15°........... 0.974
Optical Rotation................. +1° 32′
Refractive Index at 20°.......... 1.5040
Acid Number................... 5.6
Ester Number.................. 8.5
Solubility...................... Soluble in 0.5 vol. of 90% alcohol; cloudy with more.

As regards the chemical composition of the *new* matico oils, they do not contain matico camphor. Depending upon the botanical origin of the leaves, the chief constituents of the new oils, according to Thoms,[4] are

[2] "Die Ätherischen Öle," 3d Ed., Vol. II, 467.
[3] *Ibid.*, 468.
[4] *Arch. Pharm.* **242** (1904), 328; **247** (1909), 591.

apiole and *dillapiole* (cf. Vol. II of this work, pp. 537, 539), phenolic ethers that occur also in the oils of parsley seed and dill seed, respectively. Another phenolic ether, viz., *asarone* (cf. Vol. II of this work, p. 535), *limonene*, small quantities of *palmitic acid*, and traces of *phenols* have also been observed in the oil.

Continuing his work on matico oil, Thoms examined oils that had been distilled from leaves of well-established botanical origin. None of these oils is now produced commercially, but for scientific interest they will be discussed briefly.

OIL OF *Piper Angustifolium* RUIZ ET PAVON

Investigating an oil which had been distilled from the leaves of true *Piper angustifolium* Ruiz et Pavon, Thoms[5] observed the presence of *asarone* m. 60°–61°, *cineole* and *terpenes* (about 10 per cent) which, however, were not identified. Thoms did not succeed in establishing the presence of apiole and dillapiole (see above).

OIL OF *Piper Angustifolium* Var. *Ossanum* C. DC.

In an oil distilled from the leaves of *Piper angustifolium* var. *ossanum* C. DC. (yield 0.87 per cent), Thoms[6] noted only traces of phenol ethers. The oil contained probably *camphor* and *borneol*.

OIL OF *Piper Camphoriferum* C. DC.

Distilling the leaves of *Piper camphoriferum* C. DC., Thoms[7] obtained 1.11 per cent of an essential oil (d_{20} 0.9500, α_D +19° 21′) in which he noted the presence of *d-camphor, borneol, terpenes, acids, phenols*, and *a sesquiterpene alcohol* (α_D +5° 0′, n_D 1.50208).

OIL OF *Piper Lineatum* RUIZ ET PAVON

Distilling the leaves of *Piper lineatum* Ruiz et Pavon, Thoms[8] obtained 0.44 per cent of an oil, the principle fraction of which boiled between 140°–160° (at 15 mm.) and consisted chiefly of *sesquiterpenes*. Thoms could not identify any camphor or phenol ethers in the oil.

---

[5] *Pharm. Ztg.* **49** (1904), 811. *Arbeiten aus dem pharmazeutischen Institut der Universität Berlin* **2** (1905), 121.

[6] *Apoth. Ztg.* **24** (1909), 411. *Arch. Pharm.* **247** (1909), 591.

[7] *Ibid.*   [8] *Ibid.*

OIL OF *Piper Acutifolium* RUIZ ET PAVON VAR. *Subverbascifolium*

Distilling leaf material consisting of a mixture of *Piper acutifolium* Ruiz et Pavon var. *subverbascifolium*, and some *Piper mollicomum* Kunth., and *Piper asperifolium* Ruiz et Pavon, Thoms [9] obtained 0.8 per cent of an oil (d_{20} 1.10, α_D +0° 24′, $[\alpha]_D$ +0° 21.8′, methoxy content 21.8–22.1 per cent, content of acids plus phenols 1.5 per cent), in which he noted the presence of α-*pinene, sesquiterpenes, dillapiole* and *dillisoapiole* (the latter after distillation over metallic sodium).

OIL OF BETEL

Essence de Betel   *Aceite Esencial Betel*   *Betelöl*
Oleum Foliorum Betle

The preparation used by the natives of India, Malaya, Indonesia, and other parts of tropical Asia for betel-chewing consists of a thin slice of areca nut (*Areca catechu* L.), with a dash of slaked lime, the mixture wrapped in a fresh leaf derived from *Piper betle.*

Piper betle L. (fam. *Piperaceae*) is a vine allied to that yielding common pepper. Its leaves, commonly known as betel leaves ("Sirih" in Malayan), contain an essential oil of spicy and burning flavor. The oil has been distilled on several occasions for experimental purposes, but is not produced commercially. The yield of oil from dried leaves ranges from 0.6 to 1.8 per cent, being highest with young leaves.

Physicochemical Properties.—Oil of betel is a yellowish-brown liquid with a burning sharp flavor and an odor reminiscent of creosote and tea.

According to Gildemeister and Hoffmann,[1] the specific gravity of the few betel oils that have been examined ranged from 0.958 to 1.057; the optical rotation was either slightly dextro- or slightly laevorotatory.

Chemical Composition.—The chemical composition of betel oils varies with the origin of the leaves. Some oils contain up to 55 per cent of phenols. The chief constituent, characteristic of all betel oils, is *chavibetol;* some, but not all, contain also *chavicol.*

[9] *Ibid.*
[1] "Die Ätherischen Öle," 3d Ed., Vol. II, 474.

The presence of the following compounds has been observed in betel leaf oils of various origin:

Cineole

Cadinene

Caryophyllene

Methyleugenol

Chavibetol ("Betel Phenol")

Chavicol

Allylpyrocatechol

All of these compounds are fully described in Vol. II of the present work.

Use.—As was mentioned above, oil of betel is not produced commercially.

CHAPTER V

ESSENTIAL OILS OF THE PLANT FAMILY *ANACARDIACEAE*

OIL OF *SCHINUS MOLLE* L.

Introduction.—*Schinus molle* L. (fam. *Anacardiaceae*), the so-called California pepper tree, is an evergreen, up to 20 ft. in height, and native to the American tropics, where it grows wild. Because of its graceful hanging branches, feathery foliage, and yellow fragrant flowers, which grow in terminal panicles, *Schinus molle* is cultivated as an ornamental garden tree in many warm sections of America, Southern Europe, and North and South Africa. The berries (fruit) are aromatic and possess a sweetish, spicy and sharp pepper-like flavor, reminiscent also of elemi (phellandrene). In Greece the berries serve for the preparation of certain beverages, and are often employed as a substitute for true pepper. The berries, as well as the leaves, contain a volatile oil which can be isolated by steam distillation. According to Gildemeister and Hoffmann,[1] the berries yield from 3.35 to 5.2 per cent of essential oil.

When World War II brought about a shortage of true pepper from the Malayan Archipelago, attempts were made in the United States to substitute the oil derived from *Schinus molle* for true pepper oil. In general, the *Schinus molle* trees grow scattered in many places, including parks and gardens. Collection of the leaves and berries, therefore, requires much labor. In Guatemala, e.g., the Indians cannot be induced to collect the material unless they are paid sufficiently attractive wages. In some instances, natives started cutting branches from trees on public land, but were arrested and jailed because they had not procured the necessary permit from local authorities. To produce the oil in large enough quantities, it would be necessary to grow large stands of trees on private land, obviously a long and costly task. Nevertheless, efforts are being made in Central America to produce oil of *Schinus molle*. Berries have been shipped to New York for distillation, and growers in Guatemala are trying to produce the oil also locally. There has also been sporadic production of oil of *Schinus molle* during the last few years in Mexico State (Mexico), annual totals, however, varying widely. With a steady demand supplies from Mexico could undoubtedly be stabilized.

---

[1] "Die Ätherischen Öle," 3d Ed., Vol. III, 206.

OIL FROM THE BERRIES

Physicochemical Properties.—Schimmel & Co.[2] examined four oils of *Schinus molle* derived from the berries, and reported the properties listed below. No origin is indicated for oil (I); oils (II) and (III) came from Mexico; oil (IV) was of Spanish origin:

	I	*II*	*III*	*IV*
Specific Gravity at 15°............	0.850	0.8600	0.8320	0.8561
Optical Rotation...	a_D^{17} +46° 4'	+42° 30'	+60° 40'	+62° 42'
Refractive Index at 20°.............	...	...	...	1.47877
Acid Number......	...	...	0.4	...
Ester Number......	...	25.2	9.3	...
Ester Number after Acetylation......	...	56.5	14.0	...
Solubility..........	Soluble in 3.3 and more vol. of 90% alcohol	In the beginning clearly soluble in 98% alcohol; very turbid on addition of 2 vol. and more alcohol	Clearly soluble in 0.5 and more vol. of 95% alcohol. Turbid in 5 vol. of 90% alcohol, but opalescent with more	Soluble in about 4 vol. and more of 90% alcohol, with slight turbidity

Oil (III) had an odor reminiscent of water fennel oil. It contained a high percentage of phellandrene and traces of phenols (carvacrol?). Oil (IV) also contained much phellandrene and only traces of phenols.

Two oils of *Schinus molle* distilled by Fritzsche Brothers, Inc., New York, from imported berries (yield 7.2 per cent and 5.5 per cent) were light-colored liquids with an odor reminiscent of pepper and elemi. The oils had the following properties:

Specific Gravity at 15°/15°........	0.856	0.866
Optical Rotation.................	+41° 17'	+26° 55'
Refractive Index at 20°...........	1.4790	1.4818
Acid Number...................	2.4	2.5
Ester Number...................	10.5	8.3
Ester Number after Acetylation....	32.3	66.3
Solubility.....................	Hazy in 10 vol. of 90% alcohol	

[2] *Ber. Schimmel & Co.,* April (1897), 49; April (1908), 124; (1926), 112.

Chemical Composition.—The chemical composition of the oil derived from the fruit of *Schinus molle* was investigated by Spica,[3] and Gildemeister and Stephan,[4] who reported the presence of the following compounds:

α-Pinene. Passing gaseous hydrogen chloride through the lowest boiling fractions of the oil, Spica obtained a solid monohydrochloride m. 115°, which indicates the presence of *α*-pinene. In the opinion of Gildemeister and Stephan, the quantity of pinene present in the fraction b. 170° of the oil which they investigated, could not be more than 0.5 per cent, because these authors obtained only a very small quantity of a nitrosochloride.

Phellandrene. The bulk of the oil boiled between 170° and 174° (α_D +60° 21') and gave a strong phellandrene reaction with sodium nitrite and glacial acetic acid. From the optical behavior of the various crystallization products of the nitrite, it appears that the phellandrene present in oil of *Schinus molle* is a mixture of much *d*- and very little *l*-phellandrene. According to Wallach,[5] the phellandrene consists chiefly of *d*-*α*-phellandrene, but perhaps some *β*-phellandrene is also present.

Carvacrol. Extracting the phenols from the oil with alkali solution, Spica obtained a phenol, the nitroso compound of which melted at 156°. Spica considered this phenol to be thymol. On the other hand, Gildemeister and Stephan could not identify any thymol in their oil, but they established the presence of carvacrol, which they characterized by means of its isocyanate compound m. 140°.

Sobrerol(?). In the fraction b. 180°–185°, which had been exposed to light for a prolonged period, Spica observed a crystalline substance m. 160°, which he thought to be sobrerol (pinol hydrate—cf. Vol. II of this work, p. 713). This, however, is unlikely, since pinol hydrate melts at 131° (Gildemeister and Stephan).

OIL FROM THE LEAVES

Physicochemical Properties.—Schimmel & Co.[6] examined an oil (I) distilled in Mexico from the leaves of *Schinus molle*, while Roure Bertrand Fils [7] reported on the properties of three oils which had been produced in Algeria (II and III) and in Grasse (IV):

	I	*II*	*III*	*IV*
	From Leaves	*From Leaves*	*From Leaves and Twigs*	
Specific Gravity at 15°........	0.8583	0.8658	0.8634	0.8696
Optical Rotation.............	+44° 50'	+65° 20'	+50° 54'	+46° 13'
Refractive Index at 20°.......	1.47665	...	...	...
Ester Number...............	7.2	3.4	5.5	8.2
Ester Number after Acetylation ...		40.4	29.4	43.4

[3] *Gazz. chim. ital.* **14** (1884), 204.
[4] *Arch. Pharm.* **235** (1897), 589.
[5] *Nachr. K. Ges. Wiss. Göttingen* (1905), Heft 1, 2. *Chem. Zentr.* (1905), II, 674.
[6] *Ber. Schimmel & Co.*, April (1908), 124.
[7] *Repts. Roure Bertrand Fils*, April (1909), 36. Laloue, *Bull. soc. chim.* [4], **7** (1910), 1107.

Chemical Composition.—According to Schimmel & Co.,[8] the oil derived from the leaves of *Schinus molle*, like that from the berries, contains large quantities of *phellandrene*.

OIL FROM LEAVES AND BERRIES

Steam-distilling 13 kg. of leaves and attached green berries of South African *Schinus molle*, Brückner van der Lingen [9] obtained 64 cc. of a colorless oil which, on exposure to air and light, was oxidized and polymerized into a gluey mass. The oil had these properties:

Specific Gravity.................... 0.8486
Optical Rotation................... +68° 24′
Refractive Index at 20°............ 1.4732
Acid Number...................... Below 0.1
Ester Number 46.7
Ester Number after Acetylation...... 115.8
Solubility........................ The solution in 90% alcohol suddenly turned turbid when absolute alcohol was added

The same author reported the presence of α-*phellandrene* (about 26 per cent), and *carvacrol* in the oil. The α-phellandrene was identified by means of its nitrosite m. 105°–115°, the carvacrol through its phenylurethane m. 140°.

Over a number of years, the laboratories of Fritzsche Brothers, Inc., New York, have examined a number of *Schinus molle* oils, produced in various parts of Central America, probably from leaves and berries, and found the following properties:

Specific Gravity at 15°/15°...... 0.831 to 0.867
Optical Rotation............... +40° 30′ to +59° 1′
Refractive Index at 20°......... 1.4720 to 1.4831
Saponification Number.......... 5.6 to 23.6
Solubility..................... Hazy to turbid in 10 vol. of 90% alcohol

Use.—The oil may be used in all kinds of spicy flavor compositions.

SUGGESTED ADDITIONAL LITERATURE

Achille Cremonini, "Investigations on the Fruits of *Schinus molle*," *Ann. chim. applicata* **18** (1928), 361.

[8] *Ber. Schimmel & Co.*, April (1908), 124.
[9] *Perfumery Essential Oil Record* **21** (1930), 154.

OIL OF MASTIC
(Oil of Mastiche)

Mastic is the air-dried resinous exudation from *Pistacia lentiscus* L. (fam. *Anacardiaceae*), a shrub or small evergreen tree, seldom more than 12 ft. in height, and native to the countries bordering the Mediterranean. The mastic originates in special oleoresin reservoirs located in the inner bark of the trunk and branches. It is obtained by making longitudinal incisions at close intervals, with a knife, from the base of the trunk up to the thicker branches. The juice slowly exudes, and either hardens in tears on the bark, or drops to the ground, where it is collected from canvas or from the earth. It solidifies in the form of irregular masses. The tears are most esteemed; they are, in fact, the only form of mastic recognized by the National Formulary. The mastic imported into the United States and Great Britain originates chiefly from Chios, an island in the Grecian Archipelago, where the tree is cultivated.

On distillation, mastic yields from 0.7 to 1, on rare occasions up to 3, per cent of an essential oil.

Physicochemical Properties.—Oil of mastic is a colorless liquid with a pronounced balsamic odor, reminiscent of the gum. According to Gildemeister and Hoffmann,[1] its properties vary within the following limits:

Specific Gravity at 15°...........	0.857 to 0.903
Optical Rotation................	+22° 0' to +35° 0'
Refractive Index at 20°..........	1.468 to 1.476
Acid Number...................	Up to 5
Ester Number...................	2.5 to 19
Ester Number after Acetylation...	14 to 32
Solubility......................	Only sparingly soluble in 4 to 10 vol. of 90% alcohol. Up to 5 vol. of 95% alcohol are required for solution, but the solution is usually not clear, turning opalescent to turbid on further addition of alcohol

Genuine mastic oils produced in Seillans (Var), France, under the author's supervision exhibited these properties:

Specific Gravity at 15°......	0.868 to 0.887
Optical Rotation..........	+24° 18' to +30° 25'
Refractive Index at 20°.....	1.4697 to 1.4760
Acid Number.............	0.46 to 5.9

[1] "Die Ätherischen Öle," 3d Ed., Vol. III, 202.

Ester Number............. 11.0 to 16.6
Solubility................. Soluble in 0.5 vol. of 95% alcohol,
                          opalescent to cloudy with more

Chemical Composition.—The chemical composition of oil of mastic was investigated by Flückiger,[2] and by Schimmel & Co.,[3] who found that oil of mastic contains *d-α-pinene,* and a small percentage of optically inactive *α-pinene.* Aside from pinene, another terpene appears to be present in the oil, but it has not yet been characterized.

Production and Use.—Only very small quantities of oil of mastic are produced commercially. The oil finds limited use in flavoring compounds, particularly in certain liqueurs.

SUGGESTED ADDITIONAL LITERATURE

G. Reinboth, "Producing Oils and Resins in Italy," *Farbe u. Lack* (1938), 364. *Chem. Abstracts* **32** (1938), 9532.

[2] *Arch. Pharm.* **219** (1881), 170.
[3] Cf. Gildemeister and Hoffmann, "Die Ätherischen Öle," 3d Ed., Vol. III, 203.

CHAPTER VI

ESSENTIAL OILS OF THE PLANT FAMILIES *SANTALACEAE* AND *MYOPORACEAE*

OIL OF SANDALWOOD EAST INDIAN

Essence de Santal des Indes   *Ostindisches Sandelholzöl*
Aceite Esencial Sandalo Indias Orientales   *Oleum Ligni Santali*

Botany and Occurrence.—The common term "Sandalwood" refers to trees belonging to several plant families and genera. There was much controversy among taxonomists over the classification of the various types of sandalwood trees before their identity was clearly established. (Cf. the monograph on "Oil of Sandalwood, West Indian," Vol. III of this work, p. 385, and the following monograph on "Oil of Sandalwood, Australian" in the present volume, p. 187.)

By far the most important sandalwood oil is that distilled from the roots and heartwood of *Santalum album* Linn. (fam. *Santalaceae*) an evergreen tree almost entirely confined to the forests of Mysore (southern India) and closely adjacent districts. This—the East Indian sandalwood oil—differs in chemical composition and odor from other, so-called sandalwood oils, which are derived from entirely different trees.

According to Sastry,[1] sandalwood as such is used for religious purposes in India by the Hindus and Parsis; as perfume materials, the wood as well as the oil have been favorites from time immemorial. History tells us of caravans passing over the deserts of Persia, Arabia, and Asia Minor carrying Indian sandalwood to Egypt, Greece, and Rome. It has been perhaps one of the most precious perfumery materials from antiquity down to modern times, and its popularity has shown no signs of waning.

Santalum album L. is a native of the highlands of southern India and the Malayan Archipelago. It occurs ferally in open dry places, preferably at altitudes ranging from 2,000 to 3,000 ft. It is also planted (by seed), particularly in the State of Mysore, the principal producing region of the wood and essential oil. The tree, which may attain a height of 60 to 65 ft., is actually an obligate hemiparasite plant. Soon after germination of the seed, the roots attach themselves to those of nearby grasses, herbs, bushes, and undergrowth in general, obtaining food by means of the haustorium, and finally causing the host plants to perish. As the sandalwood tree grows, essential oil develops in the root and heartwood. (In this connection it should be mentioned that occasionally a tree may reach physiological maturity without forming any heartwood.) The trees reach full maturity at an age of sixty to eighty—or even more—years. The strict forestry laws enforced in the State of Mysore do not permit felling of a sandalwood tree before it is dead. Trees younger than thirty years are therefore never cut unless they show signs of the dreaded spike disease (see below). The trunks are sawed into segments about 3 ft. long, and the bark and soft sapwood are removed. The roots are treated similarly. The remaining pieces of heartwood represent the scented sandalwood, so highly valued for the carving of all kinds of art objects, for use as incense, and for distillation of the fragrant essential oil.

Total Production of Wood and Oil.—*Santalum album* L. grows wild not only in South India but also in certain parts of the Malayan Archipelago, particularly on the dry island of Timor. Wood of this origin is often referred to as "Macassar Sandalwood"; it is sold at yearly auctions in Kupang, on the island of Timor. No oil is produced locally on the island. Prior to World War II, exports of sandalwood from Indonesia varied between 80 and 150 metric tons per year. In 1938, for example, almost 150 tons were exported, 3 tons of which went to Singapore, about 12 tons to Holland (probably for distillation purposes), while the bulk was shipped to Hong Kong for use in incense and for wood carving.

As regards India, the producing regions lie in the mountainous highlands of the south, comprising parts of Coorg, Madras, and particularly Mysore, the world's greatest sandalwood producer. In the State of Mysore the sandalwood belt reaches from the southern tip about 300 miles to Dharwar in the north, and from Coorg in the west approximately 250 miles to Kuppam in the east. About 2,000 English tons of sandalwood, or more than 75 per cent of the total production of India, come from Mysore every year. Nevertheless, up to the outbreak of World War I distillation of the oil from the wood had been carried out almost exclusively by European essential oil houses. This situation has been completely revised since World War I when lack of shipping space prevented export of the bulky wood to Europe. Except for very small quantities of wood purchased by European interests at Madras and Coorg auctions, 95 per cent of the East Indian sandalwood oil is now distilled by Indian interests. At present the government of Mysore holds the key position in this industry, controlling seven-eighths of the world's production of wood, and distilling over 75 per cent of the world's output of oil. On account of tariff considerations the government of Mysore has arranged to get a portion of their wood distilled in New York, the oil being sold in the United States and Canada. The yearly average of wood thus distilled abroad on the Mysore Government account is about 500 tons.

(Left) Production of balsam copaiba in the lower Amazon Basin, State of Pará, Brazil. A hole is drilled into the lower part of the trunk, from which the balsam will flow. (*Right*) Production of balsam Peru in El Salvador, C. A. "Cascara Process": Incisions are made in the bark of a tree, after it has been scorched with a torch flame. (*Bottom Left*) Balsam Peru, El Salvador, C. A. "Trapo Process": Burlap is hammered into the incisions to absorb the flowing balsam. (*Bottom Right*) Balsam Peru, El Salvador, C. A. "Trapo Process": A bundle of burlap soaked with balsam and removed from the incisions. *Photos Fritzsche Brothers, Inc., New York.*

(*Top Left*) Production of cananga oil in the Bantam region of Java (Indonesia). Collection of the flowers with long hooked bamboo poles. (*Top Right*) Cananga oil, Bantam region. A small boy climbing into the tree with a long bamboo pole to reach the flowers. (*Bottom Left*) Cananga oil, Cheribon region of Java (Indonesia). The harvesters climb up the tall trees on a sort of bamboo ladder. (*Bottom Right*) Cananga oil, Cheribon region. A wealthy Chinese broker who purchases the bulk of the oil from the native distillers. *Photos Fritzsch Brothers, Inc., New York.*

According to Sastry,[2] the total amount of sandalwood distilled varies from 2,500 to 3,000 tons a year, depending on the demands for sandalwood oil and other economic factors. Accurate statistics are not easily available.

A published account [3] indicates the following quantities of distillation material processed in 1927–28 in India (Mysore and Bangalore factories), and in New York:

Class of Wood	India (in tons)	New York (in tons)	Total (in tons)
Roots.............	306	. . .	306
Billets.............	240	299	539
Jajpokal..........	408	75	483
Sandal chips......	78	. . .	78
Milwa chilta......	612	. . .	612
	1,644	375	2,021

(Figures approximate)

From this wood the two government factories in Mysore distilled 167,720 lb. of sandalwood oil; 45,840 lb. were distilled in New York. The average yield of oil varied from 105 to 110 lb. per ton of wood.

Habitat.—According to information gathered by the author [4] while surveying production of sandalwood oil in the south of India, *Santalum album* L. thrives on well-drained, loamy soil. It grows also on laterite, but not waterlogged ground, and preferably on slopes of hills exposed to the sun. It requires a minimum of 20 to 25 in. of rainfall per year; more than 80 in. is harmful. The finest wood, as regards odor, grows in the driest regions, particularly on red or stony ground. On rocky ground the tree often remains small, but gives the highest yield of oil. It grows most abundantly in the dry-deciduous belt along the banks of the Cauvery River, among the mountains which run through the State of Mysore from north to south. Trees more than thirty years old may have a circumference of from 18 to 38 in. The bark and the sapwood are odorless; the roots and heartwood contain the essential oil.

As has been pointed out, *Santalum album* is a parasite which attaches its roots to those of neighboring plants. In this respect the tree has certain preferences, its most suitable hosts being *Cassia siamea, Pongamia glabra* and *Lantana acuminata*. Less suited are *Bassia latifolia, Dalbergia, Sissoo* and *Fiecus* species. Experiments regarding the most suitable host

[2] *Ibid.*
[3] E. J. Parry, "Sandalwood Oil," a monograph published by the government of Mysore.
[4] The author is greatly indebted to Mr. S. G. Sastry, former Director of Industries and Commerce in Mysore, now retired in Bangalore, and to Dr. M. N. Ramaswami, Mysore Government Sandalwood Oil Factory, Mysore, for their kindness in correcting the present monograph, and bringing all information up-to-date.

plants are still being carried out by the forest service of Mysore State. If all neighboring hosts were removed, the roots of *Santalum album* would stretch out farther and farther in search of sustenance, until, finding none, the tree would perish.

Propagation.—In years past the sandal tree occurred mainly in the wild state, but reckless exploitation and the danger of losing the great revenue from forests induced the Mysore Government to restrict the cutting, and to propagate large numbers of trees every year (by sowing). At one time propagation of the sandal tree was considered a difficult matter, but the problem has been much simplified, thanks to the experimental work of the forestry department of Mysore.

A certain amount of propagation takes place by means of birds. The blue-colored fruit of the tree is juicy and sweet and much liked by birds, which eat the outer fleshy part and drop the hard seed. Falling on suitable soil, this seed germinates, taking root and, unless destroyed by its natural enemies, grows into a tree. In past years this was the only way the sandal tree was propagated.

Today most sandal trees, at least in the State of Mysore, are propagated by sowing. The seed is placed directly into the ground, because the delicate nature of the root, which must prey on a host root, does not permit transplanting of the young trees from a nursery. The sowing is done about two weeks before the arrival of the monsoon rains, usually at the end of April. Since not all seeds germinate, five seeds are planted in a semicircle and surrounded by a number of host plants. A great enemy of the sandal tree is the common rat, which burrows into the ground for food and eats the seed. Therefore, the seed is given a protective coating of red lead paste prior to planting. The young plants grow very slowly. Foraging goats frequently attempt to feed on it and may destroy it, but the dense, thorny foliage of the surrounding host plants protects the young and tender sandal tree.

The gravest peril to the sandal tree is the so-called spike disease. This is extremely dangerous because of its contagious nature, and also because no remedy has as yet been found. It represents a kind of "plant consumption," the nature of which is not fully understood. It is possible to infect a healthy sandalwood tree by grafting a piece of wood from a diseased tree upon it; soon after, characteristic symptoms of the spike disease can be noticed—a sickly appearance, a grayish color, and under-development of the leaves, which remain small and curl up. Perhaps the spike disease is carried from one tree to another by insects. Whenever the forest service notices any sandal tree attacked by the disease, they immediately order it quarantined and destroyed. This is done by cutting a ring around the tree and painting the uncovered wood with a solution of arsenic trichloride in hydrochloric acid, or with a solution of sodium arsenate. The tree then dies in three to four weeks and the germs of the disease are destroyed. Experiments have proved that oil distilled from such treated trees contained no traces of arsenic.

Under normal conditions young sandal trees grow slowly, gradually developing a core of heartwood which, together with the root, contains the much valued essential oil. It is estimated that this core should be at least 3 in. thick before exploitation of the tree becomes worth while. To attain a thickness of 3 in. requires more than thirty years. The soft, white outer layer, the so-called sapwood, contains no essential oil and is therefore removed and discarded. The older the tree, the thicker its core of heartwood and the more value it has as fragrant sandalwood. Since the roots consist chiefly of heartwood, they are richest in essential oil. According to the classification established in 1898 there exist 18 grades of sandalwood, ranging from the heartwood down to the chips and sawdust. The grading refers to the physical condition of the wood, not its quality. An exact classification is given by Gildemeister and Hoffmann.[5] Today this grading has lost much of its former significance, probably because the bulk of sandalwood is now used in the government distillery of Mysore State for distillation of the oil. There is quite a sharp line of demarcation between the soft, white wood and the darker heartwood. The thickness and appearance of the heartwood, which are functions of age, are also considered. The darker the wood the higher is the oil content. Trees from mountainous, rocky, and dry soil develop the hardest wood and the greatest amount of oil. The thickest blocks of sandalwood are too valuable for distillation because they can be used more advantageously for wood carvings, or for cabinet making, but all the lower grades may serve for distillation, the only difference then being in regard to yield of oil. The roots yield up to 10 per cent of oil, while the intermediary layers between soft and hard wood give only about 2½ per cent. The yield may be as low as 1 per cent and even less, in chips which contain much sapwood. Chips and sawdust are frequently employed in incense.

Sandalwood Production in Mysore State.—Years ago, the government of Mysore enacted laws to stop the reckless exploitation of the sandalwood forests which threatened this source of revenue to the country. The entire sandalwood industry of the State of Mysore is now organized on the basis of a strict government monopoly. Laws and regulations governing the production and handling of sandalwood are equaled in strictness only by those which govern the mining of gold in other countries. Sandalwood and

[5] "Die Ätherischen Öle," 3d Ed., Vol. II, 504.

its oil represent gold to Mysore, favored as the state is with such an abundance of this tree.

In Mysore every single sandal tree belongs to the government, whether it be grown on public, private, or temple grounds. The law obliges every proprietor of land on which a sandal tree grows to give this tree the utmost care. No individual is permitted to cut a tree, possess, transport, or export sandalwood, or distill the wood without permission. Every single piece of wood which may serve for carving, for example, must carry the government stamp, certifying that it has been purchased from the government and that it is not stolen, or "bootleg" wood. Only dead or fully grown trees which show signs of death from old age may be felled, after permission has been granted by a government forest official. He, too, is subject to fines should he allow the felling of a healthy tree, unless it is done for experimental or scientific purposes. Similar regulations apply to the transport of sandalwood, with the result that today the Mysore Government has at its disposal an abundance of old sandal trees for the distillation of oil or for the export of wood. In addition, the Mysore forestry service has embarked on a long-range program of replanting a certain acreage with sandal trees each year, so that the future supply of this valuable wood may be assured; in fact, it should be almost inexhaustible. Only 15 miles outside of the State of Mysore one sees large tracts of land covered with young sandalwood trees. The reforestation program of the government is carried out thoroughly and efficiently, and taking into consideration all kinds of possible adversities—weather, for instance, which might in some years prevent the development of newly planted trees. By strict laws regulating cutting and transporting, the sandalwood remains government property throughout the various phases from sowing of the seed to export of the wood, or marketing of the oil distilled in the government-owned factory.

The actual felling of the tree is not done by the government; the forest officials merely mark the trees which may be cut. The work is then assigned to the lowest bidding party who enters contracts to cut and transport the wood to the government stations. Contractors comprise organized groups of woodsmen whose expenses consist chiefly of labor charges and maintenance costs for bullocks and carts. Since the government pays the contractors only by weight of material received, woodcutters would prefer to fell only the trunks, the pulling of the root being a much more difficult job. However, the forestry service requires that for each trunk the corresponding root, too, must be delivered and that the wood must be shipped in long pieces. Too much sawdust would be lost if the wood were cut into small segments in the forests. It is also unlawful to remove the soft, white sapwood on the spot, and to haul only the heartwood to the government stations, because the stacking up in the forests of the valuable and less bulky heartwood is risky. Hence the transported sandalwood must have a layer of at least ½ in. of sapwood around the heartwood. This measure has been enacted chiefly to provide a certain protection against theft at night or en route; moreover, it protects the valuable core of sandalwood from mechanical injury during the transport.

The felled sandalwood trees and the pulled roots are transported first by bullock carts through the more inaccessible sections of the forests, subsequently by motor trucks over the highways, and finally by railroad to the warehouses of the forest service. These latter are distributed throughout the producing regions. Because of possible theft, guards patrol the forests, and every forest official has the right to stop and search any suspicious vehicle. Sandalwood readily betrays itself by its pronounced odor, for which reason some thieves try to hide the stolen wood in sacks of coriander seed or onions. Such tricks, however, are well known to the forestry inspectors. Guards patrol also the state line of Mysore to prevent the smuggling out of stolen wood.

The first step in the operations performed in government warehouses is the removal of the outer layer of sapwood by crews of men and women. All work is carried out under strict supervision, and only a minimum of waste is permitted in the sawing and chipping of the wood. Chips and sawdust are carefully collected and used for distillation, or sold for use in incense. Finally, the forestry department sells the wood to the Mysore Government Sandalwood Oil Factory, a mere book transaction because all the wood belongs to the government. Wood which is not used for distillation is sold or auctioned off, and transported by rail to Bombay or other ports, from where it is shipped to all parts of the world.

The State of Mysore annually produces about 2,000 tons of sandalwood, the bulk of which serves for distillation of oil in the Mysore Government factory; 10 or 12 per cent of the tonnage is sold and exported chiefly for wood carving and use in incense. Throughout India, sandalwood powder is burned in religious rites. The Hindus cremate their dead and use sandalwood in the burial pyres, while the fire-worshiping Parsis also employ much sandalwood powder in incense. The above-mentioned figure of 2,000 tons is only an average, because the government restricts cutting according to actual requirements for wood and oil, which can be anticipated quite closely by years of experience in marketing the oil throughout the world.

Wood and Oil Production in Coorg and Madras.—Next to the State of Mysore the most important producers of sandalwood in India are the neighboring Province of Coorg and the Presidency of Madras. In Coorg and Madras the cutting of sandalwood is also regulated, but much less strictly than in Mysore. Any trees growing on private land are the property of the owner and may be cut and sold freely. Auctions of sandalwood are held regularly in Fraserpett, Tirupatur, and Satayamangalam. However, these auctions have lost much of their former importance, because very little oil is produced today outside of India. (The one exception is the oil distilled in New York by the sole agents of the Mysore Government Sandalwood Oil Factory, from wood shipped from Mysore. This is done merely because of United States tariff considerations. Sandalwood enters the United States free of duty, whereas the oil is subject to an import tax.) Essential oil houses in Europe and in the United States determined to distill their own oil would have to import the wood either from Coorg or Madras, or from the island of Timor in the Malayan Archipelago (see above).

There are two small sandalwood distilleries in Kuppam (Madras Presidency) which have been operating for more than twenty years, one factory at Bombay, one at Kanauj, one at Mettur, and one at Karaikal. The average output of oil in the Mysore factory is about 120,000 lb. per year, the maximum capacity being 50 per cent more. The output of all the other factories in India is about 40,000 to 50,000 lb. a year.

Distillation.—Sastry [6] has given a detailed account of the distillation of sandalwood, as carried out today in India. The following description is based chiefly upon his experience as Director of Industries and Commerce, Mysore, and as Director of the Mysore Government Sandalwood Oil Factory.

The distillation of sandalwood oil in India has been carried on from very ancient times by the so-called water distillation method (cf. Vol. I of the present work, pp. 112, 120, 142). It consists in soaking the raw material in water in a copper vessel and heating it on an open fire. The vapors from the body of the still are conducted through a bamboo or copper pipe to receivers which are kept in cold water, the latter being renewed frequently. Even today such methods of distillation are practiced on the Malabar Coast and also in some places in northern India, notably in Kanauj. If the method is conducted with care and vigilance, it is still possible to obtain a high quality sandalwood oil, even though the yield may be somewhat low. Many old-fashioned *attar* distillers in Kanauj still prefer sandalwood oil prepared by the old-fashioned water distillation method. This oil is claimed to possess a finer odor than the oil produced in modern steam stills.

In modern operation the raw material consists of billets and roots of sandalwood and also a quantity of chips, the latter being a mixture of both the heartwood and the sapwood. The yield of oil varies, the roots giving the highest percentage and the chips the least. The average yield from good billets and roots ranges between 4.5 to 6.25 per cent.

[6] *J. Sci. Ind. Research* **3**, No. 2 (1944), 1.

The wood is first of all fed to a chipping machine consisting of a rapidly rotating disc on which are mounted 6 knives radially; the wood pressed against these knives gets reduced into a coarse powder. The latter is next fed to disintegrators which will reduce the wood into a finer state of division. In some of the factories, the chipper is avoided and manual labor employed to chop the wood by means of hatchets and adzes into small-sized sticks or chips. These are then fed into disintegrators. The powder that comes from the disintegrator is carefully sieved and remixed with the object of obtaining a powder that will not "pack in" too tightly in the stills but will form a uniform porous body of material which admits of the passage of steam easily over the entire mass of powder to be distilled.

The stills, usually made of copper, rarely of iron, are provided with goosenecks to conduct the vapor of the oil and steam to tin-lined tubular condensers. The receivers, also tin-lined, are constructed on the principle of the old Florentine flasks. The size and shape of the stills vary, but the standard equipment is a still holding a charge of $\frac{3}{4}$ to 1 ton of powdered wood, the latter being generally placed on a perforated false bottom. There is usually a little space on the top of the "burden" in the still. The height of the still is about 25 per cent more than the diameter. Stills with a height very much greater than the diameter are also in use but require a careful distribution of the powder inside, as otherwise the steam will have a tendency to force an easy passage without spreading itself uniformly over the entire mass of the powder.

Distillation generally requires 48 to 72 hr. Each factory, according to its economic conditions, has its own end point of distillation. The latter is stopped when the yield of oil ceases to be economical. The distillate collects in the receivers, the crude sandalwood oil floating on the surface. This is generally skimmed off with the help of shallow ladles and then put into a separating funnel where oil and water layers are further separated. The crude oil is stored in a separate vessel and allowed to stay in this condition for some time when a little scum with suspended woody matter comes to the top. The oil is then carefully filtered.

Distillation is usually conducted with low pressure steam, anywhere between 20 to 40 lb. High-pressure steam has been recommended for sandalwood oil distillation on the score of a slightly higher yield of the oil and a slight saving in time. Obviously, high-pressure steam means higher temperature and the delicacy of the odor of some of the constituents will suffer at higher temperatures. Orthodox opinion favors low-pressure distillation, some factories in England and America using steam only at 10-lb. pressure, even today.

In many factories in India adopting the method of steam distillation mentioned above, the oil does not receive any further treatment. At best, it is warmed once again in order to drive out most of the moisture contained in the oil and filtered once again. Such oil is marketed straight away as "Perfumery grade of oil." It has been stated that many of the *attar* manufacturers in India have no objection to the use of this oil in their preparations. Some of this oil is also exported to western countries where it is further rectified. There have been instances where by careful distillation of selected wood the oil obtained in the above manner comes within the range of pharmacopoeial standards, but there is no certainty about this.

If the oil is to conform to the pharmacopoeial standards of the different countries of the world, it should be carefully analyzed in the laboratory and further rectified. Very often this necessitates a distillation with superheated steam and final heating in a steam-jacketed vacuum still to remove the last traces of water, and also vacuum filtration. The oil finally obtained has a pale yellow color, is optically clear, and possesses the characteristic odor of sandalwood. By adopting the above methods it will be easy to produce a standard product day after day, with a certainty and a guarantee that the product will strictly conform to the rigid pharmacopoeial standards.

The first modern sandalwood oil factory to be established in India was at Bangalore during the middle of World War I, and this was followed by another distillery at Mysore. The Bangalore factory was closed down after some time and all the operations are now concentrated in the city of Mysore. This plant is complete in every branch of mechanical and chemical engineering equipment. It has an up-to-date control laboratory. The factory makes its own condensers in the copper smithy· section of the workshop. All the repairs and replacements are looked after by this workshop which has also a small foundry and a modern tin-can-making plant attached to it. The Mysore plant is the biggest and the most completely equipped sandalwood oil distillery in the world. During days of heavy production about 1,500,000 gal. of water circulate in the factory; the warm water flowing from the condensers is used again after it has been cooled in a sprinkling system and stored in large tanks. The exhausted wood is dried and used as fuel.

Packing.—The Mysore factory has set up a standard of packing sandalwood oil, and today more than 90 per cent of the oil on the world market is packed according to these specifications. This consists of a tin containing 25 lb. net of oil, after due allowances for ullage, each lot having a reference number. On the Indian market, the following other packings are also recognized: tins containing 10 and 5 lb.; aluminum bottles containing 1 lb., 8 oz., 4 oz., and 1 oz. For retail trade in India, 1-oz. aluminum containers are very popular and nearly 3 to 4 lakhs of these containers are being used in the trade every year.

Physicochemical Properties.—The volatile oil derived from the roots and heartwood of *Santalum album* L. is a somewhat viscous, yellowish liquid of peculiar, "heavy," sweet, and very lasting odor, characteristic of East Indian sandalwood.

Gildemeister and Hoffmann [7] reported these properties for East Indian sandalwood oil:

Specific Gravity at 15°............	0.973 to 0.985
Optical Rotation................	−16° 0′ to −21° 0′; in exceptional cases lower rotations have been observed
Refractive Index at 20°..........	1.504 to 1.509
Acid Number...................	0.5 to 8.0 (see below)
Ester Number..................	3.0 to 17.0
Ester Number after Acetylation...	Not less than 196
Total Alcohol Content, Calculated as Santalol...................	Not less than 90% (see below)
Solubility at 20°................	Soluble in 3 to 5 vol. and more of 70% alcohol; in 5 to 6 vol. and more of 69% alcohol; in 6 to 7 vol. and more of 68% alcohol (see below)

On these properties Gildemeister and Hoffmann [8] elaborate as follows:

Oils produced by the old-fashioned method of water distillation may exhibit a higher acid number than indicated above. The acid number can be reduced by fractionation of the oil, and by removal of the foreruns which contain teresantalic acid.

The higher its santalol content, the greater is the value of an oil. Oils of best quality exhibit a santalol content of 94 per cent and even more.

In the solubility test the solution must remain clear on further addition of solvent. Turbidity, however, is not necessarily proof of adulteration; it may be caused by the presence of decomposition or resinification products. These may be the result of improper distillation, or of age, the oil losing its originally normal solubility under the influence of air and light.

Oils produced by the old method of water distillation in primitive native stills often exhibit a dark color, an abnormally high specific gravity, and occasionally an empyreumatic odor. East Indian sandalwood oils with a high specific gravity have been known for a long time.[9] Warming sandalwood oil with water of 50° for ten days, Conroy [10] reported that the specific gravity increased from 0.975 to 0.989.

Genuine East Indian sandalwood oils examined by Fritzsche Brothers, Inc., New York, in the course of years had properties varying within the following limits:

---

[7] "Die Ätherischen Öle," 3d Ed., Vol. II, 508.   [9] *Chemist Druggist,* May 26, 1894.
[8] *Ibid.*   [10] *Ibid.*, August 19, 1893.

Specific Gravity at 25°/25°....... 0.969 to 0.975, sometimes as high as 0.979
Optical Rotation................ −15° 0′ to −19° 20′
Refractive Index at 20°.......... 1.5052 to 1.5077
Ester Content, Calculated as Santa-
  lyl Acetate.................... 1.6 to 5.4%
Total Alcohol Content, Calculated
  as Santalol.................. 90.3 to 97.4%
Solubility...................... Soluble in 3.5 to 5 vol. and more of 70% alcohol

A type sample of an oil produced and examined in the Mysore Government Sandalwood Oil Factory (April 1944) exhibited these values: [11]

Specific Gravity at 15.5°............ 0.9782
Optical Rotation at 20°............. −17° 6′
Refractive Index at 20°............. 1.5068
Ester Content, Calculated as Santalyl
  Acetate........................ 2.5%
Free Alcohol Content, Calculated as
  Santalol........................ 91.2%
Solubility......................... Soluble in 5 vol. of
  70% alcohol

Adulteration.—Present specifications of the various pharmacopoeias make it quite difficult to adulterate East Indian sandalwood oil to any considerable extent, without seriously affecting its properties. Formerly, oils of cedarwood, guaiac wood, West Indian, West Australian and South African sandalwood, and oils of copaiba balsam and gurjun balsam were used for this purpose. Addition of cedarwood oil or gurjun balsam oil increases the optical rotation, and decreases the specific gravity and solubility of the sandalwood oil. Copaiba balsam oil acts similarly, except that it usually decreases the rotation slightly. West Indian sandalwood oil is dextrorotatory and quite insoluble in dilute alcohol. West Australian sandalwood oil has a lower specific gravity, a lower laevorotation, and if not rectified, a lower ester number after acetylation, and poorer solubility than the East Indian oil. Its odor differs from that of the East Indian oil. Addition of most of the above-named oils to an East Indian oil will lower the santalol content of the latter, which should be 90 per cent or more in high-grade oils. To simulate a high alcohol content, synthetics and aromatic isolates such as terpineol, benzyl alcohol, or geraniol are occasionally used. They can be detected by acetylation of the oil; benzyl acetate and geranyl acetate will betray themselves by a fruity note in the odor of the acetylized oil. The presence of terpinyl acetate can be proved by fractional saponification of the acetylized oil for 1 hr. and 2 hr. (cf. Vol. I of this work, p. 336). A suspected oil should be submitted to fractional distillation *in vacuo;* any adulterants boiling lower than santalol can easily be separated.

[11] Sastry, *J. Sci. Ind. Research* **3**, No. 2 (1944), 1.

According to Sastry,[12] white paraffin oil has lately been observed as an adulterant of East Indian sandalwood oil. It can be detected by the so-called "Oleum Test" described in Vol. I of this work, p. 332.

A careful odor test is indispensable in the analysis of East Indian sandalwood oil.

Chemical Composition.—The chemical composition of the volatile oil derived from the root and heartwood of *Santalum album* L. has been the subject of numerous investigations. Details regarding the various constituents identified in the oil will be found in the respective monographs contained in Vol. II of the present work. The chief constituent, amounting to 90 or more per cent of the oil, is *santalol,* a mixture of two primary sesquiterpene alcohols $C_{15}H_{24}O$, viz., α- and β-santalol, in which the α- form predominates.

The presence of the following compounds has been reported in East Indian sandalwood oil:

Isovaleraldehyde. In the foreruns of the oil. Identified by means of the thiosemicarbazone m. 49°–53°. Aside from isovaleraldehyde, the oil contains other aldehydes b. 50°–130° (Schimmel & Co.[13]).

Santene. Observed by Müller [14] in the low boiling fractions of the oil (cf. Vol. II of this work, p. 79).

Nortricycloekasantalane(?). In the foreruns of the oil Schimmel & Co.[15] noted the presence of a completely saturated hydrocarbon $C_{11}H_{18}$, b. 183°, d_{15} 0.9133, d_{20} 0.9092, α_D −23° 55′, n_D^{20} 1.47860, which is probably identical with Semmler's nortricycloekasantalane. At room temperature the hydrocarbon was not affected by potassium permanganate solution.

l-Santenone. Müller [16] first reported that the foreruns b. 180°–200° of East Indian sandalwood oil contain a ketone, the semicarbazone of which melted at 224°. Later, Schimmel & Co.[17] proved that this ketone is identical with the π-norcamphor of Semmler and Bartelt,[18] and with the santenone of Aschan [19] (cf. Vol. II of this work, p. 438).

Santenone Alcohol (Santenol). In the fraction b. 196°–198°. (For details see Vol. II of this work, p. 229.)

Teresantalol. In the fraction b. 210°–220°. (For details see Vol. II, p. 257.)

Nortricycloekasantalal. In the fraction b. 222°–224°. First isolated by Schimmel & Co.[20] (Cf. Vol. II, p. 350.)

---

[12] *Ibid.*
[13] *Ber. Schimmel & Co.,* October (1910), 98.
[14] *Arch. Pharm.* **238** (1900), 366.
[15] *Ber. Schimmel & Co.,* October (1910), 102.
[16] *Arch. Pharm.* **238** (1900), 372.
[17] *Ber. Schimmel & Co.,* October (1910), 98.
[18] *Ber.* **40** (1907), 4465; **41** (1908), 125.
[19] *Ber.* **40** (1907), 4918.
[20] *Ber. Schimmel & Co.,* October (1910), 103.

Santalone. In the fraction b. 213°–216°. (For details see Vol. II, p. 494.)

A Ketone(?). The same fraction contains another ketone $C_{11}H_{16}O$, which is probably isomeric with santalone. Semicarbazone m. 208°–209°, oxime m. 97°–99°.

α- and β-Santalene. The occurrence of sesquiterpenes in East Indian sandalwood oil was first reported by von Soden and Müller.[21] Shortly afterward Guerbet [22] showed that these hydrocarbons are two isomers, which he named α- and β-santalene (cf. Vol. II, pp. 112 ff.).

α- and β-Santalol. The chief constituent of the oil, amounting to 90 or more per cent. In the mixture of α- and β-santalol present in the oil, the α-isomer predominates. The characteristic odor, as well as the medicinal value of East Indian sandalwood oil, is a result of the presence of the santalols (for details see Vol. II, pp. 265 ff.).

In 1947, Bhattacharyya [23] suggested the following structural formula for β-santalol:

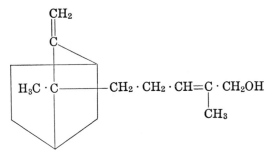

Santalal(?). According to Chapoteaut,[24] and Guerbet,[25] the aldehyde $C_{15}H_{22}O$, obtained by oxidation of santalol, occurs in East Indian sandalwood oil as a natural constituent. However, this has not been proved conclusively.[26]

Semmler and Bode [27] oxidized santalol with chromic acid in glacial acetic acid and obtained an aldehyde $C_{15}H_{22}O$, the properties of which are described in Vol. II of the present work, p. 267.

Teresantalic Acid. First reported by Müller.[28] It occurs in the oil partly free, partly in esterified form (cf. Vol. II, p. 113).

Santalic Acid. First reported by Guerbet [29] (for details see Vol. II, p. 611).

According to Müller,[30] the oil contains still other acids which, however, have not yet been identified.

21 *Pharm. Ztg.* **44** (1899), 258. *Arch. Pharm.* **238** (1900), 363.
22 *Compt. rend.* **130** (1900), 417.
23 *Science Culture* **13** (1947), 158. *Chem. Abstracts* **43** (1949), 5382.
24 *Bull. soc. chim.* [2], **37** (1882), 303.
25 *Ibid.* [3], **23** (1900), 540, 542. *Compt. rend.* **130** (1900), 417, 1324.
26 Cf. Chapman and Burgess, *Proc. Chem. Soc.* (1896), 140. Chapman, *J. Chem. Soc.* **79** (1901), 134.
27 *Ber.* **40** (1907), 1124; **43** (1910), 1722.
28 *Arch. Pharm.* **238** (1900), 374.
29 *Compt. rend.* **130** (1900), 417, 1324. *Bull. soc. chim.* [3], **23** (1900), 540, 542.
30 *Arch. Pharm.* **238** (1900), 374.

Phenols(?), Lactones(?) and Borneol(?). Von Soden and Müller [31] noted in the oil phenols of strong and disagreeable odor and lactones of fruity odor. None of these has been identified. According to the same authors, the oil probably also contains borneol in free and in esterified form. In the opinion of Gildemeister and Hoffmann,[32] it is quite possible that the borneol-like odor noted by von Soden and Müller may have been caused by santenone alcohol, which was later detected in the oil (see above).

Use.—During the early part of the twentieth century a large portion of the sandalwood oil produced in the world was used for medicinal purposes, particularly for the treatment of certain diseases, gonorrhea among them. Today the situation is entirely changed, only about 10 per cent serving for medicinal purposes. Natives in Asia still use the oil for self-treatment. In modern medical practice sandalwood oil has been replaced by the sulfa drugs, and by certain antibiotics.

As regards the perfume, cosmetic and soap industries, East Indian sandalwood oil is one of the most important ingredients. No composition of the heavy or oriental type is complete without an ample dose of this oil. Its sweet, powerful and lasting odor makes it an excellent fixative in the scenting of soaps. About 90 per cent of the East Indian sandalwood oil produced today is used in perfumes, cosmetics and soaps.

SANDALWOOD OILS AUSTRALIAN [1]

A. OIL OF *Eucarya Spicata* SPRAG. ET SUMM.
(West Australian Sandalwood Oil)

History.—The story of Australian sandalwood oil is one of vicissitudes. The oil was first distilled experimentally in 1875 by Schimmel & Co. in Leipzig, Germany.[2] It is difficult to determine exactly when Australian sandalwood oil was placed on the market; however, fifty years ago there was sporadic production of the oil by a number of small distillers in Western Australia. At that time the oil was marketed in a crude state (prac-

---

[31] *Pharm. Ztg.* **44** (1899), 259.
[32] "Die Ätherischen Öle," 3d Ed., Vol. II, 517.
[1] This monograph by A. R. Penfold and F. R. Morrison, Museum of Applied Arts and Sciences, Sydney, Australia.
[2] Gildemeister and Hoffmann, "Die Ätherischen Öle," 3d Ed., Vol. II, 521.

tically the same condition as that in which it was obtained by distillation of the wood), with the result that the santalol content was very low.

It is of interest to record that two well-known pharmacists played a prominent part in establishing the industry in Western Australia. According to Marr,[3] a Mr. Mayhew, when President of the Pharmaceutical Society of Western Australia, spent a considerable sum of money in distilling and marketing the oil in 1894–95; the late E. J. Parry, of essential oil fame, started operations in the Mount Barker district many years ago. C. L. Braddock produced the oil from time to time between 1916 and 1923.

It was not until 1921 that systematic production and scientific control of the oil were instituted by Plaimar Ltd.,[4] Perth, Western Australia. Since then the quality of the Western Australian sandalwood oil has been greatly improved; in fact, to such a degree that the oil is now recognized as being the equivalent of the well established and better known East Indian sandalwood oil. It was the great increase in its actual santalol content, and the fact that the Western Australian oil was at least equal therapeutically to the East Indian oil that resulted in its inclusion in the British Pharmacopoeia of 1932.

The improved status of the oil has been reflected in a larger demand; production increased from 3,720 lb. in 1920 to over 120,000 lb. in 1932. The volume has since dropped, however, to about 30,000 lb. per year, largely because distillers have been unable to obtain sufficient supplies of wood.

It is difficult to discuss sandalwood oil without reference to the wood itself, for the earliest record of shipment shows that 4 tons of wood, valued at £50, were exported to the Far East in 1846. The logs (stick wood) are shipped to Eastern Asia for use in incense; for the production of sandalwood oil in Western Australia chiefly roots and butts are used.

Until the adoption of the licensing system by the Western Australian Government, large tracts of forests were denuded of sandalwood, the logs being shipped to eastern markets. At the present time sandalwood can only be collected under license. Prior to World War II, the export value of Australian sandalwood logs was about £30 per ton. The continuous demand for the wood over the years has greatly reduced the existing areas of growth, and sandalwood cutters must now travel long distances for supplies. This fact, together with an increased demand from the Far East after World War II, caused the price of the wood to rise to a figure far in excess of its value for distillation purposes. The production of the oil in Australia, therefore, declined substantially. Since the Australian oil must compete on the world market with the East Indian sandalwood oil,

[3] Private communication from Mr. H. V. Marr, Perth, Western Australia.
[4] Penfold, *Australasian J. Pharm.* **18** (1937), 154.

the price of the wood in Australia must be sufficiently low to permit economical operation.

Botany.—Australian sandalwood oil is obtained from the heartwood of a diminutive tree or large bush, of straggly habit, about 12 to 20 ft. in height, with a trunk 6 to 10 in. in diameter. It occurs in the arid regions of Western and South Australia. The tree is now classified as *Eucarya spicata* (R. Br.) Sprag. et Summ.[5] (syn. *Santalum spicatum* [R. Br.] A. DC.), fam. *Santalaceae.*

The Western Australian sandalwood tree has had very rough handling by scientific botanists. It was first known as *Santalum spicatum* or *S. zygnorum.* This was later changed to *Fusanus spicatus,* and finally to *Eucarya spicata.*[6] This last classification was accepted by the British Pharmacopoeia in 1932, but synonymized as *Santalum spicatum.* The effects of a change in botanical nomenclature upon a commercial product are far-reaching, and the treatment accorded Australian sandalwood oil offers a typical example. For some years Australian sandalwood oil suffered severely in competition with the East Indian oil, because merchants used differences in nomenclature and chemical composition for trade purposes. The complete change of name from the genus *Santalum* to the genus *Eucarya* created the erroneous impression that the Australian oil was entirely different in chemical composition and in therapeutical properties from the East Indian sandalwood oil.

Method of Production.—The commercial process has been stated to be one of solvent extraction and steam distillation of the concentrated extract. Depending upon geographical origin of the tree and method of extraction the yield of oil varies between 1.4 and 2.6 per cent. According to the British Pharmaceutical Codex of 1949, Australian sandalwood oil is obtained by distillation *and rectification* (italics by the authors) from the wood of *Eucarya spicata* Sprag. et Summ.

Physicochemical Properties.—Australian sandalwood oil is a colorless or light yellow, somewhat viscous, liquid with a strong and lasting odor characteristic of the wood.

The oil exhibits these properties:

Specific Gravity at 15.5°/15.5°... 0.970 to 0.976
Optical Rotation............... −3° 0′ to −10° 0′
Refractive Index at 20°........ 1.498 to 1.508
Free Alcohol Content, Calculated Not less than 90% (in commercial,
  as Santalol   i.e., rectified oils)
Solubility at 20°............... Soluble in 3 to 6 vol. of 70% alcohol

[5] Sprague and Summerhayes, *Kew Bull. Misc. Information* **5** (1927), 195.
[6] Penfold, *J. Proc. Roy. Soc. N. S. Wales* **70** (1936), 23.

According to certificates of analyses of recent shipments from Perth to Europe,[7] the physicochemical properties of commercial (rectified) Australian sandalwood oil vary within the following limits:

Specific Gravity at 15.5°/15.5°. . . . 0.969 to 0.9725
Optical Rotation. . . . . −3° 0′ to −5° 0′
Refractive Index at 20°. . . . . 1.505 to 1.506
Total Alcohol Content, Calculated
    as Santalol. . . . . 94.3% to 95.4%
Solubility at 20°. . . . . Soluble in 3 to 3.5 vol. of 70% alcohol

Chemical Composition.—Commercial sandalwood oil from Western Australia contains more than 90 per cent of sesquiterpene alcohols, calculated as $C_{15}H_{24}O$. These alcohols have been the subject of much controversy (see "The Fusanols," Vol. II of the present work, p. 269). Their nature was not established until 1925, when Penfold [8] identified α-*santalol* as chief constituent, by preparation of its allophanate m. 162°–163°, by oxidation with chromic acid to santalal (semicarbazone m. 230°), and by oxidation with potassium permanganate solution to tricycloekasantalic acid m. 76.5°. The actual santalol content of the crude (unrectified) oil does not fall below 60 per cent.

The constitution of the other alcohols present in the oil has not yet been elucidated (cf. Vol. II of the present work, p. 270), but present investigations by Penfold are proceeding slowly. A *primary sesquiterpene alcohol* (b_{2-3} 137°–138°, d_{15}^{15} 0.9353, $α_D$ −0° 30′, n_D^{20} 1.5034) [9] has been isolated. So far, no crystalline derivatives of this alcohol have been prepared. The same is true of a *secondary sesquiterpene alcohol* (b_1 145°–150°, d_{15}^{15} 0.994 to 0.995, $α_D$ +18° 24′ to +26° 12′, n_D^{20} 1.5100 to 1.5106), which is also under investigation.

The oil furthermore contains a small percentage of *bisabolene,* which Schimmel & Co.[10] identified in the lowest boiling fraction by means of the trihydrochloride m. 79.7° to 80.5°.

In this connection it should be noted that recently Treibs [11] reported that the last runs of the Australian, but not of the Indian, sandalwood oils contain glycols and hydroxy-oxido derivatives of the sesquiterpene group. The well-known crystalline "santal camphor" (see Volume II of the present work, p. 768) of the *South* Australian sandalwood oil (*Santalum preissianum* Miq.) is a bicyclic oxido-sesquiterpene alcohol with a long side chain, hence of a structure similar to that of β-santalol.

Use.—The earliest use of Australian sandalwood oil was for therapeutical purposes. It is still being employed medicinally, contrary to general belief.

[7] Private communication from Plaimar Ltd., Perth, Western Australia.
[8] *J. Proc. Roy. Soc. N. S. Wales* **62** (1928), 60.
[9] Unpublished data.   [11] *Chem. Ber.* **84** (1951), 47.
[10] *Ber. Schimmel & Co.* (1921), 43.

According to Marr,[12] supplies of Australian sandalwood in capsules and in 1-lb. bottles are still shipped to China, the Malayan Archipelago, and South America where it is largely used for self-medication.

The oil is employed also in perfumes, cosmetics, and for the scenting of soaps. It possesses a heavy, powerful, and very lasting odor which blends particularly well with perfume compositions of the oriental type.

B. OIL OF *Santalum Lanceolatum* R. BR.

Introduction and Botany.—The botany of this tree has been fully discussed by Bentham.[13] The tree occurs in New South Wales, Queensland, and in the northwestern part of Western Australia. The essential oil derived from the wood of this tree was used at one time to bring the optical rotation of commercial Australian sandalwood up to the requirements demanded by various pharmacopoeias. However, the entire amount of oil produced from *Santalum lanceolatum* has never exceeded 5 per cent of the total output of Australian sandalwood oil.

On steam distillation the comminuted wood yields from 2 to 2.5 per cent of oil.

Physicochemical Properties.—Oil of *Santalum lanceolatum* is of unusual chemical interest, as it differs entirely from the oil of *Eucarya spicata*. The properties of the oil vary within these limits:

Specific Gravity at 15°/15°	0.9474 to 0.9628
Optical Rotation	$-45°\ 42'$ to $-61°\ 0'$
Refractive Index at 20°	1.5068 to 1.5085
Ester Number	8.4 to 23.5
Ester Number after Acetylation	200 to 205
Solubility	Soluble in 1.3 to 1.7 vol. of 70% alcohol

Chemical Composition.—Oil of *Santalum lanceolatum* contains no santalol. The chief constituent is a primary, highly laevorotatory sesquiterpene alcohol $C_{15}H_{24}O$ (b$_5$ 163°–165°, d$_{15}^{15}$ 0.9474, α_D $-66°\ 42'$, n$_D^{20}$ 1.5074; allophanate m. 114°–115°, strychnate m. 103°–105°) which was isolated in 1928 by Penfold,[14] and later named *lanceol* by Bradfield, Francis, Penfold and Simonsen.[15] (Cf. Vol. II of the present work, p. 270.) Lanceol does not appear to be structurally related to α- or β-santalol. For details of its chemical reactions and a discussion of its probable constitution the reader

---

[12] Private communication from Mr. H. V. Marr, Perth, Western Australia.
[13] "Flora Australiensis," Vol. VI, London (1873), p. 214.
[14] *J. Proc. Roy. Soc. N. S. Wales* **62** (1928), 70.
[15] *J. Chem. Soc.* (1936), 1619.

may consult the original literature. Recently, Eschenmoser, Schreiber and Keller [16] advanced a probable structural formula for lanceol, which is a primary, monocyclic sesquiterpene alcohol containing three double bonds.

Use.—Oil of *Santalum lanceolatum* is not produced commercially at present.

C. OIL OF *Eremophila Mitchelli* BENTH.[17]

Introduction and Botany.—No account of Australian sandalwood oil would be complete without some reference to the so-called "Bastard Sandalwood" or "Buddah," *Eremophila mitchelli* Benth., a tall shrub, or small tree, belonging to the fam. *Myoporaceae*. The tree attains a height of 20 to 30 ft., with a maximum diameter of 12 in. at the butt, and bears a wealth of white, scented flowers in the spring. It is one of the strongest scented woods of Australia.

Eremophila mitchelli occurs in the arid regions of New South Wales, Queensland, and South Australia. The tree, very common in Western Queensland, is often confused with *Santalum lanceolatum*, both trees being known as sandalwood.

The stick wood (logs), after reduction to shavings, yield from 2 to 3 per cent of a dark colored viscous oil with a pleasant and characteristic odor.

Physicochemical Properties.—The properties of oil of *Eremophila mitchelli* vary within these limits:

Specific Gravity at 15°/15°...... 1.03 to 1.04
Optical Rotation.............. +6° 0' to −6° 0'
Refractive Index at 20°........ 1.5260 to 1.5384
Solubility.................... Soluble in 2 to 4 vol. of
             70% alcohol

Chemical Composition.—Oil of *Eremophila mitchelli* contains three closely related sesquiterpene ketones, which were isolated by Bradfield, Penfold and Simonsen.[18] The discovery of these three ketones in oil of *Eremophila mitchelli* is probably the first record of their occurrence in nature. The ketones are:

Eremophilone............................. $C_{15}H_{22}O$, m. 42°–43°, $[\alpha]_{5461}$ −207°
2-Hydroxyeremophilone.................... $C_{15}H_{22}O_2$, m. 66°–67°, $[\alpha]_{5461}$ +153°
2-Hydroxy-2-dihydroeremophilone.......... $C_{15}H_{24}O_2$, m. 102°–103°, $[\alpha]_{5461}$ +94°

[16] *Helv. Chim. Acta* **34** (1951), 1667.
[17] "Flora Australiensis," Vol. V, London (1870), p. 21.
[18] *J. Chem. Soc.* (1932), 2744. *J. Proc. Roy. Soc. N. S. Wales* **66** (1932), 420.

Regarding the structural formulas and details pertaining to these sesquiterpene ketones, see Vol. II of the present work, p. 451.

Use.—Oil of *Eremophila mitchelli,* although dark in color, possesses marked blending and fixative properties valuable in the scenting of soap and technical preparations. While the oil is not so useful as genuine sandalwood oil, it nevertheless has remarkable qualities, which place its value in line with those of the fixative balsams.

OIL OF SANDALWOOD AFRICAN
(*Osyris* Oil)

Little is known about the so-called "African Sandalwood Oil" and its botanical origin. About sixty years ago, Schimmel & Co.[1] distilled wood imported from Tamatave (Madagascar) and obtained 3 per cent of a ruby-red oil (d_{15} 0.969), the odor of which resembled that of the so-called West Indian sandalwood oil. The wood was perhaps identical with that of *Osyris tenuifolia* Engl. (see below), or with the so-called "Hasoranto Wood"[2] from northern Madagascar.

Distilling "East African Sandalwood Oil," probably derived from *Osyris tenuifolia* Engl. (fam. *Santalaceae*),[3] Schimmel & Co.[4] obtained 4.86 per cent of a light-brown oil with an odor similar to that of vetiver and gurjun balsam oils, but very different from that of East Indian sandalwood oil. The oil had these properties:

Specific Gravity at 15°.	0.9477
Optical Rotation.	−42° 50′
Refractive Index at 20°.	1.52191
Ester Number.	11.1
Ester Number after Acetylation. . .	72.8
Total Alcohol Content, Calculated as $C_{15}H_{26}O$.	30.5%
Solubility.	Not readily soluble; soluble in 7 to 8 vol. of 90% alcohol

[1] *Ber. Schimmel & Co.,* April (1891), 49.

[2] Sawer, "Odorographia," London (1892), Vol. I, 325.

[3] Cf. A. Engler and G. Volkens, "Über das wohlriechende ostafrikanische Sandelholz (*Osyris tenuifolia* Engl.)," *Notizblatt des Königl. botan. Gartens und Museums,* Berlin, No. 9, Aug. 7, 1897.

[4] *Ber. Schimmel & Co.,* October (1908), 111.

Haensel[5] described two so-called East African sandalwood oils of unknown botanical origin. They were probably distilled from the same plant material as the *Osyris* oils. The two oils exhibited the following properties:

	I	II
Specific Gravity at 20°..............	0.9589	0.9630
Optical Rotation...................	−40° 36′	−60° 58′
Acid Number......................	1.7	...
Saponification Number..............	17.9	8.1
Ester Number after Acetylation......	88.3	68.6

These oils contained a sesquiterpene b_{747} 263.5°–265°, d_{20} 0.9243, α_D −32° 55′, and a sesquiterpene alcohol b_{25} 186°–188°.

More recently, Schimmel & Co.[6] examined an African sandalwood oil of unknown botanical origin and found these properties:

Specific Gravity at 15°..........	0.9637
Optical Rotation................	−45° 56′
Refractive Index at 20°.........	1.50762
Acid Number..................	0.7
Ester Number.................	7.5
Ester Number after Acetylation...	203.5
Total Alcohol Content, Calculated as $C_{15}H_{26}O$..................	95.2%
Solubility.....................	Soluble in 4 vol. and more of 70% alcohol; slightly opalescent on addition of more alcohol

The oil resembled the above-mentioned *Osyris* oils.

Nothing is known about the actual production of this type of oil. According to Muller[7] no sandalwood oil is produced in Madagascar.

[5] *Chem. Zentr.* (1906), II, 1496; (1909), I, 1477.
[6] *Ber. Schimmel & Co.* (1938), 97.
[7] Private communication from Mr. Charles Muller, Ambanja, Madagascar.

CHAPTER VII

ESSENTIAL OILS OF THE PLANT FAMILY *ZYGOPHYLLACEAE*

OIL OF GUAIAC WOOD

Botany, Origin and Production.—Strictly speaking, guaiac wood is the wood of *Guaiacum officinale* L. (*lignum vitae*) and of *G. sanctum* L. However, the wood of these species contains only small quantities of volatile oil. The commercial, so-called oil of guaiac wood is obtained today by steam distillation of the heartwood of *Bulnesia sarmienti* Lor. (fam. *Zygophyllaceae*), a tree 10 to 12 ft. high with a trunk diameter of about 10 in., and often gnarled and crooked in appearance. It grows wild and very abundantly throughout the Gran Chaco, the famed "Green Hell." This is a vast region consisting mostly of waterless scrub jungle, covering parts of Paraguay and Argentina and inhabited in part by wild Indian tribes. The wood of *Bulnesia sarmienti* Lor. resembles that of *Guaiacum officinale* L. and *G. sanctum* L. in being very hard; it is used for the making of fence posts, ornamental objects, and for the extraction of its volatile oil by steam distillation. The oil was first introduced on the market in 1891. Prior to World War II limited quantities of "Palo Santo" logs (as the tree is called locally) were shipped from the Gran Chaco, chiefly to Europe, for extraction of the essential oil. In 1935, for example, exports of logs from Paraguay and Argentina amounted to 257 metric tons, cost of transport being offset, at least partially, by the high yield of oil (5 to 6 per cent) obtained in the modern distilleries of Europe and North America. In 1938 a firm in Asunción, capital of Paraguay, started production of the oil, thus effecting a considerable saving in shipping costs. In this distillery the logs of "Palo Santo" trees are sawed mechanically into blocks, and then reduced to coarse sawdust by means of hogging machines, hammers and grinders, such as are employed in paper mills. The sawdust falls into a concrete basin sunk into the ground and filled with water. Two hundred and fifty pounds of coarse sawdust absorb approximately 750 lb. of water. The wetted distillation material (25 per cent of wood and 75 per cent of water) is charged into the still. The distillery contains four stills: two small ones, holding a charge of 100 kg. each, and two large ones, holding 1,000 kg. each. Pressure in the steam generators varies from 40 to 75 lb., according to requirements. Distillation of one batch lasts 8 hr. in the small stills, and 16 hr. in the large ones. Oil of guaiac wood being a quite solid mass at room temperature (see below), the distillate is not permitted to run too cool, as this would clog the condensers. The distillation waters are discarded, not cohobated.

197

The yield of oil ranges from 2.7 to 3.0 per cent. In 1950 this plant produced about 6,000 kg. of oil.

Two other distilleries of guaiac wood, producing together about 11,000 kg. of oil per year, are located in the heart of the Gran Chaco, where the Mennonites in 1928 established a colony, which now numbers more than 8,000 people. The older of these two distilleries is situated in Filadelfia, the center of the colony, the other about 40 miles to the east of Filadelfia, right in the forests where the "Palo Santo" tree grows wild and abundantly. Because of the great number of trees still available, production of oil could be increased substantially if demand and price of the oil should warrant it. The trees are felled, and the logs pulled out of the forests by teams of oxen. Then the thick layer of bark and sapwood is removed, and the wood of the trunks and heavy branches is mechanically reduced to shavings. These are steam-distilled in quite primitive stills. Distillation of one charge lasts 12 hr.; work goes on day and night, and the yield of oil averages 3 per cent. Filter paper not being available, the warm liquid oil separated from the condensate is filtered through cotton. Obviously these two distilleries, in the heart of the Gran Chaco, can produce oil at a much lower cost than a distillery to which the logs must be hauled over long distances, out of the almost inaccessible forest region, partly by land over very poor roads, partly by river on lighters.

In this connection it should be noted that the condition of the wood appears to exert some influence upon the odor of the oil. The oil produced by the Mennonites in the growing region from freshly cut wood has an odor somewhat different from that of the oils distilled from old logs dried and aged during the long transport abroad.

Physicochemical Properties.—Oil of guaiac wood is a viscous liquid which, at room temperature, slowly congeals to a yellowish-white crystalline mass. Once solidified, it can be reliquefied at temperatures ranging from 40°–50° C. The oil has a pleasant, soft and mellow rose-like odor, resembling that of tea roses, and to a slight degree that of violets.

Gildemeister and Hoffmann [1] reported the following properties for oil of guaiac wood:

Specific Gravity at 30°..............	0.967 to 0.974
Optical Rotation....................	$-3°\ 0'$ to $-8°\ 0'$
Refractive Index at 20°.............	1.502 to 1.507
Acid Number......................	0 to 1.5
Ester Number......................	0 to 7.5
Ester Number after Acetylation......	98 to 159

[1] "Die Ätherischen Öle," 3d Ed., Vol. II, 911.

Guaiol Content.................... 42 to 72%
Solubility........................ Soluble in 3 to 5 vol.
                                   of 70% alcohol

Genuine oils of guaiac wood imported by Fritzsche Brothers, Inc., New York, from Paraguay and Brazil had properties varying within these limits:

Specific Gravity Calculated at 15°/15°... 0.974 to 0.983
Optical Rotation...................... −7° 9' to −12° 16'
Refractive Index at 20°............... 1.5040 to 1.5080
Alcohol Content, Calculated as Guaiol... 62.9 to 82.5%
Solubility at 20°..................... Soluble in 2 to 6.5 vol. and
                                       more of 70% alcohol

Distilling imported guaiac wood in Southern France, the author and his associates extended the time of distillation to 24 hr. and obtained yields of oil ranging from 4.8 to 5.37 per cent. It was found that yields up to 7.5 per cent, and a good quality of oil, could be obtained by thoroughly mixing the powdered wood with water, adding traces of sulfuric acid, and distilling the mass after it had been allowed to macerate overnight. Since the acid has a deteriorating effect upon the metal of the stills, the process is impractical unless glass-lined or wooden stills are employed.

Oils distilled in Seillans (Var), France, under the author's supervision had these properties:

Specific Gravity Calculated at 15°....... 0.973 to 0.978
Optical Rotation...................... −3° 30' to −8° 48'
Refractive Index at 20°............... 1.5050 to 1.5059
Saponification Number................. 0 to 4.2
Ester Number after Acetylation........ 135.4 to 166.1
Total Alcohol Content, Calculated as
   Guaiol............................ 59.8 to 75.2%
Solubility at 20°..................... Soluble in 3.5 to 4 vol. and
                                       more of 70% alcohol

Chemical Composition.—Relatively little is known about the chemical composition of the volatile oil derived from the wood of *Bulnesia sarmienti*, particularly in regard to its odoriferous principles. The presence of the following compounds has been reported:

Guaiol. A bicyclic, tertiary sesquiterpene alcohol $C_{15}H_{26}O$, containing one double bond. On dehydration it yields guaiene $C_{15}H_{24}$. When dehydrogenated with sulfur, guaiene gives deep blue guaiazulene $C_{15}H_{18}$ (for details re Guaiol, see Vol. II of this work, 278).

Bulnesol. Also a bicyclic, tertiary sesquiterpene alcohol $C_{15}H_{26}O$, containing one double bond (see Vol. II of this work, p. 756).

Use.—Oil of guaiac wood is an interesting perfumer's material, quite widely employed in rose compositions. Its principle merit lies in its lasting qualities, whence the use of the oil as a natural fixative of odors. Being moderately priced, oil of guaiac wood serves to good advantage in the scenting of soaps. It helps to conceal harsh notes of synthetic aromatics.

In years past, the oil was occasionally employed as an adulterant of Bulgarian and Turkish rose oils.

CHAPTER VIII

ESSENTIAL OILS OF THE PLANT FAMILY *LEGUMINOSAE*

OIL OF BALSAM COPAIBA

Essence de Baume de Copahu   *Copaivabalsamöl*
Aceite Esencial Balsamo Copaiba   *Oleum Balsami Copaivae*

BALSAM COPAIBA

Botany and Origin.—Balsam [1] copaiba is the oleoresin obtained from the trunk of various species of the genus *Copaifera* L. (fam. *Leguminosae*), tall, multi-branched trees, with a large trunk and a smooth bark. The tree, a native of the northern part of South America, grows wild in the vast forests and jungles of Brazil (Amazon Basin), and to a lesser extent in Venezuela, Guiana, and Colombia.

The oleoresin originates probably by decomposition of the cell walls of the wood parenchyma. The schizogenous cavities within the wood and pith of the tree trunk, in which the copaiba accumulates, enlarge and join, forming reservoirs of surprisingly large capacity, which hold gallons of oleoresin. On tapping of the tree, the oleoresin flows from the wound as a clear, colorless, thin liquid which, however, soon acquires a thicker consistency and a yellowish tinge.

Literature cites a number of *Copaifera* species which produce copaiba balsam, but the taxonomy of the trees that are actually exploited is still somewhat confused. According to Freise,[2] who undertook a study of the subject, the genus *Copaifera* is represented in Brazil by fifteen species occurring in all states from the extreme north to São Paulo; but only the following species are of commercial interest:

(1) *Copaifera reticulata* Ducke.

70% of all Brazilian copaiba balsam originates from this species.
(2) *C. guianensis* (Desf.) Benth.

About 10% of the exported balsam comes from this species. It grows also in Guiana.
(3) *C. multijuga* Hayne.

About 5 per cent are collected from this species.

---

[1] The term "balsam" for copaiba is a misnomer since copaiba is actually an oleoresin. However, the term balsam will be retained in this monograph because the trade has become used to it in the course of many years.
[2] *Perfumery Essential Oil Record* **25** (1934), 218.

(4) *C. officinalis* L.

Not more than 3 to 5 per cent of the exported copaiba is now derived from this species. It grows also in Venezuela, where it supplies the so-called Maracaibo balsam.

(5) *C. martii* var. *rigida* (Benth.) Ducke.

(6) *C. coriacea* (Mart.) Ktze.

The balsam from this species is exported mostly to Belem-Pará for admixture with better grades.

Of the different types of copaiba reaching the world market, the product from the middle and lower Amazon Basin is by far the most important. Commercially known as "Pará balsam," it includes also balsam from Manáos, a city located on the junction of the Amazon and the Rio Negro. Both products are shipped abroad from Belem-Pará, located in the Amazon Delta. Markets in North America, Europe and the Far East are supplied mainly with copaiba from the Brazilian states of Pará and Amazonas, and only seldom with that from Bahia.

Chief producing regions of balsam copaiba in Brazil are:

(a) State of Pará. These stretch over the lower Amazon Basin, mostly south of Belem-Pará. The trees grow in vast areas, along the larger tributaries of the Amazon, the smaller ones not yet being exploited. The southern tributaries are the Tocantins, Anapú, Pacajá, Xingú and Tapajoz. Production along some of the northern tributaries—e.g., the Parú, Jary, and Yamunda—is insignificant.

(b) State of Amazonas. As will be explained later, only the so-called insoluble balsam comes from this part of Brazil. Its insolubility probably results from soil conditions and from the very humid climate of this region. Here the balsam is collected chiefly along the banks of the Amazon and adjacent tributaries—e.g., the Madeira, Purus, Jurua, Rio Negro, and Rio Branco.

In all these areas there are no plantations, the copaiba trees growing wild in the somewhat drier sections of the jungles. Native woodsmen, financed by traders and exporters, exploit the trees. They refuse to be employed on a fixed wage basis and consider the job of day laborer beneath their dignity. Therefore, they prefer to work on their own.

It is probably safe to assume that neither North American nor European importers have ever handled a lot of copaiba balsam of definite geographical origin. None of the shippers in Belem-Pará or Manáos has any idea regarding the species from which the various small lots of copaiba originate. They receive the product in cans or sometimes in drums that have formerly served for gasoline. Indians bring them down the river after they have spent weeks or months in the wild interior collecting the balsam. The trees, which usually grow on dry land, may almost be compared with containers, being filled in the hollow center with oleoresin, which can easily be drawn from them. This is done simply by drilling two holes into the trunk, one 2 to 3 ft. above the ground, and the other 10 to 20 ft. higher. A bamboo tube, provided with a simple stopcock, is driven into the lower hole. Under normal conditions, the copaiba flows easily and as a clear liquid, particularly if the tree has not been tapped for several years. The product is then collected in cans. If the copaiba does not flow readily, a fire is made at the base of the trunk and at the elevated temperature the oleoresin will flow more easily. After completion of the tapping operation, both holes are closed with a simple plug.

The quality of copaiba drawn from each tree depends upon the botanical species of the tree, its age, the lapse of time since the previous tapping, and upon the season. On the average, a tree yields from about 17 to 18 kg. of copaiba balsam per year. In general, the trees are exploited all year around, provided production is stimulated by demand and sufficiently attractive prices. It must be borne in mind that it takes many days, and often weeks, of difficult traveling to reach the remote regions of production, and that so far away from civilization the cost of food, which must be carried along, increases enormously. The same holds true of the transport of the collected copaiba back to the trading centers.

The season of production depends, to a certain extent, upon the region and the prevailing weather. In most sections the best period for collecting the copaiba is in the dry months from July to November, because the trees then give the largest yield. During the rainy season, they contain very little oleoresin.

In the trading centers, the product is either sold to intermediaries for cash or bartered against food, clothing, and other necessities. Frequently, advances in the form of cash or food are made to the workers by traders or exporters in Belem-Pará or Manáos. Accounts are then settled by delivery of copaiba balsam and other products of the jungles to the trading posts. The crude product is subsequently shipped by boat down the rivers to Manáos or Belem-Pará, the two principal exporting centers. Exporters thus receive many small lots, all of different origin, color, and solubility. The exporters remove water, particles of bark and other foreign matter and assort the lots according to solubility and color.

Shippers in Brazil differentiate between two qualities of copaiba balsam:

1. The so-called soluble copaiba: one part must be clearly soluble in 5 to 6 parts of 95 per cent alcohol and must remain clearly soluble.

2. The so-called insoluble copaiba: not clearly soluble in 95 per cent alcohol.

The State of Pará produces both soluble and insoluble copaiba; the State of Amazonas supplies only insoluble copaiba. Sixty per cent of the exports from Pará are of the soluble, and about 40 per cent of the insoluble, type. Thus Brazil supplies much larger quantities of insoluble copaiba than of soluble copaiba. The difference in price usually varies from 10 to 50 per cent.

Solubility of the copaiba has no relation to its color and specific gravity; nor is solubility a criterion for purity and odor of the product. The solubility simply offers a means of distinguishing between the two commercial types of copaiba. Exporters in Brazil apply no other tests. No attention is paid to specific gravity or other properties, which appear to vary with the trees exploited, their age, the region of production, the time of collection, and other factors. The odor of the different lots is quite similar, at least in regard to the pure, unadulterated product. Some attention is paid to the color, because certain countries specify a light color, while others are less particular in this respect. To summarize, exporters roughly grade and ship their lots according to solubility and color, conforming to the requirements of the country of destination. In most cases, the exporters do not bulk the many small lots into large lots, but simply select a number of cans according to the specifications of the consumers abroad.

The United States buys insoluble copaiba chiefly, and pays little attention to the color. Great Britain insists upon light or amber colored, soluble copaiba. Japan and Germany, before the last war, bought both the soluble and insoluble product.

Exports of Balsam Copaiba from Brazil.—According to a report of the United States Department of Commerce, Office of International Trade,[3] Washington, D. C., Brazil exported the following quantities of balsam copaiba during the years following World War II:

Year	Metric Tons
1946	81
1947	94
1948	47

Physicochemical Properties of the Oleoresin.—Copaiba balsam is a pale-yellow to yellowish-brown, more or less viscous liquid, some lots exhibiting a slight greenish fluorescence. It possesses a peculiar, aromatic odor. On exposure to air, the oleoresin acquires a dark brown color, a more viscous consistency and a higher specific gravity. When spread over a large surface it dries and turns brittle, the change being caused partly by evaporation

[3] Private communication from Mr. T. W. Delahanty, Chief, Consumers Merchandise Branch.

(*Top*) Production of cananga oil in the Bantam region of Java (Indonesia). Primitive direct fire stills serve for the distillation of the flowers. (*Bottom*) Cananga oil, Bantam region. View of primitive condensers and oil separators. *Photos Fritzsche Brothers, Inc., New York.*

(*Top*) Production of ylang ylang oil in Nossi
(Madagascar). A branch of an ylang ylang
bearing flowers. (*Left*) Native girls with
vested ylang ylang flowers in baskets (Nossi-
(*Bottom*) A large ylang ylang distillery in No
Bé. Native girls carrying the harvested flow
to the distillery in the morning. *Photos Fritz*
Brothers, Inc., New York.

and partly by polymerization of the essential oil. On aging of the oleoresin, the resinous, nonvolatile portion increases.

Four samples of genuine copaiba balsam collected by the author during a visit to the Amazon Basin had the following properties:

	Insoluble	Soluble	Dark	Pale
Specific Gravity at 25°/25°	0.973	0.962	0.961	0.975
Optical Rotation.........	−21° 20′	−20° 20′	−20° 18′	−25° 20′
Refractive Index at 20°...	1.5117	1.5099	1.5101	1.5118
Acid Number............	74.2	67.2	67.1	79.8
Saponification Number...	81.2	71.9	73.7	86.8
Evaporation Residue.....	61.1%	50.0%	49.7%	58.9%
Color..................	Brown with greenish fluorescence	Reddish-brown	Reddish-brown	Yellow with greenish fluorescence

Freise [4] examined samples of copaiba from the eight most important species of *Copaifera*, viz., *reticulata, guianensis, multijuga, officinalis, martii* var. *rigida, coriacea, martii* Hayne, and *glycycarpa*, and observed considerable variations even in samples from the same species. The range of properties determined by Freise was:

Specific Gravity at 15°/4°...... 0.888 to 1.004
Optical Rotation at 20°........ −14° 45′ to −37° 10′
Acid Number................ 48 to 86
Saponification Number........ 44 to 92
Essential Oil Content......... 26 to 63%

Numerous lots of copaiba balsam imported by Fritzsche Brothers, Inc., New York, from the States of Pará and Amazonas in Brazil had properties varying within these limits:

Specific Gravity at 25°/25°...... 0.922 to 0.982
Optical Rotation.............. +20° 0′ to −23° 0′
Refractive Index at 20°........ 1.5022 to 1.5122
Acid Number................. 20 to 72.1
Saponification Number......... 33.6 to 87.5
Evaporation Residue........... 27.1 to 55.6%

The wide varieties in these properties, particularly in regard to the optical rotation, must be explained by the fact that each can received from Brazil represented a lot originating from a different region, as was explained above. Every lot was tested for adulteration with gurjun balsam and with African copaiba, but none gave a positive reaction.

In general it can be said that copaiba balsam with a high specific gravity and high evaporation residue is preferable for employment as a fixative for

---

[4] *Perfumery Essential Oil Record* **25** (1934), 218. Cf. Deussen, *Scientia Pharm.* **10** (1939), 69. *Chem. Abstracts* **33** (1939), 5122.

soap perfumes; this type of copaiba possesses a smoother and more lasting odor than copaiba with a low specific gravity and a low evaporation residue. The latter contains more volatile oil and, therefore, lends itself better to distillation purposes.

Adulteration of the Balsam.—Copaiba balsam is occasionally adulterated with fatty oils, turpentine oil, kerosene, or rosin.

As regards rosin, the reader will find a test for its detection in Vol. I of this work, p. 334. It should be pointed out, however, that an infallible test for rosin has not yet been developed.

A test for kerosene and mineral oils in general is described in Vol. I of this work, p. 332 (Oleum Test). Another test for paraffin oils in copaiba balsam is given in the National Formulary.

The National Formulary also describes a test for the presence of turpentine or fatty oils. On evaporation of an accurately weighed sample of copaiba balsam in a shallow dish on a water bath, no odor of turpentine oil should develop. The evaporation resin remaining should not be less than 27 per cent of the weighed balsam. When cooled, the resin must be hard and brittle. A further test for fatty oils may be found in Vol. I of the present work, p. 344 (Fatty Oils). It should be mentioned in this connection that, according to information gathered by the author in the producing regions of the Amazon Basin, the natives often employ the oil expressed from the seed of the tree *Carapá guianensis* (fam. *Meliaceae*) for adulteration of copaiba balsam.

Formerly copaiba balsam was frequently adulterated with gurjun balsam and with the so-called African copaiba. This, however, is rarely, if ever, encountered today, because these two products are not always easily available; moreover, their prices are too high to permit use in copaiba balsam as an adulterant. Tests for the presence of African copaiba and gurjun balsam in copaiba balsam are given in the National Formulary. One drop of nitric acid is added to 3 cc. of glacial acetic acid, and to this mixture are added 4 drops of volatile oil obtained by steam distillation of the copaiba balsam. No reddish or purplish color should appear on shaking of the mixture. Otherwise presence of gurjun balsam is indicated. As regards adulteration with African copaiba: the volatile oil obtained by steam distillation of true copaiba balsam does not boil below 250°, and has an optical rotation of not less than −7°; African copaiba (and its volatile oil) usually exhibit dextrorotation. Therefore, addition of African copaiba to true copaiba affects the optical rotation of the volatile oil derived from the latter.

Chemical Composition of the Oleoresin.—Copaiba balsam contains substantial quantities of a volatile oil, the balance being resinous substances and small quantities of acids. The resinous matter is hard, brittle, translucent, greenish-brown, almost odorless and tasteless. Relatively little is known about its composition except that it contains several resinoic acids. Most probably the resinous matter is the result of oxidation and polymerization of the oil in the plant cells. The higher its content of oil, the less resinous substances the copaiba contains.

OIL OF COPAIBA

For the industrial isolation of the volatile oil from the oleoresin, the so-called Pará copaiba (from the Brazilian States of Pará and Amazonas) is preferable to the product from Maracaibo (Venezuela) because the former contains up to 85 per cent of volatile oil. The Maracaibo product (which originates from *Copaifera officinalis*) is more viscous and, upon distillation, yields only 35 to 58 per cent of volatile oil.

Physicochemical Properties of the Oil.—Oil of balsam copaiba is a colorless, yellowish, or slightly bluish liquid possessing a somewhat peppery and aromatic odor, characteristic of the oleoresin itself. Oils distilled from copaiba balsams of different origin exhibit widely varying physicochemical properties, particularly in regard to the optical rotation. Gildemeister and Hoffmann [5] reported properties for several types of oil; we shall quote only those of oils which are of commercial importance today.

	Oils from *Pará Copaiba*	*Oils from* *Mandos Copaiba*	*Oils from* *Maracaibo Copaiba*
Specific Gravity at 15°	0.886 to 0.910	0.9036 to 0.9095	0.900 to 0.905
Optical Rotation	$-7°\,0'$ to $-33°\,0'$	$-16°\,4'$ to $-32°\,28'$	$-2°\,30'$ to $-14°\,0'$
Refractive Index at 20°	1.493 to 1.502	1.50045 to 1.53012	About 1.498
Acid Number	0 to 1.9	0.3 to 0.6	0.9 to 1.0
Ester Number	0 to 4 (in 1 case 13)	0.9 to 6.4	1 to 1.6
Solubility in 95% Alcohol	Soluble in 5 to 6 vol.	Soluble in 6 vol.	Soluble in 5 to 6 vol.

Oils distilled by Fritzsche Brothers, Inc., New York, from imported Pará and Amazonas copaiba had properties varying within these limits:

Specific Gravity at 15°/15°	0.901 to 0.907
Optical Rotation	$-10°\,30'$ to $-14°\,22'$
Refractive Index at 20°	1.4958 to 1.4990
Saponification Number	Up to 1.5

Chemical Composition of the Oil.—The early investigations on the chemical composition of oil of copaiba balsam, carried out by Blanchet,[6] Sou-

---

[5] "Die Ätherischen Öle," 3d Ed., Vol. II, 864.
[6] *Liebigs Ann.* **7** (1833), 156.

beiran and Capitaine,[7] Posselt,[8] Strauss,[9] Brix,[10] and Levy and Englaender,[11] require no discussion here, because they did not lead to any conclusive results. It was only in 1892 that Wallach [12] succeeded in identifying caryophyllene as the chief constituent of the volatile oil derived from copaiba balsam. Since then, the presence of the following compounds has been reported in the oil:

α- and β-Caryophyllene.   Treating the fraction b. 250°–270° with sulfuric acid and glacial acetic acid, Wallach [13] obtained caryophyllene hydrate m. 96°.   He arrived at the conclusion that caryophyllene, which also occurs in oil of clove, is the principal constituent of copaiba balsam oil.   In 1910, Deussen and Hahn [14] prepared the nitrosochloride and the nitrosate of caryophyllene from copaiba balsam oil, and declared the optically inactive modification of this sesquiterpene to be α-caryophyllene.   Schimmel & Co.[15] found that the oil also contains optically active β-caryophyllene, an observation later confirmed by Deussen and Eger,[16] who identified β-caryophyllene in copaiba balsam oils of various origin.   (Regarding caryophyllene, see Vol. II of this work, p. 99).

l-Cadinene.   In their investigation of copaiba balsam oil, Schimmel & Co.[17] also obtained a dihydrochloride m. 113°–115°, identical with l-cadinene dihydrochloride. This led Schimmel & Co. to the conclusion that l-cadinene is a constituent of the oil.

Other Sesquiterpenes(?).   The same workers also expressed the opinion that copaiba balsam oil contains still other sesquiterpenes, which were not identified, however.

Examining an oil derived from Surinam copaiba balsam, van Itallie and Nieuwland [18] noted the presence of l-cadinene or of another sesquiterpene yielding l-cadinene dihydrochloride m. 116°–117°, $[\alpha]_D$ −36° 5′ (in chloroform solution).   In addition, they isolated a sesquiterpene alcohol $C_{15}H_{26}O$, m. 113°–115° which, on dehydration with formic acid, gave a sesquiterpene b_{759} 252°, d_{15} 0.952, α_D −61° 42′, n_D^{15} 1.5189.   Van Itallie and Nieuwland did not succeed in identifying caryophyllene, but expressed the opinion that their oil contained other sesquiterpenes, aside from the l-cadinene mentioned above.

[7] Ibid. **34** (1840), 321.
[8] Ibid. **69** (1849), 67.
[9] Ibid. **148** (1868), 148.
[10] Monatsh. **2** (1881), 507. Cf. Umney, Pharm. J. [3], **24** (1893), 215.
[11] Liebigs Ann. **242** (1887), 189. Ber. **18** (1885), 3206, 3209.
[12] Liebigs Ann. **271** (1892), 294.
[13] Ibid.
[14] Chem. Ztg. **34** (1910), 873.
[15] Ber. Schimmel & Co., October (1910), 177.
[16] Liebigs Ann. **388** (1912), 136. Chem. Ztg. **36** (1912), 561.
[17] Ber. Schimmel & Co., October (1910), 177.
[18] Arch. Pharm. **242** (1904), 539; **244** (1906), 161.

Use.—Copaiba was once used quite extensively in all kinds of medicinal preparations, internal as well as external. Today it is employed chiefly as an odor fixative in the scenting of soaps and technical preparations. It blends well with the ionones and with oil of cedarwood.

The volatile oil derived from the oleoresin is a good diluent of low-priced perfume compositions; its soft odor has a tendency to mellow and harmonize harsh notes of synthetic aromatics.

SUGGESTED ADDITIONAL LITERATURE

F. W. Freise, "Certain Clarifications Respecting Brazilian Copaiba Balsam," *Süddeut. Apoth. Ztg.* **77** (1937), 11. *Chem. Abstracts* **31** (1937), 2351.

O. R. Gottlieb and A. Iachan, "Study of Copaiba Balsam," *Rev. quim. ind. Rio de Janeiro* **14**, No. 163 (1945), 20. *Chem. Abstracts* **40** (1946), 5207.

OIL OF BALSAM COPAIBA AFRICAN

The African copaiba balsam, also known as Illurin balsam, originates from Nigeria and adjacent parts of West Africa. According to Umney,[1] Solereder,[2] and Engler and Prantl,[3] the tree which produces the oleoresin, is probably *Oxystigma mannii* Harms (*Hardwickia*(?) *mannii* Oliv.), fam. *Leguminosae*. African copaiba balsam, which was once handled via the London market, is now rarely met with, and few, if any, commercial lots are encountered.

Physicochemical Properties of the Oil.—On steam distillation, the oleoresin yields from 37 to 57 per cent of volatile oil, for which Gildemeister and Hoffmann,[4] reported these properties:

Specific Gravity at 15°........	0.917 to 0.930
Optical Rotation.............	+5° 45′ to +39° 0′
Refractive Index at 20°.......	1.50574 to 1.50811
Acid Number................	0.5 to 9.3
Ester Number..............	0 to 5.6
Ester Number after Acetylation (One Determination)........	10

[1] *Pharm. J.* [3], **22** (1891), 449; **24** (1893), 215.
[2] *Arch. Pharm.* **246** (1908), 72.
[3] "Die natürlichen Pflanzenfamilien," Suppl. to the 2nd to 4th Parts (1897), 195.
[4] "Die Ätherischen Öle," 3d Ed., Vol. II, 870.

Solubility.................... Soluble in 98% alcohol; slightly
opalescent on addition of 2
and more vol. Soluble in 10
vol. of 95% alcohol with opalescence

Chemical Composition of the Oil.—The following compounds have been
reported in the volatile oil derived from African copaiba balsam:

β-Caryophyllene. Very small quantities (less than 1 per cent) were found in the
lowest boiling fraction b_{15} 128°–129.5° by Deussen [5] and Schimmel & Co.[6]

d- and l-Cadinene. The last-named authors [7] also noted that the fraction b. ~240°
consists chiefly of d-cadinene $[\alpha]_D$ +60° 40'. As regards the l-cadinene, it was
identified by von Soden,[8] and Schimmel & Co.[9]

Copaene. Investigating the fraction b. 246°–251°, Schimmel & Co.[10] noted the presence of a sesquiterpene $[\alpha]_D$ −13° 21', which yielded a dihydrochloride m. 117.5°–
118° identical with that of l-cadinene. Shortly afterward, however, Semmler and
Stenzel [11] determined that the sesquiterpene in question was not l-cadinene, but
copaene (cf. Vol. II of this work, p. 122).

Use.—The oil derived from African copaiba balsam was at one time used
as an occasional adulterant of the oil from Brazalian copaiba balsam.

OIL OF BALSAM PERU

Essence de Baume de Pérou   *Perubalsamöl*
Aceite Esencial Balsamo Peru   *Oleum Balsami Peruviani*

Balsam Peru

History.—The designation "Balsam Peru" is misleading, inasmuch as it
implies that the balsam originates from Peru, whereas in reality it is produced chiefly in El Salvador. The misnomer dates from early colonial days,
when products from the Pacific coast of Spanish America were assembled
in Callao, Peru, shipped to Panama, transported across the Isthmus, and
transshipped to Europe. In those days balsam was transported by land
from Salvador across the Isthmus and loaded into ships carrying Peruvian

[5] *Liebigs Ann.* **388** (1912), 142.
[6] *Ber. Schimmel & Co.,* April (1914), 44.
[7] *Ibid.*
[8] *Chem. Ztg.* **33** (1909), 428.

[9] *Ber. Schimmel & Co.,* October (1909), 31.
[10] *Ibid.,* April (1914), 44.
[11] *Ber.* **47** (1914), 2555.

merchandise to Europe. Moreover, during the Spanish rule in Central and South America, El Salvador was for a time under the jurisdiction of the viceroy of Peru, and merchandise originating from El Salvador could only be exported under his seal. These facts led to the erroneous impression in Europe that the balsam originated in Peru.

Botanical Origin.—According to official standard works (United States Pharmacopoeia and British Pharmacopoeia), Peru balsam is the exudation from the trunk of *Myroxylon pereirae* (Royle) Klotzsch (fam. *Leguminosae*). Years ago Harms [1] classified the tree as *Myroxylon balsamum* (L.) Harms var. *β-pereirae* (Royle) Baill. (see the monograph on "Oil of Balsam Tolu.").

Geographical Origin.—The principal areas producing Peru balsam are located northwest of San Salvador, the capital of El Salvador. Rather limited in area, the "Balsam Coast" consists of a mountainous district covered with tropical forests. Chief producing centers lie around San Julian, Isuhatan, Teopotepeque, Chiltinyan, and Izalco. Balsam trees occur also in other sections, but balsam from these regions is of such thick consistency that it can hardly be purified. They are not, therefore, exploited. Small quantities of Peru balsam have been produced in Guatemala, particularly during World War II, but this type has not attained much commercial importance.

The so-called Nicaraguan balsam must not be confused with Peru balsam, as it possesses different physicochemical properties. The Nicaraguan product appeared on the market some years ago as a low-priced substitute for the Peru balsam. When some tobacco manufacturers started to use it for the flavoring of certain types of tobacco, the price of the Nicaraguan balsam increased to such an extent that it could no longer be employed. Annual postwar exports of this type of balsam averaged 35,000 lb.

Growth of the Trees.—The trees grow wild, single or in groups, in altitudes up to 1,000 ft. They usually attain a height of 45 to 60 ft., and grow from one-half to more than one yard in diameter. Trees from twenty-five to thirty years old may be tapped, but they yield a very poor quality of balsam. Balsam productivity increases as the trees grow older, reaching a maximum at about sixty years.

Producing Season.—The highest yield is obtained during the dry and hot season, because at such times the trees develop more balsam. The quality, too, is better. The dry season lasts from the cessation of the rains (usually the end of November) until the middle of June, when the heavy rains start again.

Production of balsam during the rainy season amounts to only about two-

[1] *Notizbl. bot. Gart. Berlin-Dahlem* **5** (1908), 95.

thirds of that during the dry season. In the flowering season (April) the trees also seem to yield less balsam than at other times.

Production of Peru Balsam.—Two processes are in use today:

1. The so-called *cascara* or "bark process." Here (as well as in the other process described below), the tree must first be scorched by making a fire at its base. After ten minutes of scorching, the tree is left standing for about eight days, when the bark becomes so soft that pieces 1 to 2 ft. long and about 1 ft. wide can be cut out with a machete. Subsequent incisions are made either on the opposite side of the tree or above the first incision so that the tree will not be damaged too much. If made one above the other, the incisions become longer and longer. Supported by ropes and pulleys, the workers then climb high up—quite a dangerous task—and scorch the tree with torches.

The removed bark is crushed and subsequently pressed in primitive presses sheltered beneath a shed, hot water being poured over the bark several times, to soften the balsam and facilitate its flow. After the liquid is cooled, the balsam, which is heavier than water, settles at the bottom of the container and can be separated. The supernatant layer of water contains only very little balsam.

2. The so-called *pañal* (cloth) or *trapo* process. This is simply a continuation of the *cascara* process. After the bark has been removed by incisions, burlap is calked into the incisions to absorb the balsam exuding from the wound. After eight or ten days, these pieces of burlap are saturated with balsam and can be removed. They are boiled in water for about 1 hr., and subsequently pressed in the above-described presses, which also serve for the bark. The balsam separates on the bottom of the container and can easily be obtained by drawing off the water.

The *trapo* process yields about two to four times as much balsam as the *cascara* process, because new incisions for obtaining new pieces of bark should be made only every two months, while the same incisions can be treated every two weeks by the *trapo* process as long as the trees are in full vigor, i.e., from forty to sixty or more years old. However, it is advisable to treat one incision only twice and then make a new incision above the previous one, or on the opposite side of the trunk.

Properly worked and tapped twice a month, or twenty-four times a year, a healthy, fully grown tree not less than forty to fifty years old yields, on each extraction, from 4 to 6 oz. of balsam. Trees over sixty years old produce up to 8 oz. Trees younger than twenty-five years give a poor yield of balsam, of inferior quality.

Thus it appears that an average yield of balsam is 4 to 6 oz. with each extraction. Assuming that a tree is tapped twenty-four times a year, the total yearly yield would be from 96 to 144 oz. per tree. However, during the rainy season the tree "cools off," and the tapping cannot be done twice a month. Eighteen tappings per year per tree would be a safe estimate.

Purification of the Balsam.—After separation from the aqueous layer, the balsam is purified by boiling for 2 or 3 hr. above an open fire. This causes the water still present in the balsam to evaporate. Scum, dirt particles, and impurities are skimmed off, and the product is strained while still hot. The final balsam must flow clear and easily.

Obviously, the heating of the balsam with open fire is a rather crude practice, and has some deteriorative effect upon the odor of the final product, the more volatile constituents being driven off. Purification by more advanced methods produces a balsam superior to that regularly shipped from El Salvador—as has been proved in the laboratory. Unfortunately, competition in balsam Peru is so keen that the article can scarcely stand the added expense of purification outside of the country of origin.

Twenty pounds of balsam obtained by the *cascara* process are usually mixed with about one hundred pounds of *trapo* balsam. Since thinner, more liquid balsam is considered to be of higher quality than thick balsam, the *trapo* product is preferred to the *cascara;* the former has a lower specific gravity and a higher oil content.

Economic Set-up.—The trees are generally owned by landed proprietors who, during periods of low balsam prices, share the cost of production with the peons on an equal basis, all actual working expenses being carried by the peons. The lots produced on the various estates are sold, usually through middlemen, to exporters in San Salvador.

Exports of Balsam Peru from El Salvador.—According to a report of the United States Department of Commerce, Office of International Trade,[2] Washington, D. C., the total exports of balsam Peru prior to 1940 varied between 100,000 and 150,000 lb. annually; about 50 per cent of this was shipped to the United States. From 1940 to 1944 the exports averaged 218,000 lb. per year; 77 per cent of this was shipped to the United States. In 1947 the exports were 212,898 lb.; of this 77,218 lb. went to the United States. In 1948 El Salvador exported 106,274 kg.

Physicochemical Properties of the Balsam.—Peru balsam is a viscous, dark brown, transparent liquid which does not harden on exposure to air. The balsam possesses a pleasant, soft and sweet, very lasting odor, slightly reminiscent of vanilla. Peru balsam is soluble in 95 per cent alcohol, in chloroform, and in glacial acetic acid, with opalescence; only partly soluble

[2] Private communication from Mr. T. W. Delahanty, Chief, Consumers Merchandise Branch.

in ether and in petroleum ether. Shipments of genuine Peru balsam from El Salvador examined by Fritzsche Brothers, Inc., New York, had properties varying within these limits:

Specific Gravity at 25°/25°.......	1.152 to 1.170
Acid Number...................	56.2 to 83.0
"Cinnamein" Content............	47.0 to 58.7%, seldom below 50%
Saponification Number of the "Cinnamein"......................	230.5 to 240.0
Solubility......................	Soluble in high-proof alcohol, chloroform and glacial acetic acid, with not more than opalescence. Only partly soluble in ether and petroleum ether

Bennett [3] reported these properties for Peru balsam:

Specific Gravity at 15.5°.........	1.140 to 1.160
Refractive Index at 25°..........	1.588 to 1.595
Refractive Index at 25° of the extracted "Cinnamein"..........	1.575 to 1.582
Solubility......................	Not clearly soluble in 90% alcohol, a flocculent, waxy precipitation finally settling out on standing

Numerous papers have been published on various methods of assaying the balsam, but the scope of the present work does not permit their discussion, particularly in view of the fact that many of these methods have little, if any, value. Suffice it to refer the reader to the official tests of the United States Pharmacopoeia. The "cinnamein" content (see below) is determined by extraction with ether. It ranges from 50 to 60 per cent by weight, calculated upon the original balsam. In balsams of good quality, the "cinnamein" content should not be below 50 per cent. The analysis of Peru balsam also includes determination of the saponification number of the extracted "cinnamein," which should range from 230 to 240.

Occasionally Peru balsam is adulterated with rosin. A test for the detection of rosin will be found in Vol. I of the present work, p. 334. It should be pointed out, however, that an infallible rosin test has not yet been developed.

Other adulterants of Peru balsam are fatty oils and all kinds of low-priced gums. The rigid specifications of the United States Pharmacopoeia, however, prohibit addition of appreciable amounts of adulterants. A good criterion is the specific gravity which at 25° should be between 1.15 and 1.17.

Producers in the interior of the "Balsam Coast" claim that occasionally honey is used as an adulterant, but this too can easily be detected. The exporters in San Salvador usually do not submit the incoming lots of bal-

[3] *Perfumery Essential Oil Record* **19** (1928), 423.

sam to any chemical tests because the price of the product does not warrant analysis in special test laboratories. The exporters simply examine the balsam according to color, transparency, and flow, and whenever necessary a lot is reworked and purified by boiling it over an open fire, and eliminating moisture and mechanical impurities.

Chemical Composition of the Balsam.—Peru balsam is a complex mixture consisting of about 25 to 30 per cent of resin and 60 to 65 per cent of essential oil [4] (see below). The resinous part of the balsam is composed chiefly of an alcohol, viz., peru-resinotannol, which occurs in the balsam partly free, and partly esterified with cinnamic acid and benzoic acid.

OIL OF BALSAM PERU

Extraction of the Oil.—The oil of balsam Peru, in commerce at one time referred to as "cinnamein," possesses such a high boiling point that it cannot be isolated from the balsam by steam distillation. In technical practice the oil is, therefore, extracted from the balsam by means of volatile solvents such as petroleum ether or benzene, after neutralization of the free acids. The yield of oil in commercial production ranges from about 43 to 55 per cent, the average being about 50 per cent.

Physicochemical Properties of the Oil.—Oil of balsam Peru is a reddish-brown, slightly viscous liquid, possessing a pleasant, warm, balsamic, sweet and lasting odor. Gildemeister and Hoffmann [5] reported these properties for the oil:

Specific Gravity at 15°..........	1.102 to 1.122
Optical Rotation...............	Slightly dextrorotatory up to $+2°\ 30'$
Refractive Index at 20°.........	1.570 to 1.580
Acid Number..................	24 to 52
Ester Number.................	200 to 250
Solubility....................	Miscible with 90% alcohol in any proportion; sometimes soluble up to 1 vol. of 90% alcohol. On dilution the solution exhibits slight opalescence either immediately or after standing

Numerous lots of oil isolated by Fritzsche Brothers, Inc., New York, from pure Peru balsam (origin El Salvador) had properties varying within the following limits:

Specific Gravity at 15°/15°.......	1.102 to 1.110
Optical Rotation...............	$+1°\ 0'$ to $+1°\ 40'$
Refractive Index...............	1.5676 to 1.5728, usually above 1.5700
Acid Number..................	30.0 to 44.8

[4] "United States Dispensatory," 24th Edition (1947), 859.
[5] "Die Ätherischen Öle," 3d Ed., Vol. II, 880.

Saponification Number........... 246.5 to 267.9
Solubility...................... Soluble in 0.5 vol. of 90% alcohol and more

Chemical Composition of the Oil.—Oil of balsam Peru consists chiefly of "cinnamein," the latter being largely a mixture of esters in which benzyl benzoate and benzyl cinnamate predominate. The proportions of these esters appear to vary in oils derived from balsam of different origin. The following compounds have been reported in the oil:

Benzyl Benzoate. An oil examined by Tschirch and Trog [6] consisted chiefly of benzyl benzoate.

Benzyl Cinnamate. On the other hand, in several oils investigated by Thoms,[7] benzyl cinnamate appeared to predominate.

Peruviol(?). The same author also reported the presence of an alcohol $C_{13}H_{22}O$, b[7] $139°–140°$, $d_{17.5}$ 0.886, α_D $+13°$ 0', esterified probably with a dihydrobenzoic acid. The freed alcohol exhibited a pleasant odor, reminiscent of honey and narcissus.

d-Nerolidol. A more thorough investigation of peruviol by Schimmel & Co.,[8] however, showed that this alcohol consists to the greater part, if not entirely, of nerolidol (cf. Vol. II of this work, p. 260).

Farnesol. Reported as constituent of Peru balsam oil by Elze.[9]

Vanillin. Identified by Schmidt.[10]

Cinnamyl Alcohol(?). Delafontaine [11] expressed the opinion that cinnamyl alcohol occurs in oil of balsam Peru, but according to Gildemeister and Hoffmann [12] this is probably not the case.

Cinnamyl Cinnamate. The presence of this ester in oil of balsam Peru has been reported in literature.[13] However, no experimental data have been presented to this effect.

Stilbene(?). Also reported many years ago by Kachler,[14] but presence unlikely (Gildemeister and Hoffmann).

The fact that the above-mentioned compounds occur in the balsam (which is a pathological exudate of the tree) raises the question whether all of them are pathological products of the tree. This problem was recently studied by Naves [15] who examined wood of *Myroxylon pereirae* Klotzsch,

---

[6] *Arch. Pharm.* **232** (1894), 70.
[7] *Ibid.* **237** (1899), 271.
[8] *Ber. Schimmel & Co.,* April (1914), 75.
[9] *Chem. Ztg.* **34** (1910), 857.
[10] *Tagebl. d. Naturforscher-Vers. Strassburg* (1885), 377.
[11] *Z. Chem.* [2], **5** (1869), 156.
[12] "Die Ätherischen Öle," 3d Ed., Vol. II, 881.
[13] "United States Dispensatory," 24th Edition (1947), 859.
[14] *Ber.* **2** (1869), 512.
[15] *Helv. Chim. Acta* **30** (1947), 275, 278; **31** (1948), 408. *Perfumery Essential Oil Record* **39** (1948), 280.

imported from the province of Sonsonato in El Salvador. One log was derived from a healthy tree, the other from a tree that had been used for the production of balsam. Examination under the microscope showed that the brown heartwood is surrounded by a wide whitish zone, viz., the sapwood. The oil glands in the heartwood are impregnated with essential oil, whereas those in the sapwood are empty. The sapwood, whether of healthy trees or of trees used for the production of balsam, contains very little essential oil. The oil is localized in the heartwood.

Distilling the wood of the healthy and that of the wounded tree, Naves obtained 0.736 and 1.08 per cent, respectively, of essential oil. The two types of oil had the following composition:

	From Healthy Trees (%)	From Wounded Trees (%)
Cadinenes	28 to 30	6 to 8
l-Nerolidol, free	50 to 52	68 to 70
l-Cadinol, free	12 to 14	7 to 9
Alcohols, esterified	About 2	About 6

Thus the wood of the healthy tree yielded less essential oil and much less nerolidol than the wood of the tree used for the production of balsam. As a result of his observations, Naves suggested that the wood of trees exhausted of balsam be used in Salvador for recovery of its oil by distillation. The essential oil thus obtained is very rich in *d*-nerolidol, a highly esteemed perfume material that imparts "sleekness" and smooth tonalities to otherwise harsh compositions.

Continuing his work, Naves [16] also investigated the essential oil derived from the *fruit* of *Myroxylon pereirae* Klotzsch and found that it contains a laevorotatory bicyclic sesquiterpene. On treatment with gaseous hydrogen chloride, this sesquiterpene yielded *l*-cadinene dihydrochloride; on hydrogenation the sesquiterpene gave a hydrocarbon which appeared to be *trans*-tetrahydrocadinene. The sesquiterpene was accompanied by an isomer that readily oxidized or polymerized, and by a small quantity of sesquiterpene alcohols.

More recently Naves and Ardizio [17] found that the oil derived from the *leaves* of the tree contains neither nerolidol nor farnesol; these two sesquiterpene alcohols, therefore, occur only in the wood of the tree.

Resinoid of Peru.—On repeated extraction with volatile solvents, removal of the insoluble portions by filtration, and concentration of the filtrates (removal of the solvent), Peru balsam yields the so-called resinoid

[16] *Helv. Chim. Acta* **32** (1949), 2180.   [17] *Ibid.* **33** (1950), 169.

of Peru, a very viscous, dark brown mass of pleasant odor and high fixation value. According to Naves and Mazuyer,[18] the yield of resinoid on extraction with benzene varies between 80 and 86 per cent.

Naves and Mazuyer reported the following properties for resinoids of Peru, obtained by extraction with alcohol (I), and by extraction with benzene (II):

	I	II
Specific Gravity at 15°	1.140 to 1.170	1.103 to 1.126
Optical Rotation	Up to +2° 20'	Up to +2° 44'
Refractive Index at 20°	1.590 to 1.599	1.568 to 1.582
Acid Number	55 to 84	22.4 to 48.6
Ester Number	168 to 220	198 to 260

The resinoid prepared by extraction with alcohol offers the advantage of solubility in high-proof alcohol and a very lasting odor.

Use.—Peru balsam is used mainly in pharmaceutical preparations—dressing of wounds, e.g. The oil and the resinoid are employed chiefly in perfumes, soaps and cosmetics (see the monograph on "Oil of Balsam Tolu").

OIL OF BALSAM TOLU

Essence de Baume de Tolu   *Tolubalsamöl*
Aceite Esencial Balsamo Tolu   *Oleum Balsami Tolutani*

BALSAM TOLU

Botany and Origin.—The taxonomy of the tree from which balsam Tolu is derived has been the subject of considerable controversy. Official standard works (United States Pharmacopoeia and British Pharmacopoeia) now classify the tree as *Myroxylon balsamum* (L.) Harms (fam. *Leguminosae*). Tschirch [1] expressed the belief that the trees yielding balsam Tolu and balsam Peru are merely physiological forms of *Toluifera balsamum* L. Harms [2] stated that balsam Tolu is derived from *Myroxylon balsamum* (L.) Harms var. *α genuinum* Baill., and balsam Peru from *Myroxylon balsamum* (L.) Harms var. *β pereirae* (Royle) Baill.

---

[18] "Les Parfums Naturels," Paris (1939), 295.
[1] "Die Harze und Harzbehälter," Vol. 2, 2d Ed. (1906), 1204, 1208.
[2] *Notizbl. bot. Gart. Berlin-Dahlem* **5** (1908), 95.

Myroxylon balsamum L., a tall tree with a spreading crown, differs from the Peru balsam tree only by its habit of branching about 15 m. aboveground. It occurs wild in the great forests of the province Tolu, along the Magdalena and Cauca rivers in Colombia. The balsam forms in trunk tissues as the result of injuries. To stimulate the formation of balsam, Indians lacerate the bark of the trunk by making deep V-shaped incisions with machetes. The balsam is collected in small cups, which are inserted in slight excavations beneath the point of the two vertical incisions meeting at the lower end. As many as twenty cups have been observed at a time on one tree. The workers go from tree to tree, emptying the cups into bags of rawhide slung over the back of a donkey. In these skin vessels the product is then brought to the shipping ports.[3] Collection of the balsam continues all year around, except for the period of heavy rains, when the forests become impassable. The balsam is packed in tin cans and exported mainly from Barranquilla.

Total exports of balsam Tolu per year amount to about 80,000 kg. net.

There exists only one quality of balsam Tolu; various lots differ merely in color, which ranges from light yellow to reddish black. No price differential is made because of color.

Physicochemical Properties of the Balsam.—Tolu balsam is a brown or yellowish-brown, semisolid or plastic mass with a pleasant, aromatic, lasting and sweet odor, reminiscent of vanilla. The consistency of the balsam depends upon its age and the temperature. On standing, it turns into a plastic solid, softening at about 30° and melting between 60° and 65°. With age the balsam becomes brittle. Tolu balsam is soluble in 95 per cent alcohol, in ether, and in chloroform, but nearly insoluble in petroleum ether.

Shipments of genuine Tolu balsam imported by Fritzsche Brothers, Inc., New York, from Barranquilla had properties varying within the following limits:

Acid Number	102 to 158
Saponification Number	154 to 236.7 (one lot as low as 150)
Solubility	Soluble in high proof alcohol, chloroform and ether

These properties apply to both the crude and the purified balsam, because the limits for the two types overlap and nothing would be gained by listing them separately. Purification, in this case, simply means removal of dirt, wooden particles, and insects, but no special treatment. It should be pointed out here that in the producing regions it is almost impossible to

[3] "United States Dispensatory," 24 Edition (1947), 1229.

remove all such foreign material, and the presence of minute amounts does not violate the usual standards set up in the United States.[4]

Many cans of Tolu balsam contain small quantities of water which rise to the top and should be taken into consideration when drawing a sample of balsam for analysis. Standards and tests are described in the United States Pharmacopoeia. Balsam Tolu is occasionally adulterated with rosin. A test for the detection of rosin in Tolu balsam will be found in Vol. I of the present work, p. 334. It should be mentioned, however, that an infallible rosin test has not yet been developed. Much work has been done on the analysis of Tolu balsam, particularly by van Itallie and his collaborators,[5] by Rosenthaler,[6] and Bennett.[7] For details the reader is referred to the original literature.

The chemical composition of Tolu balsam is complex with the result that the literature contains contradictory statements on the subject.[8] To summarize investigations covering a period of more than fifty years: the balsam, like Peru balsam, appears to contain a mixture of resinous substances, benzyl benzoate, some benzyl cinnamate, a small percentage of essential oil, and traces of vanillin. The proportion of resinous matter is larger than in Peru balsam, amounting to 75 or 80 per cent. On saponification, the chief constituent of Tolu balsam, the so-called "tolu-resin," yields an alcohol —viz., tolu-resinotannol—and cinnamic and benzoic acids. The "tolu-resin" is easily soluble in 95 per cent alcohol, acetone, or glacial acetic acid, insoluble in petroleum ether. (Petroleum ether dissolves from 2 to 10 per cent of the balsam; the soluble portions consist of cinnamic and benzoic esters, vanillin, and perhaps some terpenes.[9])

According to the British Pharmacopoeia, 1948 (p. 84), Tolu balsam should contain "not less than 35 per cent, and not more than 50 per cent, of total balsamic acids, calculated with reference to the dry alcohol-soluble matter."

OIL OF BALSAM TOLU

Distillation.—Steam distillation of Tolu balsam yields from 1.5 to 7 per cent of a volatile oil which is lighter or heavier than water, depending upon the quality of the live steam employed. Distillation with low-pres-

---

[4] "Vegetable drugs are to be as free as practicable from insects or other animal life, animal material, or animal excreta."

[5] *Pharm. Weekblad* **56** (1919), 1185; **62** (1925), 510, 893.

[6] *Pharm. Ztg.* **73** (1928), 837.

[7] *Perfumery Essential Oil Record* **19** (1928), 464.

[8] Tschirch and Oberländer, *Arch. Pharm.* **232** (1894), 559. Cocking and Kettle, *Pharm. J.* [4], **47** (1918), 40. Rollett and Schneider, *Monatsh.* **55** (1930), 151.

[9] Burger, *Riechstoff Ind.* **7** (1937), 135.

sure steam results in oils lighter than water; superheated steam or steam of high pressure gives better yields and oils of higher specific gravity.

Physicochemical Properties of the Oil.—Oil of balsam Tolu is a somewhat viscous liquid with an agreeable, lasting and sweet odor, slightly reminiscent of hyacinth. Gildemeister and Hoffmann [10] reported the following properties for oil of balsam Tolu:

Specific Gravity at 15°........	0.945 to 1.09
Optical Rotation..............	−1° 20′ to +0° 54′
Refractive Index at 20°.......	1.537 to 1.560
Acid Number.................	5 to 34
Ester Number...............	153 to 208
Solubility...................	Usually not clearly soluble in 80% alcohol. Soluble in 1 vol. of 90% alcohol, opalescent to turbid with more; separation of floccules in the latter case

Oils of balsam Tolu distilled by Fritzsche Brothers, Inc., New York, had properties varying within these limits:

Specific Gravity at 15°/15°....	0.907 to 1.016
Optical Rotation.............	−1° 5′ to −4° 0′
Refractive Index at 20°.......	1.5075 to 1.5347
Acid Number.................	23.8 to 80.0
Saponification Number........	85.9 to 171.1
Solubility...................	Soluble in 0.5 vol. and more of 90% alcohol

Chemical Composition of the Oil.—Early investigations [11] into the chemical composition of the volatile oil derived from balsam Tolu, which were carried out about a century ago, gave only vague and partly contradictory results. They need not be discussed here. The following compounds are probably present in the oil:

Phellandrene(?). In the fraction b. 170° Kopp [12] observed a terpene with an odor reminiscent of elemi gum. It was apparently phellandrene.

Benzyl Benzoate and Benzyl Cinnamate. Years ago Busse [13] identified benzyl benzoate and benzyl cinnamate in the balsam itself. Most probably these esters are present also in the distilled oil, because the oil exhibits a high ester number and, on saponification, yields crystalline acids (apparently benzoic acid and cinnamic acid).

Farnesol. According to Elze,[14] oil of balsam Tolu also contains traces of farnesol. Recently confirmed by Naves.[15]

[10] "Die Ätherischen Öle," 3d Ed., Vol. II, 878.
[11] Deville, *Ann. chim. phys.* [3], **3** (1841), 151. *Liebigs Ann.* **44** (1842), 304. Scharling, *ibid.* **97** (1856), 71.
[12] *J. pharm. chim.* [3], **11** (1847), 425.   [14] *Chem. Ztg.* **34** (1910), 857.
[13] *Ber.* **9** (1876), 830.   [15] *Helv. Chim. Acta* **32** (1949), 2181.

Submitting Tolu balsam to dry distillation, Dupont and Guerlain [16] obtained substantial quantities of benzoic acid and cinnamic acid, as well as guaiacol, creosol, and perhaps 4-hydroxy-3-methoxy-1-ethylbenzene. Carbon dioxide and hydrogen developed during distillation.

In order to compare the chemical composition of the oil derived from balsam Tolu (which is a pathological exudate of the tree) with that of the essential oil contained in the *wood* of a healthy tree, Naves [17] imported two logs of *Myroxylon balsamum* L. from San Juan de Uraba, one of the principal producing regions of Tolu balsam in Colombia. One log came from a healthy tree, the other from a tree which had served for the production of balsam. Naves found that the oil is present chiefly in the red heartwood of the trunk. The wood of the tree that had been wounded contained more oil than the wood of the healthy tree. However, in regard to chemical composition the oils from the two trees were identical. Naves reported the presence of the following compounds in the oil derived from the wood of *Myroxylon balsamum* L.:

l-Cadinol. The chief constituent.

d-Cadinene and Other Sesquiterpenes.

Farnesol. Small quantities.

Nerolidol. Traces (only one-fifteenth of the quantity of farnesol).

Resinoid of Tolu.—On repeated extraction with high-proof alcohol, removal of any impurities present by filtration, concentration of the filtrates (removal of the alcohol), Tolu balsam yields the so-called resinoid of Tolu, a very viscous, dark mass of pleasant odor and high fixation value. According to Naves and Mazuyer,[18] the yield of resinoid varies between 60 and 66 per cent. The resinoid has an acid number ranging from 112 to 154, and an ester number varying between 53 and 72. The resinoid contains from 2.5 to 12 per cent of essential oil. The resinoid offers the advantage of being a clear, viscous, and water-free liquid.

Use.—Tolu balsam itself is used as a feeble expectorant in cough mixtures and as an inhalent in cases of obstinate catarrh.

The volatile oil derived from the balsam, on the other hand, is employed chiefly in perfumes, cosmetics and soaps. Its pleasant, aromatic, and hyacinth-like odor blends well with certain floral, and even better with

[16] *Compt. rend.* **191** (1930), 716.
[17] *Helv. Chim. Acta* **32** (1949), 2181.
[18] "Les Parfums Naturels," Paris (1939), 298.

oriental, types of compositions. The oleoresin is an excellent odor fixative, and at the same time imparts warm tonalities to a perfume compound.

SUGGESTED ADDITIONAL LITERATURE

C. T. Bennett, "Balsam of Tolu," *Perfumery Essential Oil Record* **19** (1928), 464.

OIL OF CABREUVA

Myrocarpus frondosus Fr. Allem. and *Myrocarpus fastigiatus* Fr. Allem. (fam. *Leguminosae*), the so-called "Cabreuva," are tall trees attaining a height of 12 to 15 m. The cabreuva tree grows wild in many parts of southern Brazil (States of São Paulo and Rio de Janeiro), in Paraguay, and northern Argentina.

The wood of these trees is very hard and is used extensively for the construction of houses and ships. In the State of São Paulo, the waste shavings and sawdust are occasionally employed for the extraction of the essential oil. When lacerated by incisions, the trunks of the live trees produce a pathological exudate closely resembling the balsam Peru. Because of this similarity, the balsam of the cabreuva tree has been known for centuries as "Baume du Pérou brun" or "Baume du Pérou rouge en coques."[1] It was described in old European pharmacopoeias, but is now used only by natives in Brazil to heal wounds and ulcers and to obviate scars.

The heartwood of the trees contains an essential oil of pleasant odor, which can be isolated by steam distillation. It was distilled experimentally for the first time many years ago by Peckolt.[2] Lately substantial quantities of the so-called cabreuva oil have been produced commercially, particularly in the State of São Paulo. According to information gathered by the author in the producing regions, the distillation of 700 kg. of shavings and sawdust requires about 24 hr. The yield of oil averages 1.5 per cent.

Physicochemical Properties.—Oil of cabreuva is a somewhat viscous liquid of pleasant, balsamic, very lasting odor, slightly reminiscent of rose and sandalwood.

[1] Schaer, *Arch. Pharm.* **247** (1909), 176. Tschirch and Werdmüller, *ibid.* **248** (1910), 431.
[2] *Katalog der National-Ausstellung in Rio de Janeiro* (1866), 48.

A genuine oil produced under the author's supervision in Mattão, São Paulo, had these properties:

Specific Gravity at 15°	0.893
Optical Rotation	+8° 0'
Refractive Index at 20°	1.4845
Acid Number	0
Ester Number	0
Ester Number after Acetylation	144
Boiling Range at 6 mm	124°–138°
Solubility	Soluble in 6 vol. and more of 70% alcohol

An oil of cabreuva examined by Naves [3] exhibited the following properties:

Specific Gravity at 20°/4°	0.8875
Optical Rotation	+8° 20'
Refractive Index at 20°	1.48322
Ester Number	2.2
Alcohol Content, Calculated as $C_{15}H_{26}O$ (Cold Formylation)	82.9%

On fractionation, this oil yielded about 80 per cent of nerolidol.

Chemical Composition.—The chemical composition of cabreuva oil was investigated by Naves [4] who reported the presence of these compounds in the oil:

Tetrahydro-Δ^3-*p*-toluic Aldehyde. Very small quantities only. Identified by oxidation to tetrahydro-Δ^3-*p*-toluic acid m. 99°.

p-Methylacetophenone. Also in very small quantities (cf. Vol. II of this work, p. 476).

l-1-Methyl-4-acetyl-1-cyclohexene (*p*-Methyl-3-tetrahydroacetophenone). In very small quantities only (cf. Vol. II, p. 482).

d-Nerolidol. The chief constituent (about 80 per cent) of the oil. The nerolidol present in cabreuva oil is a mixture of two isomers (cf. Vol. II, p. 260).

Farnesol. Aside from *d*-nerolidol, the oil contains about 2.5 per cent of farnesol (cf. Vol. II, p. 258).

Other Sesquiterpene Alcohols(?). The nerolidol and farnesol are accompanied by a mixture of laevorotatory sesquiterpene alcohols $C_{15}H_{26}O$. One of them is *bisabolol*, another appears to be a bicyclic sesquiterpene alcohol of the cadalene group, and a third one an azulogenic sesquiterpene alcohol.

[3] *Helv. Chim. Acta* **30** (1947), 277.
[4] *Ibid.* **30** (1947), 275, 278; **31** (1948), 44. Cf. *ibid.* **31** (1948), 408; **32** (1949), 2181. *Perfumery Essential Oil Record* **38** (1947), 191; **39** (1948), 280.

A comparison of the chemical composition of cabreuva oil with that of bois de rose oil shows that cabreuva oil may be considered a sesquiterpene homologue of bois de rose oil (Naves):

Cabreuva Oil	*Bois de Rose Oil*
Nerolidol	Linaloöl
Farnesol	Geraniol-Nerol
Bisabolol	Terpineol

Use.—Oil of cabreuva is an excellent odor fixative in perfumes, cosmetics and in the scenting of soaps. The chief use of the oil, however, is as a starting material for the isolation of farnesol (Patent L. Givaudan et Cie., S.A. Geneva, Switzerland).

CONCRETE AND ABSOLUTE OF CASSIE

There are two species of *Acacia* (fam. *Leguminosae*) which have been used for the commercial extraction of their natural flower oils, viz., *Acacia farnesiana* Willd., and *A. cavenia* Hook. et Arn. According to some botanists,[1] the latter is only a variety of the former. Of the two, *Acacia farnesiana* has always been preferred because it yields a flower oil with a much finer odor than that from *Acacia cavenia*. As a matter of fact, the latter is no longer produced in substantial quantities, having been replaced almost entirely by *A. farnesiana*. For the sake of completeness, however, we shall now discuss the two species in order.

I. *Acacia Farnesiana*

Acacia farnesiana Willd., our common sweet acacia, in France called *Cassie ancienne*, is a thorny, much branched shrub, up to 10 ft. in height, and bearing round, yellow-red flower beads in small clusters. It grows spontaneously in many tropical and warm countries. In Southern France, Syria, Egypt, Algeria, and lately in Morocco, it is cultivated for the treatment (extraction) of its flowers, which contain a most agreeable perfume, reminiscent of violet and raspberry. *Acacia farnesiana*, however, is quite a sensitive plant, easily damaged by frost or drought, and requires great care in cultivation. On the French Riviera, for example, where it was

[1] Cf. C. Wehmer, "Die Pflanzenstoffe," Jena, 2d Ed., I (1929), 492.

once grown extensively, it has repeatedly suffered from attacks by cold weather; therefore plantings here are gradually disappearing, being replaced by those in North Africa. Another reason lies in the development of the French Riviera into a fashionable resort and tourist center; cassie plantations formerly located near Cannes (Le Cannet, Vallauris, Mougins, Saint-Laurent-du-Var) have been parceled off for the building of villas. Labor, too, has become too expensive on the glamorous Côte d'Azur—and the cultivation, particularly the harvesting, of *Acacia farnesiana* requires a great deal of labor. The plant is covered with numerous large thorns, the flowers are hard to reach, and the picking represents a difficult, if not painful, task.

In Southern France, branches of *Acacia farnesiana* are often grafted upon *A. cavenia,* which is a much hardier and frost-resistant plant. Any branches damaged by frost are trimmed off and replaced by new graftings from *A. farnesiana.* Two years after such operation the branches will again bear a normal flower harvest. When in full production, and depending upon its size, a bush yields from 1 to 2, in exceptional cases up to 5, kg. of flowers per year. The harvest begins at the end of September and lasts until the end of December. The flowers are collected twice every week. Toward the end of February the branches are usually pruned, in order to keep the bush within a size that facilitates harvesting of the flowers during the next season.

In Egypt, which now produces substantial quantities of concrete of cassie, conditions for the growth of *Acacia farnesiana* are much more favorable than in Southern France. Attacks by frost are practically unknown, labor is low priced and ample, the harvesting of the flowers being done by native children—and they are numerous. It means no hardship to them; on the contrary, it brings them much desired pocket money. According to observations made by the author when he visited Egypt some years ago, the cassie plantations are located not far from Cairo. For propagation, seed is planted in a nursery on fertile, black soil. This requires much experience and attention; otherwise the bulk of the seed will not develop. The young plants must be shaded and watered most carefully. They remain in the nursery for almost two years. When they appear to have stopped growing, the young plants are ready for transplanting into the field. For this purpose furrows are dug sufficiently deep to hold the plants without necessity of trimming their root system. The plants are set out 2½ to 6 ft. apart, in rows 14 to 28 ft. apart. The field has to be amply manured and frequently irrigated. The amount of water, however, must be controlled carefully. Provided the seed has been raised in fertile, black soil and has remained in the nursery for two years, the young plants will withstand transplanting without too high mortality. In the first year after transplanting the bushes will develop a few flowers; after the third year a bush will yield from 300 to 1,000 g. of flowers. The buds appear in September, but the harvest does not start before October and lasts until April, with a marked decline of productivity, however, in December and January. In April the branches are pruned (while they still contain some flowers) in order to keep the bushes within a size that allows the children to reach all flowers by hand. Four to six weeks after pruning the branches again attain a length of 1 m. For this reason the trimming has to be done quite closely, particularly after the fourth year, when the bushes have attained full growth.

As regards production in France, about 35 metric tons of flowers were processed in 1900. Since then the quantity has declined steadily and at present amounts to only 2 metric tons per year. Egypt and Algeria each produce 3 or 4 tons of flowers every year for extraction purposes. Morocco has only recently started to develop plantations of *Acacia farnesiana*, and substantial quantities of flowers may be expected in the future.

In the author's experience, about 200 to 250 kg. of *Acacia farnesiana* flowers are required to yield 1 kg. of concrete. According to Naves and Mazuyer,[2] the yield of concrete usually varies within 0.50 and 0.70 per cent, but may be as high as 0.82 per cent, in exceptional cases. The flowers collected at the end of the season (late spring) in general give a higher yield of concrete than those gathered in the fall.

Physicochemical Properties.—Concrete of *Cassie ancienne* is a solid, waxy, dark brown mass which, on exposure to air and light, becomes lighter. Walbaum and Rosenthal[3] reported a congealing point of 46.5° and a saponification number of 103.6 for a concrete.

On steam distillation, the concrete yields from 6.5 to 9 per cent of a volatile oil, which, however, is not produced commercially but is used only for analysis of the concretes or absolutes (cf. Vol. I of this work, p. 213, 215). Naves, Sabetay and Palfray,[4] reported the following properties for four steam-volatile oils which they had obtained from a concrete of Egyptian origin:

	I	II	III	IV
Specific Gravity at 15°.....	1.032	1.037	1.029	1.043
Optical Rotation..........	+0° 55'	+0° 20'	+1° 10'	−0° 15'
Refractive Index at 20°....	1.5045	1.5082	1.5018	1.5120
Acid Number.............	5.60	3.70	4.20	6.30
Ester Number............	154.0	148.0	166.0	182.0

[2] "Les Parfums Naturels," Paris (1939), 198.
[3] *Ber. Schimmel & Co., Jubiläums-Ausgabe* (1929), 193.
[4] *Perfumery Essential Oil Record* **28** (1937), 336.

On treatment with high-proof alcohol in the usual way, concrete of *Cassie ancienne* yields from 30 to 35 per cent of alcohol-soluble absolute.

An absolute of *Cassie ancienne* from Grasse (I), described by Naves and Mazuyer,[5] and absolutes from Liguria, Italy (II), prepared by Rovesti,[6] had the following properties:

	I	II
Specific Gravity at 15°	0.988	1.020 to 1.070
Optical Rotation	...	0° up to $-3°0'$
Refractive Index	...	1.514 to 1.521
Acid Number	62.8	18 to 55
Ester Number	33.5	97 to 243

Chemical Composition.—About fifty years ago the chemists of Schimmel & Co.[7] investigated an absolute of maceration which they had obtained by treating 115 kg. of cassie pomade from India three times with alcohol. Concentration of the alcoholic washings and steam distillation of the residual absolute yielded 315 g. of a dark oil which, on rectification with steam, gave 197 g. of oil; d_{15} 1.0475, α_D $\pm 0°$, n_D^{20} 1.51331, sap. no. 176. Schimmel & Co. reported the presence of the following compounds in the volatile oil thus obtained:

Benzaldehyde. Semicarbazone m. 214°.

Anisaldehyde. Oxidation to anisic acid m. 180°.

Decylaldehyde. Semicarbazone m. 97°.

Cuminaldehyde. Semicarbazone m. 200°.

A Ketone(?). With a menthone-like odor; b. 200°–205°, d_{15} 0.9327, α_D $-3°$ 50'; semicarbazone m. 177°–178°.

A Ketone(?). With a violet-like odor; b_{15} 133°; *p*-bromophenylhydrazone m. 103°–107°. This ketone appears to be very important for the odor of the cassie flower oil.

p-Cresol. Small quantities.

Methyl Salicylate. The oil obtained by Schimmel & Co. from the Indian pomade contained 11 per cent of methyl salicylate. A wax-free extract of *Cassie ancienne* examined by von Soden [8] contained 30.9 per cent of esters, calculated as methyl salicylate.

Benzyl Alcohol. Identified by means of the phenylurethane m. 177°, and of the acid phthalic ester m. 105°–106°.

[5] "Les Parfums Naturels," Paris (1939), 198.

[6] *Profumi ital.* **3** (1925), 277.

[7] Gildemeister and Hoffmann, "Die Ätherischen Öle," 3d Ed., Vol. II, 859. Cf. Walbaum, *J. prakt. Chem.* [2], **68** (1903), 235.

[8] *J. prakt. Chem.* [2], **69** (1904), 270.

Geraniol(?) and Linaloöl(?). Presence probable, but not definitely established by Schimmel & Co.

Farnesol. Reported in the high boiling constituents by Haarmann and Reimer.[9]

More recently, La Face [10] investigated an absolute of *Acacia farnesiana* (prepared from a genuine Calabrian concrete; yield of absolute 36 per cent), and found that the absolute contained no eugenol and no methyl salicylate —important constituents of the natural flower oil derived from Roman cassie (*Acacia cavenia*—see below).

La Face proved the presence of the following compounds in the natural flower oil obtained by volatile solvent extraction of *Acacia farnesiana:*

Benzyl Alcohol.

Linaloöl.

α-Terpineol.

Nerolidol or Farnesol (or both), partly esterified.

A Mixture of Cresols.

A Mixture of Ethyl Phenols.

Hydroxyacetophenone.

A Ketone with an ionone configuration (probably optically active α-ionone).

Coumarin.

Butyric Acid.

Palmitic Acid, free (about 50 per cent of the free acids present in the absolute).

Benzoic Acid.

Salicylic Acid.

n-Eicosane.

According to La Face, the perfume of the absolute is due chiefly to the odor complex cresol, ethyl phenol, and coumarin.

Use.—Absolute of *Cassie ancienne* is a valuable adjunct in the creation of high-grade perfumes of the French type. It imparts elegant notes to floral as well as oriental scents. The absolute has to be dosed carefully and well blended in order to obtain the desired effects. It harmonizes particularly well with violet, orris, and the ionones.

[9] German Patent No. 149,603 (1902). Cf. *Chem. Zentr.* (1904), I, 975.
[10] *Helv. Chim. Acta* **33** (1950), 249.

II. *Acacia Cavenia*

Acacia cavenia Hook. et Arn. is a large shrub with stout spines, often grown in hedges. It was introduced into Southern France from Italy; hence the name "Roman cassie" or *Cassie romaine*. The flowers resemble those of *Acacia farnesiana*, but their perfume is somewhat harsher, more spicy, and less pleasing than that of the flowers from *Acacia farnesiana*. For this reason, perfumers prefer the latter, and, when buying cassie flowers, the manufacturers in Grasse attempt to obtain only those of *Acacia farnesiana*. This, however, is not always possible and for a long time it has been the practice of the growers to supply a mixture of two parts of *Acacia farnesiana* and one part of *Acacia cavenia*. The price of the former has usually been about twice that of the latter. Because of the development of real estate in the vicinity of Cannes, the plantations of *Cassie romaine* are now gradually disappearing, and it will not be long before flowers from this plant are no longer available. The manufacturers in Grasse do not regret this development. In general, *Acacia cavenia* is a hardier plant than *Acacia farnesiana* and does not suffer so much from attacks by frost and drought. It requires much less care than *Acacia farnesiana*. The flowers appear in September and can be harvested up to May. There is a lull in December and January, and a new bloom in spring.

Formerly the natural perfume was isolated from the flowers by means of maceration with hot fat (cf. Vol. I of this work, p. 198), but this process has been gradually replaced by extraction with petroleum ether.

According to the author's experience, about 200 kg. of flowers are required to yield 1 kg. of concrete. Naves and Mazuyer [11] reported yields of concrete ranging from 0.55 to 0.88 per cent, in most cases from 0.60 to 0.84 per cent. Flowers harvested in October give a better yield of concrete than those harvested in spring (0.60 to 0.92 per cent against 0.49 to 0.56 per cent).

Physicochemical Properties.—Concrete of *Cassie romaine* is a solid, waxy mass of brown color. Naves and Mazuyer [12] reported for a concrete of French origin: m. 52°–53°, acid number 32.4, ester number 46.2.

On steam distillation, the concrete yields from 7 to 10.4 per cent of a volatile oil which, however, is not a commercial article. Naves, Sabetay and Palfray [13] reported the following properties for two steam-volatile oils derived from two concretes of *Cassie romaine:*

[11] "Les Parfums Naturels," Paris (1939), 200.
[12] *Ibid.*
[13] *Perfumery Essential Oil Record* **28** (1937), 336.

	I	II
Specific Gravity at 15°	1.031	1.028
Optical Rotation	±0°	+0° 14'
Refractive Index at 20°	1.5120	1.5140
Acid Number	11.2	22.0
Ester Number	96.0	112.0

On treatment with high-proof alcohol, in the usual way, the concrete yields from 30 to 35 per cent of an alcohol-soluble absolute. An absolute of *Cassie romaine* of French origin was examined by Naves and Mazuyer: [14]

Specific Gravity at 15°	0.984
Optical Rotation	+8° 20'
Refractive Index at 20°	1.5037
Acid Number	33.6
Ester Number	49

Chemical Composition.—Walbaum [15] investigated a steam-volatile oil of *Acacia cavenia* which he had obtained by steam distillation of a petroleum ether concrete. Treatment of two samples with dilute sodium hydroxide solution gave these portions:

	Per Cent	Per Cent
Phenolic Constituents	5.5	3
Nonphenolic Constituents	8.8	4.5
Salicylic Acid	1.1	1.66

The percentages are calculated upon the concrete.

In the volatile oil Walbaum identified the following compounds:

Eugenol. Characterized by means of its benzoyl compound m. 69°–70°. The phenols present in the oil contained at least 90 per cent of eugenol.

Methyleugenol. Oxidation to veratric acid m. 178°.

Methyl Salicylate. Saponification of the ester gave salicylic acid m. 156°.

Benzaldehyde. Semicarbazone m. 214°.

Anisaldehyde. Semicarbazone m. 203°–204°. Oxidation to anisic acid m. 184°.

Decyl Aldehyde(?). Presence probable.

A Ketone(?). Of violet-like odor.

Benzyl Alcohol. Phenylurethane m. 77°–78°.

Geraniol. Diphenylurethane m. 81°.

Linaloöl(?). Presence probable.

[14] "Les Parfums Naturels," Paris (1939), 201.
[15] *J. prakt. Chem.* [2], **68** (1903), 235. *Ber. Schimmel & Co.*, October (1903), 14.

Use.—When it was still produced in substantial quantities, absolute of *Cassie romaine* served as a lower priced substitute, and occasionally as an adulterant, of absolute of *Cassie ancienne* (see above).

CONCRETE AND ABSOLUTE OF MIMOSA

The name mimosa popularly applies not only to species of the true genus *Mimosa,* of which there are many, but to certain of the genus *Acacia* as well. As a matter of fact, the natural flower oils commercially known as concrete and absolute of mimosa are derived not from any true mimosa, but from *Acacia decurrens* var. *dealbata* (syn. *A. dealbata* Lk.) fam. *Leguminosae.* This tree was introduced to Southern France during the first half of the nineteenth century from Australia, where it is known as "Silver Wattle." Another so-called mimosa is *Acacia floribunda* Willd., cultivated widely in Southern France almost exclusively for the florist trade because of its long-lived flowers. *Acacia decurrens* var. *dealbata,* on the other hand, serves both for the extraction of the oil and for the cut flower trade.

Most mimosas prefer siliceous soils, and do not grow well on calcareous ground. Large forests and groves of mimosa extend along the French Riviera, in the Éstérelle and Maures Mountains and in the massif of Taneron (Vallauris, Supercannes, La Croix-des-Gardes, La Bocca, Théoule, Auribeau, Pégomas, etc.). From February to March these mimosa trees, with their golden-yellow flowers, afford a magnificent spectacle. The bulk of the harvested flowers goes to the florists in Paris and other cities of Central Europe and England; but toward the end of the blooming season, when the flowers are fully developed, and contain a maximum of perfume, large quantities are processed in Grasse for the extraction of the natural flower oil. The quantities of mimosa flowers (*Acacia decurrens* var. *dealbata*) treated for this purpose have varied from 50 to 100 metric tons per year. Steam distillation giving no results, the flowers are extracted with petroleum ether (see Vol. I of this work, p. 200), yielding concrete of mimosa, which can be transformed into an alcohol-soluble absolute. In the author's experience, 180 to 200 kg. of mimosa flowers yield 1 kg. of concrete which, in turn, gives 0.18 to 0.20 kg. of absolute. According to Naves and Mazuyer,[1] the yield of concrete ranges from 0.70 to 0.88 per cent. The perfume is contained exclusively in the flowers. Carefully culled flowers gave as much as 1.06 per cent of concrete against 0.77 per cent from unselected material. The concrete yields from 20 to 25 per cent of absolute.

Physicochemical Properties.—Concrete of mimosa is a solid, waxy mass of light brown color. Its perfume is not strongly characteristic of the live flowers; its odor slightly recalls beeswax. (The concrete derived from *Acacia decurrens* var. *dealbata* exhibits a "by-note" of violet, that from *Acacia floribunda* possesses a phenolic note, reminiscent also of ylang ylang.)

Three concretes described by Naves and Mazuyer[2] (I, II, and III), and by Sabetay and Trabaud[3] (IV) had these properties:

	I	II	III	IV
Melting Point	48°	51°	54°	52.5°
Acid Number	12.4	16.1	9.8	14
Ester Number	32.4	26.2	38.8	38

According to Naves and Mazuyer, the concrete contains from 4.1 to 6.2 per cent of steam-volatile oil. Sabetay and Trabaud reported a yield of 3 per cent of steam-volatile oil. Co-distillation with ethylene glycol *in vacuo* (see Vol. I of this work, p. 215) gives 4.7 per cent of volatile oil.

Absolute of mimosa is a viscous, syrupy liquid of yellow-brownish color, with an odor more typical of the live flowers than is that of the concrete.

Naves and Mazuyer[4] (V), and Sabetay and Trabaud[5] (VI) reported the following values:

	V	VI
Specific Gravity at 15°	1.002	0.9797
Optical Rotation	+22° 40'	+15° 40'
Refractive Index at 20°	1.5184	1.5175
Acid Number	8.9	28.2
Ester Number	60.4	42.3

On co-distillation with ethylene glycol at reduced pressure, the absolute yielded 14.5 per cent of volatile oil; on distillation with steam, 15.1 per cent of oil (Sabetay and Trabaud). These volatile oils, however, are not commercially available. They are prepared only for the purpose of checking the purity of a concrete or absolute. For this reason they merit a short

[1] "Les Parfums Naturels," Paris (1939), 248.
[2] *Ibid.*
[3] *Perfumery Essential Oil Record* **31** (1940), 120.
[4] "Les Parfums Naturels," Paris (1939), 249.
[5] *Perfumery Essential Oil Record* **31** (1940), 120.

description in these pages. Two volatile oils described by von Soden [6] (I), and Naves, Sabetay and Palfray [7] (II) had these properties:

	I	II
Specific Gravity at 15°................	0.816	...
Optical Rotation....................	Inactive or slightly laevorotatory	+0° 35'
Refractive Index at 20°...............	...	1.4812
Acid Number........................	12	3.6
Ester Number.......................	20.5	22

Naves, Sabetay and Palfray obtained their oil by vacuum distillation with superheated steam in a specially constructed apparatus.[8]

Sabetay and Trabaud [9] submitted mimosa concrete to steam distillation *in vacuo,* and obtained 3 per cent of a limpid volatile oil with a waxy, fatty, honey-like and "green" odor. The oil exhibited the following properties:

Specific Gravity at 15°/15°..........	0.8100
Optical Rotation...................	−0° 30'
Acid Number......................	14.1
Ester Number.....................	18.2
Ester Number after Formylation......	21.2
Methoxy Groups (Zeisel)............	1 g. required 0.90 cc. of N/10 AgNO$_3$
Phenols...........................	Traces present
Sulfur and Nitrogen................	Absent
Acetylation in Pyridine Solution......	Primary and secondary alcohols absent

The typical mimosa odor persisted after the neutralization of the free acids, after oximation. On the other hand, the odor underwent a total change after saponification, the mimosa fragrance giving way to a strong "green" odor of legumes. The really odorous constituents formed only a small fraction both in the product obtained by vacuum distillation and in the concrete. The major portion of the oil consisted of hydrocarbons (see below).

Chemical Composition.—The chemical composition of mimosa flower oil was first investigated by Walbaum and Rosenthal,[10] who identified *anisaldehyde* by means of its semicarbazone m. 204°–205°. A more thorough in-

[6] *J. prakt. Chem.* **110** (1925), 276.

[7] *Perfumery Essential Oil Record* **28** (1937), 336.

[8] Cf. Naves and Mazuyer, "Les Parfums Naturels," Paris (1939), 173.

[9] *Perfumery Essential Oil Record* **31** (1940), 123.

[10] *Ber. Schimmel & Co., Jubiläums-Ausgabe* (1929), 194.

vestigation was carried out by Sabetay and Trabaud [11] (see above) who reported the presence of the following compounds:

Hydrocarbons(?). In crystalline form. Not further identified.

A Hydrocarbon(?). In crystalline form. The properties resembled those of 1-hexadecene $C_{16}H_{32}$.

An Alcohol(?). With a "green" odor, reminiscent of cucumber and ambergris.

Palmitic Aldehyde. Observed for the first time in a natural flower oil. Identified by the oxime m. 86°, and semicarbazone m. 105°–106°.

Enanthic Acid. In the free state. Amide m. 95°.

Anisic Acid. In free state, m. 185°.

Acetic Acid, Palmitic Acid, and Anisic Acid. In ester form.

Phenols(?). Small quantities only, with an odor of leather reminiscent also of pepper.

Use.—Absolute of mimosa is a very interesting and useful natural flower oil that merits much wider application in fine perfumery than it finds at present, particularly in the United States. French perfumers have long recognized its value as an excellent blender and "smoothing agent" for synthetics and as an effective fixative in high-grade perfumes. The odor of absolute of mimosa is mild and smooth; it blends well into scents of floral, as well as of oriental, character. It is one of the lower priced natural flower oils, and therefore can be quite freely employed.

SUGGESTED ADDITIONAL LITERATURE

R. Arnaud, "Culture des Plantes à Parfums en France," *Ind. parfum.* **4** (1949), 370.

CONCRETE AND ABSOLUTE OF *SPARTIUM JUNCEUM* L.
(Concrete and Absolute of *Genêt*)

There are several species of *Spartium* (fam. *Leguminosae*), popularly known as "Spanish Broom," or "Weavers Broom." In the essential oil industry the most important species is *Spartium junceum* L., known in France as "Genêt d'Espagne." In the south of France it grows wild and

[11] *Perfumery Essential Oil Record* **31** (1940), 123.

abundantly over wide areas, particularly in the Départements Alpes-Maritimes, Var, Basses Alpes, Drôme, and Vaucluse. The plant prefers rocky argillocalcareous soil, and is often found in the company of lavender. The flowers, which appear in June and July, are of beautiful yellow-golden color and exhale a delightful perfume reminiscent of orange blossoms and grape. Bees are greatly attracted to these flowers, with the result that the honey from certain sections of southern France exhibits a decided flavor of *genêt*.

In the Grasse region substantial quantities of the flowers of *Spartium junceum* L. are extracted with petroleum ether. Prior to World War II the quantity of flowers processed reached as much as 50 metric tons yearly; much larger quantities could be processed if demand should warrant it. In the extraction of these flowers it is of fundamental importance to treat them as quickly as possible, because the perfume of the flowers deteriorates rapidly after picking.

In the author's experience, about 1,200 kg. of flowers are required to yield 1 kg. of concrete which, in turn, gives 0.30 to 0.35 kg. of alcohol-soluble absolute. Naves and Mazuyer [1] reported yields of concrete ranging from 0.09 to 1.18 per cent, the concrete giving from 30 to 40 per cent of absolute.

The concrete is a solid, waxy mass of dark brown color, with a most peculiar, "heavy," sweetish and honey-like odor.

For the absolute, a viscous, dark-brown oil of equally peculiar odor, Walbaum and Rosenthal [2] (I), and Sabetay and Igolen [3] (II) reported these properties:

	I	II
Acid Number	33.6	98.9
Ester Number	85.9	81.0

Steam-distilling a concrete and an absolute according to the method of Sabetay (see Vol. I of this work, p. 215), Sabetay and Igolen obtained 2 and 4.1 per cent, respectively, of volatile oil. The oil from the concrete exhibited the following values:

Refractive Index	1.4712
Methoxy Content (Zeisel's Method)	4.40%

An examination of this oil led Sabetay and Igolen [4] to the conclusion that the flowers of *Spartium junceum* owe their perfume primarily to free acids (chiefly caprylic acid), to phenols with a leathery and peppery odor, aldehydes (chiefly aliphatic), terpenes with a pinene-like odor, and to esters

[1] "Les Parfums Naturels," Paris (1939), 213.
[2] *Ber. Schimmel & Co., Jubiläums Ausgabe* (1929), 201.
[3] *Ann. chim. anal.* **27** (1945), 224.   [4] *Ibid.*

of formic, acetic, and higher aliphatic acids, the alcohols of which exhibit a "green" odor.

Absolute of *genêt* is an interesting flower oil which has to be employed with great skill and discretion to obtain the desired effect. It blends particularly well with absolute of orange flowers (in which it is occasionally used as an adulterant). In the hands of an expert perfumer, absolute of *genêt* imparts remarkable effects to compositions of the heavier type.

CONCRETE AND ABSOLUTE OF *GENISTA TINCTORIA* L.

Genista tinctoria L., syn. *G. sibirica* (fam. *Leguminosae*), the common "Dyers-Greenweed," or "Woodwaxen," grows wild and abundantly in many dry locations, with a mild climate, in Europe and Western Asia. It has been naturalized in North America. Extracting the flowers with low boiling petroleum ether, Treff, Ritter, and Wittrisch,[1] in Germany, obtained 0.161 per cent of a concrete, which yielded 53 per cent of an alcohol-soluble absolute. On steam distillation the concrete gave 2.26 per cent of a yellowish volatile oil with these properties:

> Specific Gravity at 15°.............. 0.9335
> Optical Rotation................... −9° 10′
> Acid Number..................... 18
> Ester Number.................... 35
> Ester Number after Acetylation....... 156.0

The volatile oil had a "heavy" and "green" odor.

According to the author's knowledge, the concrete and absolute of *Genista tinctoria* are not produced on a commercial scale.

CONCRETE AND ABSOLUTE OF LUPINE

The common lupine, *Lupinus luteus* L. (fam. *Leguminosae*), is widely cultivated in Germany. The yellow flowers exhale a strong odor. Extracting the flowers (without the stalks) with petroleum ether, Treff, Ritter,

[1] *J. prakt. Chem.* **113** (1926), 360.

and Wittrisch [1] obtained 0.205 per cent of a concrete which, on treatment with alcohol in the usual way, yielded 60.4 per cent of an alcohol-soluble absolute. The concrete contained 0.95 per cent of a steam-volatile oil with these properties:

Specific Gravity at 15°...............	0.900
Optical Rotation...................	+7° 30′
Acid Number.....................	38
Ester Number....................	31
Ester Number after Acetylation.......	143

The odor of the volatile oil was penetrating, sweet, and herb-like.

Nothing is known about the chemical composition of the lupine flower oil.

According to the author's knowledge, the oil is not produced on a commercial scale.

[1] *J. prakt. Chem.* [2], **113** (1926), 359.

CONCRETE AND ABSOLUTE OF WISTARIA

Extracting the flowers of *Wistaria sinensis* (Sims) Sweet (fam. *Leguminosae*), called "Glycine" in French, Sabetay [1] obtained 0.197 per cent of a dark brown concrete with these properties: m. 47°, acid number 20, ester number 19.3.

Nothing is known about the chemical composition of this flower oil. The oil is not produced on a commercial scale.

[1] *Ind. parfum.* **5** (1950), 86.

CHAPTER IX

ESSENTIAL OILS OF THE PLANT FAMILY *HAMAMELIDACEAE*

OIL OF STYRAX
(Oil of Storax)

Essence de Styrax   *Aceite Esencial Estoraque*   *Storaxöl*
Oleum Styracis

Styrax, or storax, is an aromatic balsam formed and exuded by the styrax tree when the sapwood is injured. The pathological (rather than physiological) exudate congeals upon exposure to air. It is collected for commercial purposes, being widely employed in perfumery, for the scenting of soaps, and in certain medicinal preparations.

There exist on the market two types of styrax, viz., the so-called Levant or Asiatic styrax, and the American styrax, which since World War I has become of much more importance than its Old World prototype. The latter is inferior in quality to the American styrax.

I. Levant or Asiatic Styrax

The oriental sweet gum *Liquidambar orientalis* Mill. (fam. *Hamamelidaceae*) is a native of Asia Minor, in the southwestern parts of which it forms large forests. The tree grows to a height of 20 to 40 ft., in some cases 90 ft.; the smooth leaves are shiny bright green on the upper surface, and pale underneath. The tree grows wild in colonies, the most important of which are located in swampy valleys near the sea, between Makri and Giova. These styrax forests belong to the Turkish government or to private individuals; they are exploited by groups of woodsmen who sell their output to the owners of the forest or to exporters. One foreman and four workers can treat about five hundred trees daily. According to a government report,[1] styrax is collected from the time the trees are three to four years old, beginning early in May when they are in full foliage.

The surface is prepared by removing the outer layer of bark and sapwood from opposite sides of the tree. The exposed surfaces are then scraped at intervals of several days, particularly in August and September. Collection is suspended in the rainy season, which starts in mid-November. Thus the trees are exploited for seven or eight months every year, until the trunks are whittled down from the opposite sides to about 2-in. thickness. They are then left to recuperate for three or four years, treated again, and finally

[1] *L'Agronomie Coloniale* **15** (1926), 165. *Perfumery Essential Oil Record* **18** (1927), 222.

felled for firewood. The shavings collected each day are boiled in water and filtered through coarse cloth bags with the aid of a press. An emulsion of resin and water exudes from the press and is collected. The resin rising to the surface is separated and, while still containing 25 to 30 per cent water, is sold to dealers who remove most of the water and dirt, occasionally also adulterating it with pine resin or oil of turpentine. In normal times the balsam is then sent to Constantinople, Smyrna, and other ports of the Levant, and from these to Marseilles. In 1934, the United States imported more than 35,000 lb. of Levant styrax from Turkey, Italy, and France. Prior to World War I, Anatolia produced from 100,000 to 120,000 kg. of styrax annually (Rollet[2]).

According to Jeancard,[3] young trees yield only a few hundred grams of balsam, older ones from 6.4 to 7.7 kg. At the beginning of the season 10 kg. of shavings are required per kilogram of balsam, in the summer only 5 to 6 kg. During hot spells tears of white styrax can be obtained; their odor is superior to that of ordinary styrax. Quite different from the yields reported by Jeancard are those claimed by Berkel and Hus:[4] 40 to 50 g. of the purified resin from trees forty to fifty years old, 150 to 180 g. from those eighty to ninety years old, 250 to 400 g. from trees one hundred and thirty to one hundred and fifty years old.

Physicochemical Properties of Crude Levant Styrax.—Levant styrax is a semiliquid, sticky, opaque mass of grayish to grayish-brown color which deposits, on standing, a heavy, dark brown layer. It possesses a peculiar, characteristic odor and a sharp, somewhat spicy taste. When heated moderately, the mass melts. Crude Levant styrax contains a considerable amount of water and foreign matter, which must be removed in order to clarify the product.

A number of methods[5] have been devised for testing styrax for purity and for the presence of adulterants such as rosin, kerosene, oil of turpentine, fatty oils or exhausted styrax (from which the volatile oil has previously been removed, or partly removed, by steam distillation). Ahrens[6] based his method of analysis upon the partial solubility of styrax in petroleum ether. He also developed a direct test for the amount of water present in the balsam. Bohrisch[7] applied the method of Hill and Cocking[8] for the quantitative determination of the cinnamic acid in styrax. Another method of

[2] *Parfumerie moderne* **29** (1935), 473.
[3] *Ibid.* **18** (1925), 73.
[4] *Ankara Yüksek Zir. Enstitüsü Derg.* **3** (1944), 9. *Chem. Abstracts* **41** (1947), 1466.
[5] Tschirch and van Itallie, *Arch. Pharm.* **239** (1901), 506. Spokes, *J. Am. Pharm. Assocn.* **9** (1920), 1055.
[6] *Z. öffentl. Chem.* **18** (1912), 267. *Apoth. Ztg.* **27** (1912), 651.
[7] *Pharm. Zentralhalle* **61** (1920), 275.
[8] *Chemist Druggist* **80** (1912), 412.

assaying the content of cinnamic acid has been recommended by van Itallie and Lemkes.[9] The most important principles for the testing of styrax have been described by Anselmino, Seitz and Bodländer.[10] Examining samples of pure styrax from Asia Minor, they found properties varying within these limits:

	Acid Number	Saponification Number	Ester Number	Ester Number : Acid Number
Authentic Crude Styrax..	45 to 61	125 to 147	79 to 92	1.4 to 1.9
Calculated upon Styrax Freed from Water.....	64 to 80	178 to 195	107 to 122	...

These investigators also established the following limits for pure styrax from Asia Minor:

	Per Cent
Loss on Drying	25.5 to 32.5
Water Content	22.3 to 31.5
Alcohol-soluble Portion	64.8 to 72.9
Alcohol-insoluble Portion	1 to 3
Total Cinnamic Acid, Calculated upon Water-free Styrax	21.3 to 25.89
Total Cinnamic Acid	14.6 to 19.4
Free Cinnamic Acid	0.08 to 4.43
Phenol Content	19.93 to 29.42

When examining a sample of styrax according to the methods developed by the above-mentioned authors, the procedures suggested by them must be followed strictly; otherwise divergent results will be obtained. In the United States the most widely used method of analyzing styrax is that official in the United States Pharmacopoeia.

Shipments of crude Levant styrax examined in the laboratories of Fritzsche Brothers, Inc., New York, had properties which varied within these limits:

Loss on Drying	7.0 to 20.0%, in exceptional cases up to 28.0%
Alcohol-soluble Residue	70.0 to 92.5%
Alcohol-insoluble Residue	2.0 to 8.0%
Acid Number of Styrax Purified	51 to 66
Saponification Number of Styrax Purified	168 to 194
Color	Brown
Appearance	Opaque, semisolid

Crude styrax should be free of rosin. A test for the presence of rosin, which is occasionally added as an adulterant, will be found in Vol. I of this work, p. 334.

---

[9] *Pharm. Weekblad* **55** (1918), 142.   [10] *Arb. Reichsgesundh.* **57** (1926), 162.

The percentage of alcohol-soluble and alcohol-insoluble residue in Levant crude styrax differs considerably from that of American crude styrax. The variation probably results from the age of the Asiatic product, which is usually examined much longer after production than the American styrax. During the long transport from Asia Minor, styrene, an important constituent of the balsam, polymerizes; the resulting polymer is insoluble in alcohol. Therefore, certain allowances should be made in evaluating these factors. In this connection it should also be pointed out that old lots of crude styrax, in which polymerization has taken place to its fullest extent, are most suitable for purification because it is easy to separate the polymerized, alcohol-insoluble portions, and no further polymerization will take place in the purified product.

Incidentally, the technique of sampling styrax is of great importance in evaluating a lot. It is necessary to stir the product thoroughly so that the water will be evenly distributed. If this is not done, the properties obtained will be quite abnormal, and an actually normal product may be rejected.

Physicochemical Properties of Purified Levant Styrax.—For the preparation of *purified* styrax, the crude product is dissolved in 95 per cent alcohol, the insoluble residue is filtered off, and the alcoholic filtrate concentrated. The resulting concentrate, a viscous mass yellow to brown in color, represents the purified styrax. It may not always be clear, as its appearance depends greatly upon the styrene content of the crude styrax employed for purification. The purified styrax, prepared commercially on a large scale, is not necessarily identical with that described in the U.S.P. test for alcohol-soluble residue, the latter forming only a part in the analysis of Levant and American styrax.

Commercially purified Levant styrax, prepared by Fritzsche Brothers, Inc., New York, had the following properties:

> Acid Number. . . . . 53 to 59
> Saponification Number. . . . . 178 to 188
> Color. Brown
> Appearance. . . . . Clear, viscous, liquid

Chemical Composition of Styrax.—Styrax (Levant and American) contains the following compounds:

"Storesin." There has been much confusion about the so-called "storesin," an alcoholic resin which was discovered in 1877 by von Miller, and which Tschirch and van Itallie [11] declared to be a single compound $C_{16}H_{26}O_2$, m. 161°–162°.

In 1916, Henze [12] found that "storesin" m. 156°–161°, can be separated into 5, if not 6, different substances. According to Henze, "storesin" actually consists of isomeric acids $C_{20}H_{30}O_2$, identical with pimaric and abietic acids, which are con-

---

[11] *Arch. Pharm.* **239** (1901), 533.   [12] *Ber.* **49** (1916), 1622.

tained as such, and free, in styrax; the remainder consists of 3 or 4 substances, chiefly a ketone and an easily esterified alcohol.

In 1944, Berkel and Hus [13] reported that the original formula $C_{36}H_{58}O_3$ assigned to "storesin" by von Miller in 1877 is erroneous, the correct formula being either $C_{30}H_{46}O_3$ or $C_{30}H_{48}O_3$. Pure storesin prepared by Berkel and Hus melted at 157°–167°, $[\alpha]_D$ +43° 0'. These two authors arrived at the conclusion that storesin contains a phenolic OH group, but no primary or secondary alcohol group.

The United States Dispensatory [14] still describes storesin, the most abundant constituent of styrax, as an alcohol resin $C_{36}H_{55}(OH)_3$. It is present in two forms, designated respectively α and β, both free and in the form of a cinnamic ester. These together make up from $\frac{1}{3}$ to $\frac{1}{2}$ of the resin. Storesin is an amorphous substance m. 168°, readily soluble in petroleum benzin.

Cinnamic Acid. Another important constituent, present in styrax both in free and in ester form. The content of free cinnamic acid ranges from 5 to 15 per cent, but may be as high as 23 per cent.

Cinnamyl Cinnamate ("Styracine"). Styrax contains from 5 to 10 per cent of this ester which can be isolated by repeated extraction with ether, benzene, or alcohol, after the cinnamic acid has been separated from the resin. Cinnamyl cinnamate is volatile only with superheated steam; it crystallizes in tufts of long rectangular prisms m. 44°, which frequently do not form readily. Saponification with concentrated solutions of potassium hydroxide yields potassium cinnamate and cinnamyl alcohol (formerly called styrone). Cinnamyl alcohol is not a constituent of liquid styrax.

Phenylpropyl Cinnamate. Amounting to about 10 per cent of the balsam.

Ethyl Cinnamate. Present in small quantities only.

Benzyl Cinnamate. Also in small quantities only.

Styrene (Styrol). This hydrocarbon C_8H_8 occurs in the resin as a liquid, and also in polymeric form as a solid. The liquid form, called styrene or stryol, is a colorless, mobile oil, possessing the characteristic odor and taste of styrax. Styrene is phenylethylene, $C_6H_5 \cdot CH{=}CH_2$. When heated, styrene polymerizes to metastyrene, a colorless, transparent solid, which unlike styrene, is not soluble in alcohol or in ether.

"Styrocamphene"(?). Liquid styrax also contains about 0.4 per cent of a pleasant-smelling laevorotatory oil of the empirical molecular formula $C_{10}H_{16}O$ or $C_{10}H_{10}O$.

Vanillin. Present in traces.

Oil of Levant or Asiatic Styrax

When distilled with steam, Levant styrax yields only about 0.5 per cent of a volatile oil; with superheated steam more than 1 per cent. By treating the resin with caustic soda, and subsequent distillation, much higher yields can be obtained, but these oils differ in chemical composition from normal

---

[13] *Ankara Yüksek Zir. Enstitüsü Derg.* **3** (1944), 9. *Chem. Abstracts* **41** (1947), 1466.
[14] "United States Dispensatory," 24th Edition (1947), 1116.

styrax oils. Because of the very low yield from Levant styrax, oil of styrax is now distilled almost exclusively from the American balsam.

Physicochemical Properties.—The volatile oil possesses a light yellow to dark brown color and a pleasant, but peculiar odor. Its properties vary greatly according to the method of distillation employed. The specific gravity of the oil is below 1 if the hydrocarbons predominate, and higher than 1 if the oil contains a high percentage of alcohols and cinnamic esters.

Gildemeister and Hoffmann [15] reported these properties for the oil distilled from Levant styrax:

Specific Gravity............. 0.89 to 1.06
Optical Rotation............ −38° 0′ to +0° 30′
Refractive Index at 20°...... 1.53950 to 1.56528
Acid Number............... 0.5 to 33
Ester Number.............. 0.5 to 130
Solubility.................. Soluble in 1 vol. of 70% alcohol, opalescent on addition of 2 to 5 vol. of 70% alcohol. Soluble in any proportion of 80% alcohol, but the diluted solution often shows opalescence. Oils containing a high percentage of esters are less soluble

Chemical Composition.—The essential oil derived from styrax obviously contains only those constituents of the balsam which are carried over with the steam during distillation. Since some of the components have a high boiling point, the quantitative composition of the oil depends greatly upon the length of distillation, the temperature of the steam, and the steam pressure applied.

The chemistry of the volatile oil from Asiatic styrax was investigated by Simon,[16] Laubenheimer,[17] van't Hoff,[18] von Miller,[19] Dieterich,[20] and Tschirch and van Itallie,[21] who reported the presence of the following compounds:

Styrene. Identified more than a century ago by Simon. Dibromide m. 74°. The peculiar odor of styrax oil (reminiscent of illuminating gas) is a result of the presence of styrene, $C_6H_5 \cdot CH{=}CH_2$, b. 146° (cf. Vol. II of this work, p. 14).

Styrocamphene(?). According to van't Hoff, the optical rotation of styrax oil is due to an oxygenated compound $C_{10}H_{16}O$ or $C_{10}H_{18}O$, which he named styrocamphene.

Ethyl Cinnamate. First reported by von Miller; presence later confirmed by Tschirch and van Itallie.

Benzyl Cinnamate(?). Reported by Laubenheimer; presence doubtful.

[15] "Die Ätherischen Öle," 3d Ed., Vol. II, 786.
[16] *Liebigs Ann.* **31** (1839), 265.
[17] *Ibid.* **164** (1872), 289.
[18] *Ber.* **9** (1876), 5.
[19] *Liebigs Ann.* **188** (1877), 184.
[20] *Pharm. Zentralhalle* **37** (1896), 425.
[21] *Arch. Pharm.* **239** (1901), 506.

Phenylpropyl Cinnamate. Identified by von Miller; presence later confirmed by Tschirch and van Itallie.

Cinnamyl Cinnamate ("Styracine"). Also identified by von Miller, and later by Tschirch and van Itallie.

Vanillin. Identified by Dieterich, and later by Tschirch and van Itallie.

It should be mentioned here that oil of styrax contains not only the esters listed above, but also their parent substances, chiefly *phenylpropyl alcohol, cinnamyl alcohol,* and free *cinnamic acid.*

In an oil derived from the bark of the styrax tree, von Soden and Rojahn [22] identified *naphthalene* m. 79°.

II. American Styrax

Liquidambar styraciflua L. (fam. *Hamamelidaceae*), our common sweet-gum, or alligator tree, is a beautiful native American tree occurring along the Atlantic Coast from Connecticut southward to Central America. Fully grown, it reaches a height of 100 ft. and sometimes even 150 ft.

There exist two varieties besides the type species. *L. styraciflua* var. *mexicana* has three-lobed leaves, instead of the five- to seven-lobed leaves common to the northern *L. styraciflua* L. *L. styraciflua* var. *macrophylla* of Central America is the second variety; its leaves, also three-lobed, are much larger than those of the type species.

The wood is hard, close-grained, and of a reddish-brown color. The balsam apparently has been long used by the natives; after the conquest by Cortes it was exported to Spain in substantial quantities for use both as a perfume and as a vulnerary. The balsam, commonly known as liquidambar, copal balsam, Honduras balsam, or white Peru balsam, originates naturally as a pathological product in secretion reservoirs under the bark in old trees, being rarely found in young ones. Not all *Liquidambar* trees produce balsam; sometimes completely sound trees more than ten years old are tapped in vain. During World War I, the American product came into wider use as a result of the scarcity of the Asiatic variety, which it came to surpass in importance; now, however, the picture appears to be changing again.

American styrax originates almost exclusively from the Central American republic of Honduras. During World War II small quantities were produced in the adjacent sections of Guatemala, but owing to the low price the gum fetches on the world market, production in Guatemala has lately been discontinued.

A. Honduras Styrax.—*Liquidambar styraciflua* L. var. *macrophylla* is a tall, stately tree, reaching great height with age and a diameter up to

[22] *Pharm. Ztg.* **47** (1902), 779.

50 in. in exceptional cases. It grows wild in the enormous forests of Honduras and nearby parts of Guatemala. The trees usually occur in large stands at elevated altitudes between the tropical jungles in the valleys and the pine forests of the high mountains. A *Liquidambar* tree can readily be recognized in the fall by the color of its leaves, which turn red.

The styrax is a pathological exudate and not all trees contain it. As many as 15 trees may be examined in vain before one containing balsam is found. Small pockets develop beneath the bark, and gradually increase in size; in these the balsam accumulates. The collectors of the balsam go from tree to tree, examining each for pockets. To ascertain whether balsam is actually present in a pocket they hit the bark with a wooden mallet, the sound then indicating whether the pocket should be tapped. If the pocket is located high and not easily accessible the woodsmen are tempted to fell the tree to get at the pocket. In fact, this was the method practiced formerly. Of late, however, felling of the trees has been prohibited by the government because a felled *Liquidambar* tree represents an economic loss, as the wood cannot be hauled out of the forests. Moreover, when a *Liquidambar* tree is cut down it usually breaks at its weakest point, i.e., at the location of the pocket, and as the tree crashes down the pocket breaks open and the balsam spills on the ground, absorbing much earth, dirt, leaves and impurities. Therefore, the balsam is collected today by tapping the pockets of live trees only. For this purpose a cut is made into the pocket and a small gutter inserted, its free end leading into a receiving vessel. The viscous liquid flows very slowly. A tree yields on the average 40 to 50 lb. of balsam, in very exceptional cases up to 200 lb. Collection of the balsam goes on almost all year round, except for the rainy season from July to October, when it is impossible to penetrate into the forests.

The most difficult part of the work consists in getting the balsam out to civilization; this can only be accomplished by transporting the 50 lb. containers on the shoulders of the natives. Frequently 10 to 12 miles must be covered in this way before the remainder of the journey can be made on muleback.[23] The country is primitive, of very rugged topography; no roads exist; violent and sudden rainstorms add to the difficulties and dangers of those engaged in the enterprise. Not infrequently a mule with its load of styrax goes over a precipice, entailing considerable financial loss to the manager of the expedition. Collection of the balsam, therefore, forms one of the hardest tasks the natives are called upon to perform. In the region of Olancho, which supplies almost all of the Honduran styrax, the balsam is gathered by the Payas, a quite primitive tribe of Indians, who from the deep forests also bring snake, alligator, panther, and deer

[23] Watermeyer, *Am. Perfumer* **14** (1919), 161.

skins. After the styrax has reached the first outposts of civilization it must be transported to Tegucigalpa, capital of Honduras. Since no road exists between Olancho and Tegucigalpa this has to be accomplished by air freight; in fact, even the empty drums have to be shipped by air from Tegucigalpa to Olancho—and the freight charge is calculated by volume, not by weight of the empty drums. From Tegucigalpa the exporters then send the drums of styrax by rail or road to an Atlantic or Pacific seaport, for transshipment to the United States or Europe.

As was mentioned above, most of the Honduran styrax originates from the Olancho region. According to information gathered by the author in 1950 while surveying production of styrax in Honduras, collection of styrax in the various regions of Honduras is distributed as follows:

	Per Cent
Olancho	99.0
Comayagua	0.7
Progresso	0.2
Santa Barbara	0.1
	100.0

Total yearly production of styrax in Honduras averages about 100 drums (one drum = 520 lb.).

B. Guatemala Styrax.—During World War II small quantities of styrax were collected in areas adjoining Honduras. After the war, however, production diminished almost to the vanishing point; the following account is, therefore, given merely for completeness' sake.

According to Schaeuffler,[24] the producing regions of styrax (*Estoraque* or *Liquidambar*) in Guatemala are located in the Departments of Alta Verapaz, Zacapa, and Chiquimula. The tree from which the balsam is extracted grows in a temperate and humid climate, at altitudes ranging from 2,500 to 5,000 ft. In Guatemala *Liquidambar* trees are not planted or specially cultivated; they grow wild and usually stand in groups among longleaf pine trees. Extraction of the balsam is started when the trees are about ten years old, younger ones not yielding any balsam.

In Guatemala two methods are employed for the collection of styrax:

(a) *Box Method—"Casillas."* [25]—A groove, running diagonally through the bark and inclined inward, is cut with an ax about $\frac{1}{8}$ in. into the outer layer of the wood and at the height of approximately 2 ft. above the ground. From the upper end of this groove two or three strips are cut into the

[24] Private communication from Mr. Carlos Schaeuffler, Retalhuleu, Guatemala.
[25] Cf. Stark, *Chemurgic Digest* **1**, No. 23 (1942), 183.

sapwood in a "V" shape. These strips induce the flow of balsam, which runs along the strips into the groove from where it is removed with a spoon and transferred into a tin or any other receptacle. At first it takes two to three months, depending on weather conditions, for the groove to fill up, but later the balsam flows more freely. After approximately nine months, additional strips are peeled off and the old ones scrapped, which procedure opens up new passages for the flow of balsam. On old trees as many as six to ten grooves with their corresponding strips can be cut at different heights.

(b) *Deposit Method.*—Many *Liquidambar* trees have natural deposits of balsam beneath the bark, probably provoked by some injury to the sapwood by insects. These deposits can be detected by outgrowths on the tree trunk and also by the texture of the veins; in either case quite some experience is required to locate trees which have natural deposits. If these deposits are at a height which can be reached from the ground, a hole is drilled and the balsam which flows from the deposit is collected in a container. If the deposit cannot be reached from the ground, the tree must be felled and tapped on the ground.

The collected balsam is heated in a water bath, whereby coarse impurities, such as particles of bark, sapwood, and bugs, rise to the surface and are removed. While still hot, the balsam is roughly filtered through cheesecloth to remove the finer impurities, and then stored in five-gallon tins or iron drums.

Liquidambar trees in the region described above change their foliage once a year, from January to March. During this period no balsam flows. The most intensive flow is observed during the warm and rainy season from April to August, particularly in the month of June.

The color of raw styrax varies slightly with the altitude at which the trees are located. Experience has shown that styrax from trees at lower altitudes has a yellowish color; that from trees at higher altitudes exhibits a greenish color. There is no difference in color according to age of the tree or season. Crude Guatemala styrax is not processed locally, but sold on the market in raw condition.

In order to stimulate rapid flow of the styrax, Schaeuffler [26] applied a 40 per cent solution of sulfuric acid to the open wound. This procedure yielded a more liberal flow of resin for a few days. However, through action of the sulfuric acid, the fibers of the surrounding sapwood were completely burnt and killed. After a few days the flow stopped entirely.

[26] Private communication from Mr. Carlos Schaeuffler, Retalhuleu, Guatemala.

Physicochemical Properties of American Styrax

In the crude state, American styrax is a gray, occasionally pale brown, viscous mass, usually free flowing, but sometimes a semisolid substance, which softens on gentle warming. It often shows separation of water and contains specks of impurities or foreign matter. The crude balsam contains much less water than the corresponding Asiatic product and is, therefore, of higher quality. Crude American styrax imported from Honduras and Guatemala and analyzed by Fritzsche Brothers, Inc., New York, had properties varying within these limits:

Loss on Drying.......................... 7.0 to 27.0% (seldom over 15%)
Alcohol-soluble Residue.................. 70.8 to 95.0% (seldom less than 85%)
Alcohol-insoluble Residue................ 0.2 to 6.8% (seldom over 5%)
Acid Number of Styrax Purified........... 36 to 52.3
Saponification Number of Styrax Purified... 160 to 194

The alcohol-insoluble portions of certain lots of American styrax occasionally exceed 5 per cent. This is an indication of the age of the material and is caused by polymerization of styrene.

Commercially purified American styrax prepared by Fritzsche Brothers, Inc., New York, exhibited the following properties:

Acid Number.............. 35.5 to 47.6
Saponification Number....... 164 to 186

Purified American styrax is usually a clear, viscous liquid of brown color.

Oil of American Styrax

Depending upon its quality, the length of distillation, the temperature of the steam, and the steam pressure applied, American styrax yields from 15 to 20 per cent, occasionally even 30 per cent of volatile oil.

Physicochemical Properties.—In odor and appearance the oil from American styrax resembles that from the Asiatic product. The color ranges from pale to light yellow. Oils distilled by Fritzsche Brothers, Inc., New York, from Central American styrax had properties varying within these limits:

Specific Gravity at 15°/15°...... 0.986 to 1.008
Optical Rotation............... +0° 17' to +1° 45'
Refractive Index at 20°......... 1.5325 to 1.5415
Acid Number.................. 2.8 to 26.1
Saponification Number.......... 4.7 to 46.7
Solubility.................... Soluble in 0.5 vol. of 80% alcohol. Occasionally hazy to cloudy in 10 vol.

As regards congealing point, these oils did not readily congeal and temperatures ranging from $-10°$ to $-20°$ or even lower were required to bring about congelation.

Chemical Composition.—The chemical composition of American styrax and its volatile constituents was investigated by von Miller,[27] Tschirch and van Itallie,[28] Thoms and Biltz,[29] Hellström,[30] Burchhardt,[31] and Tschirch and Werdmüller.[32] The presence of the substances reported by these workers was later confirmed by Schimmel & Co.[33] in the *volatile oil* derived from American styrax:

Hydrocarbons(?). Among them compounds of the empirical molecular formula C_8H_8, C_8H_{10} and C_9H_{12}.

Styrene. Identified by von Miller through its dibromide m. 73° (cf. Chemical Composition of "Asiatic Styrax Oil").

Cinnamic Acid. Free and in esterified form (Schimmel & Co.).

Cinnamyl Alcohol. Free and in esterified form (Schimmel & Co.).

Phenylpropyl Alcohol. Also free and in esterified form (Schimmel & Co.).

Phenylpropyl Cinnamate. First reported by von Miller; presence later confirmed by Schimmel & Co.

Cinnamyl Cinnamate ("Styracine"). Also first reported by von Miller, and later by Schimmel & Co.

A Sesquiterpene(?). B. 261°–262° (Schimmel & Co.).

Vanillin. Presence in small quantities probable.

USE

Styrax is seldom employed today in medicinal preparations, except as a constituent of the compound tincture of benzoin.

Styrax is a valuable raw material for the isolation of cinnamyl alcohol (by saponification of cinnamyl cinnamate). Substantial quantities of styrax are employed for the scenting of soaps. Owing to its content of high boiling constituents, styrax acts as a most efficient odor fixative.

Oil of styrax is an ingredient in all kinds of perfume compounds, particularly those of oriental character.

[27] *Arch. Pharm.* **220** (1882), 648.
[28] *Ibid.* **239** (1901), 532.
[29] *Z. oesterr. Apoth.-Ver.* **42** (1904), 943.
[30] *Arch. Pharm.* **243** (1905), 218.
[31] *Schweiz. Wochschr.* **43** (1905), 238. *Chem. Zentr.* (1905), I, 1705.
[32] *Arch. Pharm.* **248** (1910), 420.
[33] *Ber. Schimmel & Co.,* April (1921), 73.

OIL OF *HAMAMELIS VIRGINIANA* L.
(Oil of Witch Hazel)

On simple steam distillation the leaves and terminal branches of *Hamamelis virginiana* L. (fam. *Hamamelidaceae*) yield not an essential oil but an aromatic water, viz., the popular witch hazel, used widely as an after-shaving lotion and in certain toilet preparations.

Little is known about the odoriferous principles of this distillate. To isolate the (concentrated) volatile oil of *Hamamelis virginiana* L., it would probably be necessary to cohobate the distillation waters (witch hazel) repeatedly, until the oil separates clearly from the water. Very probably most of the odoriferous constituents of the oil are soluble in large quantities of water, as is the case with oil of rose, which can be recovered fully only by repeated cohobation of the distillation waters.

Scoville,[1] and Jowett and Pyman [2] investigated small quantities of oil and observed the presence of an alcohol (about 7 per cent), of an ester $C_{10}H_{17} \cdot COOCH_3$ (7.3 per cent), low molecular fatty acids (0.6 per cent), high molecular fatty acids (small quantities), phenols with an odor of eugenol, a sesquiterpene $C_{15}H_{24}$, b. 259°–260°, α_D +14° 53', solid waxes and paraffins, and a compound with an odor of safrole.

SUGGESTED ADDITIONAL LITERATURE

S. L. Hilton, *"Aqua Hamamelidis* N. F. Witch Hazel Water," *J. Am. Pharm. Assocn.* **19** (1930), 232.

John Uri Lloyd and John Thomas Lloyd, "History of *Hamamelis* (Witch Hazel), Extract and Distillate," *J. Am. Pharm. Assocn.* **25** (1935), 220.

T. Ruemele, *"Hamamelis* (Witch Hazel)," *Perfumery Essential Oil Record* **41** (1950), 323.

[1] *Am. Perfumer* **2** (1907), 119.
[2] *Pharm. J.* **91** (1913), 129.

CHAPTER X

ESSENTIAL OILS OF THE PLANT FAMILY *DIPTEROCARPACEAE*

OIL OF GURJUN BALSAM

Gurjun balsam is the pathological exudation, caused by incision or scorching with fire, of the wood of several species of *Dipterocarpus* (fam. *Dipterocarpaceae*). There are about fifty species of this genus, tall and beautiful trees which grow wild in the mountain forests of India, Burma, and Indo-China. A fully grown tree yields as much as 180 liters of balsam during a summer. The balsam is obtained like gum turpentine. Large quantities are used in Indian varnishes. The trade often refers to gurjun balsam as "East Indian Copaiba Balsam," but this is a misnomer. Other vernacular terms are "Wood Oil" (in India [1]) and "Huile de Bois" (in French Indo-China).

In India and Burma there are two types of gurjun balsam, viz., the "Kanyin Oil," derived chiefly from *Dipterocarpus turbinatus* Gaertn., and the "In Oil," obtained principally from *D. tuberculatus* Roxb.[2] The former is the true gurjun balsam; "Kanyin Oils" are obtained by hacking triangular holes into the base of the live trees, scorching the opening with torches or with burning wood coal, and by collecting the exuding balsam with the aid of spoons. "In Oils" are obtained in a similar manner, but without the use of fire. The surface of the hole in the trunk is rasped repeatedly to facilitate the flow of the balsam. Yield and value of the two types of gurjun balsam are about the same.

The consistency, color, and solubility of gurjun balsam depend greatly upon the method of production. The better commercial grades are greenish-gray, slightly turbid and fluorescent in reflected light, but perfectly clear and reddish-brown in transmitted light. Odor and flavor are reminiscent of copaiba balsam.

According to Gildemeister and Hoffmann,[3] good grades of gurjun balsam exhibit these properties:

Specific Gravity.............. 0.95 to 0.97
Optical Rotation............. −23° 0′ to −70° 0′
Refractive Index at 20°....... 1.510 to 1.516

[1] Not to be confused with "Chinese Wood Oil" which is tung oil, a fast drying fatty oil obtained from the seed of *Aleurites cordata* Steud. (fam. *Euphorbiaceae*).
[2] *Oil, Paint Drug Reptr.* **73** (1908), No. 13, 41. Cf. Hanbury, *Science Papers* (1876), 118.
[3] "Die Ätherischen Öle," 3d Ed., Vol. III, 237.

On steam distillation gurjun balsam yields from 60 to 75 per cent of volatile oil.

Physicochemical Properties of Gurjun Balsam Oil.—The volatile oil derived from gurjun balsam is a yellow, somewhat viscous liquid, for which Gildemeister and Hoffmann [4] reported the following properties:

Specific Gravity................. 0.918 to 0.930
Optical Rotation............... $-35° 0'$ to $-130° 0'$
Refractive Index at 20°.......... 1.501 to 1.505
Acid Number................... 0 to 1
Ester Number.................. 0 to 8
Ester Number after Acetylation... 6 and 10 (two determinations)
Solubility...................... Not completely soluble in 90% alcohol. Soluble in 7 to 10 vol. of 95% alcohol, but not always in additional vol. of 95% alcohol

As regards the optical rotation, most gurjun balsam oils are strongly laevorotatory, few essential oils exhibiting equally high laevorotation. However, an optically inactive [5] and a strongly dextrorotatory oil [6] of gurjun balsam have also been reported.

Fractionating an oil of gurjun balsam (d_{15} 0.9255, α_D $-64° 26'$, n_D^{20} 1.50308, Acid Number 0, Ester Number 4.7) at 3.5 mm. pressure, Schimmel & Co. [7] obtained the following ten fractions:

	$b_{3.5}$	*Per Cent*	α_D
1.	101°–103°.........	11.1	$-85° 30'$
2.	103°–103.5°.......	11.1	$-83° 58'$
3.	103.5°–104°.......	8.5	$-80° 10'$
4.	104°.............	10.3	$-78° 44'$
5.	104°–105°.........	6.8	$-76° 8'$
6.	105°–106°.........	11.1	$-71° 40'$
7.	106°.............	9.4	$-66° 4'$
8.	106°–107°.........	9.4	$-58° 22'$
9.	107°–107.5°.......	10.3	$-44° 0'$
10.	107.5°–114°.......	11.1	$-21° 4'$

Of special interest are the properties of two oils which were derived from two species of *Dipterocarpus* identified with certainty and submitted to Schimmel & Co.[8] by Mr. C. G. Rogers, Conservator of Forests in Rangoon. The first balsam was obtained from *Dipterocarpus turbinatus* Gaertn. fil., and on steam distillation yielded 46 per cent of a light yellow

---

[4] *Ibid.*
[5] Tschirch and Weil, *Arch. Pharm.* **241** (1903), 382.
[6] W. Dymock, C. J. H. Warden and D. Hooper, "Pharmacographia Indica," Vol. I (1890), 193.
[7] Gildemeister and Hoffmann, "Die Ätherischen Öle," 3d Ed., Vol. III, 237.
[8] *Ber. Schimmel & Co.*, April (1913), 61. Cf. R. S. Pearson, "Commercial Guide to the Forest Economic Products of India," Calcutta (1912), 140.

oil (I) of balsamic odor. The second balsam originated from *D. turberculatus* Roxb., and on steam distillation gave 33 per cent of a yellowish-brown oil (II). The two oils exhibited these properties:

	I	II
Specific Gravity at 15°.	0.9271	0.9001
Optical Rotation.	−37° 0′	−99° 40′
Refractive Index at 20°.	1.50070	1.50070
Acid Number.	0	0
Ester Number.	1.9	0
Solubility in 95% Alcohol.	Soluble in 7 and more vol.	Soluble in 6 and more vol.

Chemical Composition of Gurjun Balsam Oil.—Oil of gurjun balsam consists chiefly of sesquiterpenes of closely similar boiling points. Fractionating an oil at atmospheric pressure (741 mm.), Schimmel & Co.[9] obtained 86 per cent of a fraction boiling between 260° and 265°, and 6 per cent of a fraction boiling from 265° to 269°.

The principal constituents of gurjun balsam oil are α- and *β-gurjunene* (cf. Vol. II of the present work, p. 119); in fact rectified gurjun balsam oil represents little more than a mixture of α- and β-gurjunene (Gildemeister and Hoffmann).

Pfau and Plattner[10] have reported that gurjun balsam oil contains a mixture of *sesquiterpenes* and *sesquiterpene alcohols* (b_{10} 127°–128°, d_{20} 0.9140, $α_D$ −179° 30′, n_D^{20} 1.5010); on dehydrogenation with sulfur or selenium, this mixture yields S-guaiazulene.

Use.—At one time, gurjun balsam oil was used quite extensively as a fixative of perfume compounds in the scenting of soaps and technical products. Occasionally the oil was also employed as an adulterant of copaiba balsam oil. The oil has now lost much of its importance.

OIL OF BORNEOCAMPHOR

Dryobalanops aromatica Gärtn., syn. *D. camphora* Coleb. (fam. *Dipterocarpaceae*), the so-called borneocamphor tree, is a tall, majestic tree, native to the northwest coast of Sumatra and to northern Borneo. The crevices and fissures in the wood of old trees occasionally contain crystalline

[9] *Ber. Schimmel & Co.*, April (1909), 51.
[10] *Helv. Chim. Acta* **19** (1936), 858.

d-borneol, which is variously known as borneocamphor, Baros camphor, Sumatra camphor, and Malayan camphor. Among the Malayan and Chinese population of the Far East, borneocamphor is highly esteemed for embalming and for ceremonial purposes. Indeed, such extraordinarily high prices are paid in the Far East for the natural *d*-borneol that none is exported to Europe or America, where it would have to compete with the very low-priced synthetic *d*-borneol. For this reason the natural product has completely lost the commercial importance which it had many years ago, before the introduction of the synthetic *d*-borneol. Borneocamphor and its corresponding oil will, therefore, be discussed here only briefly.

Aside from *d*-borneol (borneocamphor), the trees also contain an essential oil which can be isolated by distillation of the wood. It is a curious fact that only about 1 per cent of the trees contain *d*-borneol in crystalline form. The problem of whether the borneol is a natural or a pathological product of the tree has not yet been solved. Janse [1] observed that old trees are occasionally infested with larvae of insects which bore holes into the wood. Essential oil accumulates in these holes and, under favorable conditions, borneol may sublimate from the oil and form crystals in the clefts and fissures of old trees. Experiments carried out in the botanical garden of Buitenzorg [2] did not reveal the presence of crystalline borneol in the wood of felled trees, but when holes were drilled into the live trees, a white mass, consisting chiefly of borneol, accumulated in the holes.

Since only a very small percentage of trees contains crystalline borneol, the natives who collect the borneocamphor in the jungles of Borneo and Sumatra test every ·single tree by making incisions into the trunk. If the wound exhales no odor of borneol, the tree is left standing. In some instances, particularly when the trees are located near rivers, they may be felled, cut into logs, and floated to sawmills. Being resistant to the attacks of white termites, the wood is highly esteemed for construction purposes. Wood which exhibits the characteristic odor of borneol is often made into chests because the odor repels insects. The collection of borneocamphor in the wild interior of Borneo has been described in a fascinating report by Furness,[3] who spent considerable time among the head hunters of Borneo. According to Spoon,[4] exports of borneocamphor from Sumatra and Borneo declined from 853 kg. in 1920 to 558 kg. in 1927. They have further diminished since 1927. The entire trade in Borneo and Sumatra is in the hands

[1] *Annales du Jardin botanique de Buitenzorg,* Second Series, Suppl. III (1910), 947.

[2] *Verslag omtrent de te Buitenzorg gevestigde technische Afdeelingen van het Departement van Landbouw* (1905). Batavia (1906), 46, 63.

[3] Cf. Kremers, *Pharm. Rev.* **23** (1905), 7. Gildemeister and Hoffmann, "Die Ätherischen Öle," 3d Ed., Vol. III, 230.

[4] *Berichten van de Afdeeling Handelsmuseum van de Kon. Vereeniging Koloniaal Instituut* Nr. 46. *Overgedrukt uit* "De Indische Mercuur," July 24, 1929.

of Chinese, who ship the borneocamphor to Singapore, Penang, Hong Kong, and India. Despite its much higher price, the Chinese and Malayans still prefer the natural borneol to the synthetic product for use in rituals.

As was mentioned above, borneocamphor trees also contain an essential oil, the so-called borneocamphor oil. An oil from Singapore examined by Schimmel & Co. [5] had these properties:

Specific Gravity at 15°.	0.9180
Optical Rotation.	+11° 5'
Refractive Index at 20°.	1.48847
Acid Number.	5.6
Ester Number.	0
Ester Number after Acetylation.	50.5, corresponding to 17.67% of ester $C_{10}H_{17}OCO \cdot CH_3$
Solubility.	Soluble in 5 vol. and more of 90% alcohol, with slight turbidity

The oil had a dark brown color and an odor reminiscent of turpentine and borneol. Investigating the chemical composition of this oil, Schimmel & Co. found its composition to be approximately as follows:

35% Terpenes	10% Alcohols
d-α-Pinene	d-Borneol
Camphene	l-α-Terpineol
β-Pinene	
Dipentene	20% Sesquiterpenes
	35% Resin

As was explained above, natural d-borneol derived from the borneocamphor tree cannot compete with synthetic d-borneol, and is used only by the natives of the Far East for ceremonial purposes. As regards the essential oil obtained from the same tree, it offers no practical interest to the essential oil industry.

[5] *Ber. Schimmel & Co.*, April (1913), 31.

CHAPTER XI

ESSENTIAL OILS OF THE PLANT FAMILY *ANONACEAE*

OIL OF CANANGA

Essence de Cananga   *Aceite Esencial Cananga*   *Canangaöl*
Oleum Canangae

Botany and Occurrence.—*Cananga odorata* Hook. f. et Thomson, syn. *Canangium odoratum* Baill. (fam. *Anonaceae*), a tree which in the wild state may attain a height of 35 m., is probably a native of the Moluccas, and occurs wild, semiwild, and cultivated in many parts of tropical Asia, as well as on some islands of the Indian Ocean adjacent to the east coast of Africa. It was first designated *Arbor saguisan* by Ray [1] in 1704. The first correct drawings of the fruit and the inflorescence were published in 1829 by Blume,[2] who also noted that the flowers of the wild-growing tree are almost odorless.

On steam distillation the fresh flowers of the (semiwild and cultivated) tree yield an essential oil with a strong and pleasant odor, greatly appreciated and widely used in perfumery. Two distinct types of oil are derived from the flowers of *Cananga odorata*, viz., the lower priced cananga oil, and the more expensive ylang ylang oil. The former is produced almost exclusively in Java, the latter in Madagascar, the Comoro Islands, Réunion Island, and the Philippines. Oil of cananga is produced in primitive stills by natives of Java, using the flowers of semiwild trees. Oil of ylang ylang, on the other hand, is the product of meticulous distillation (fractionation) in modern stills, owned in large part by white planters; the carefully picked flowers originate exclusively from planted and pruned trees. The odor of oil of cananga is much coarser than that of oil of ylang ylang.

For a long time it was assumed that the trees, the flowers of which yield oils of cananga and ylang ylang, were botanically identical, and that the dissimilarity in odor and physicochemical properties between the two oils resulted chiefly from differences in the methods of distillation employed. (The latter is to some extent true, although other factors, such as selection of the flower material, soil, altitude, and climatic conditions, also enter the picture.) In order to ascertain whether ylang ylang oil could be ob-

---

[1] "Historia plantarum, Supplementum tomi 1 et 2. Historia stirpium insulae Luzonensis et Philippinarum a Georgio Josepho Camello," London (1704), 83. Cf. Gildemeister and Hoffmann, "Die Ätherischen Öle," 3d Ed., Vol. II, 580.
[2] "Flora Javae," Brussels (1828–9). Cf. Gildemeister and Hoffmann, "Die Ätherischen Öle," 3d Ed., Vol. II, 580.

tained from the same botanical source as cananga oil, and whether the difference between the two oils was principally a matter of distillation (as was formerly believed), Koolhaas[3] imported seed of the true ylang ylang from the Philippines. He planted these in Java, then distilled the oils from the flowers of native cananga trees and from those of the imported ylang ylang trees. Koolhaas found that the trees which yield ylang ylang oil can be distinguished botanically from those that yield cananga oil. The former he classified as *Canangium odoratum* Baill. forma *genuina*, the latter as *Canangium odoratum* Baill. forma *macrophylla*. (The more common designation *Cananga odorata* Hook. f. et Thomson is used in the present monograph.) Observations by Hischmann[4] tend to confirm the supposition that *Cananga odorata* occurs in two forms; he found that the trees grown from imported (Manila) ylang ylang seed grew only to half the size of the simultaneously planted native cananga trees. The leaves of the ylang ylang trees turned out to be smaller than those of the native cananga (Bantam) trees. The oils distilled from the flowers of the two types of trees exhibited marked differences, the oil from the ylang ylang flowers possessing a much finer odor than that from the cananga flowers.

As regards the cananga tree growing wild and semiwild in Java, it appears to exist in three varieties. Bobiloff[5] distinguishes between the cananga varieties "kerbo," "toelen," and "teri"—which would partly account for the distinctive qualities of cananga oils produced in different sections of Java.

Producing Regions.—There are two principal cananga producing regions in Java:

1. The province of Bantam, with Serang as center, in the western part of Java, about 80 km. from Batavia. Years ago practically all of the Java cananga oil was produced in this area, but by 1939 only half of Java's total output originated from Bantam, the balance coming from the province of Cheribon. Because of the cruder process of distillation employed in Bantam, the oil produced there is inferior in quality to that originating from Cheribon.

2. The province of Cheribon on the north coast of Java, about 250 km. east of Batavia. The oil produced here is better than that from Bantam, chiefly because more efficient stills and methods of distillation are employed in Cheribon.

---

[3] *Landbouw* **15** (1939), 587. Archipel Drukkerij, Buitenzorg, Java.
[4] Conversation of the author with Dr. A. Hischmann in Batavia, Java (1939). Cf. *Ber. Schimmel & Co.* (1937), 13.
[5] "De cultuur van *Cananga odorata* en de mogelijkheid der bereiding van ylang ylang olie in Nederlandsch-Indië," Teysmannia (1922).

Planting and Cultivating.—In Java the cananga tree occurs largely wild and semiwild; it is planted only occasionally. Propagation usually takes place spontaneously by means of fruit (seed) falling off the trees, or by birds eating the fruit and dropping the seed. On occasion, a native may find a spontaneously growing young plant and transplant it into his garden, without, however, giving it any special care. Rarely, and only when prices of cananga oil are very attractive, do the natives plant new trees, using for this purpose one-year-old seed. (Fresh seed does not seem to germinate well.) During the rainy season the seed is planted into holes and fertilized with cow dung. The young plants develop more or less rapidly, depending upon climate and altitude. At low altitudes the trees require only a few years to give a worth-while flower harvest, but at higher and cooler altitudes (500 m. above sea level) they grow much more slowly, and may require seven to ten years to produce flowers in quantities sufficient for harvesting. The cananga tree possesses a deep taproot; if this root encounters hard subsoil, the crown of the tree will not develop well. Trees more than twenty years old may reach a height of 90 to 100 ft. A seven-year-old tree bears from 30 to 100 kg. of flowers per year. A fully grown tree may produce up to 300 kg. of flowers, the quantity depending upon conditions of climate and altitude.

Flower Harvest.—In Java climatological conditions differ in various sections. Rainfall is local, rather than general. Since the flower harvest depends upon weather factors, it is impossible to count upon any uniform harvesting season throughout the island. In Bantam, for example, the heaviest harvest takes place from October to December (which is the main rainy season). A medium-sized harvest may be gathered from January to February, and a small one from June to July. In general, however, the natives collect the flowers whenever the trees bear a sufficient quantity to make distillation worth while. They do not care whether the flowers are green and undeveloped, or yellow and fully matured (as they should be in order to yield an oil of good quality). Always in need of money, the natives are generally unwilling to await full development of the flowers. They pick them while still green. This is one of the reasons why the odor of cananga oil does not measure up to that of ylang ylang oil.

Because of the great height of the cananga trees and the brittleness of the wood, the flower harvest represents a difficult and dangerous task. The natives, usually young village boys, have to climb high into the trees, step on the thicker, more solid branches, and with the aid of long bamboo rods tear the flowers off the outer branches. These rods are provided, at one end, with a small propeller-like, sharp-edged hook made from a bone of the water buffalo. Under the weight of the harvesters the brittle branches sometimes break, resulting perhaps in fatal accidents. To facilitate the task, especially after a rainfall, when the smooth bark is very slippery, high bamboo poles, 6-in. thick and provided with horizontal spikes, are permanently lashed to the trees. On these "ladders" the agile boys then climb into the base of the crown, from where they can reach the outer branches with their long bamboo rods. The torn-off flowers drop to the ground and are collected by other members of the harvesting crew.

Heavy tropical showers occasionally beat the flowers down prematurely and damage the crop. A prolonged dry spell may cause the flowers to shrivel and do even more harm to the crop than rain or storms.

Economic Set-up.—Most of the cananga trees belong to small-scale native growers who sell their flower harvest, through middlemen, to native or Chinese distillers. Frequently money is advanced by the middlemen, and the harvest or the trees may be mortgaged, or the trees may be rented for a certain period by the middlemen. The latter are often financed by local distillers who, in turn, depend financially upon local native or Chinese oil dealers, through whom the oil reaches the Chinese or European oil brokers in Batavia. In brief, the flower material and the oil pass through many hands before the oil finally reaches the export houses in Batavia. It is a characteristic case of the intricacies connected with any Malayan village industry, hard to understand and, for an outsider, impossible to penetrate.

Distillation in Bantam.—Distillation of cananga flowers for production of the essential oil is an old-established, typical, village and home industry, in which the entire family participates.

Prior to distillation the flowers are crushed—another reason why the odor of the cananga oils produced in the Bantam district is inferior to that of the oils from Cheribon. (As will be explained in the following monograph on "Oil of Ylang Ylang," it is very important to eliminate all damaged or crushed flowers from the distillation material; otherwise the oil will not be of good quality.) Nevertheless, the natives in Bantam insist on crushing the flowers before distillation, primarily for the reason that, with the primitive stills used there (in which condensation is insufficient), uncrushed, whole flowers do not yield all their oil.

The stills, made of copper, are small and primitive, particularly in regard to the condensers. They are usually set up in the corner of a palm-thatched, dark bamboo hut, in the backyard of the compound, behind the living quarters. The condenser consists of a straight bamboo or copper tube running diagonally through a large earthen pot filled with water. The cooling surface of the condensers in most cases is quite insufficient, and the joints are not well sealed—factors responsible for the excessively long hours of distillation required in the stills of Bantam. Whenever the cooling water in the earthen pot becomes too warm, cold water is added. The oil separator consists of a simple wine bottle, with a hole in the side, just above the bottom. It stands in a pot filled with distillation water. Only male members of the family carry out the actual distillation and handle the oil. Women and children cast an occasional glance at the flow of the distillate and the fire beneath the still, rekindling it with a piece of wood, especially at night when the distiller falls asleep. Distillation thus goes on day and night, aside from the regular house and field work. The flow of the oil being quite irregular, the quality of oil in every bottle varies. Since in the primitive equipment used in Bantam distillation has to be carried out very slowly, completion of one batch requires from 36 to 48 hr. This results in that slightly "burnt" and "leathery" odor so typical of cananga oils from Bantam. In fact, there is hardly any resemblance in odor between the oil and the living flowers.

The producers sell their oil in bottles to local native middlemen, who bulk the small lots, then filter and sell them to other dealers (often Chinese) through whom the oil finally reaches the brokers and the exporters in Batavia.

Distillation in Cheribon.—As in the Bantam region, production of cananga oil in the Cheribon district is entirely in native or Chinese hands. However, much of the Cheribon oil is produced on a "community" basis, whereas all Bantam oil is distilled by small individual operators and their families.

The stills used in Cheribon are much larger and of better construction than those employed in Bantam. They consist of galvanized iron drums holding 300 to 400 liters. The flowers can be charged and discharged through a small manhole. Several stills are mounted in a row on a brick hearth, with a fire beneath each drum (retort). The condensers used in Cheribon are larger and much more effective than those of the Bantam stills. Several condensers are usually immersed in a single water tank of about 10 cu.m. Since not all stills operate simultaneously, condensation is quite sufficient, permitting more rapid distillation than in Bantam. Completion of one batch requires only 24 hr. The flowers do not have to be crushed prior to distillation.

An additional reason for the better quality of the Cheribon oils lies in the fact that the flowers in that district are more carefully selected; whereas, in Bantam, flowers from the wild growing, so-called "Buffalo" cananga ("Karbouw") are occasionally mixed with the distillation material. These possess a somewhat harsh, rather unpleasant scent.

As a result of these factors, obviously, the Cheribon cananga oil possesses a finer odor, truer to nature, and usually fetches a slightly higher (about 10 per cent) price than the Bantam oil. As a matter of fact, the perfume of the Cheribon cananga oil almost approaches that of the lower grades of ylang ylang oil.

As regards the economic set-up of the cananga oil production in the Cheribon district, there are several distilleries, each equipped with a number of stills which the owner of the building rents out to native distillers. The owner, usually a Chinese, alone holds the license for distillation, and supervises all operations, seeing to it that the natives who lease the various stills distill properly, produce a good oil, and do not adulterate it. Often he has advanced money to the natives who, in turn, have financed the flower harvesters and the tree owners. In the final settlement of accounts the oil distilled then goes to the owner of the distillery who sells it to middlemen or brokers in Cheribon, from where the oil reaches the exporters in Batavia.

It is quite obvious that in Bantam, as well as in Cheribon, cananga oil passes through many hands before reaching the exporters in Batavia. All other assertions to the contrary, there is not one distiller in the Bantam or Cheribon districts who produces more than 100 kg. of cananga oil. There are, however, a few wealthy Chinese brokers who handle relatively large quantities of the oil, and exert a considerable influence upon its price.

Yield of Flowers and Oil.—The yield of oil varies considerably, depending upon season, condition of the flower material, distillation equipment, and care taken during operation. According to estimates, the average cananga tree produces from 125 to 150 lb. of flowers per year. From 350 to 400 lb. of flowers are said to be required to yield 1 lb. of oil.[6] From local sources in Java the author learned that the yield of oil ranges within 0.5 and 1.0 per cent. On the basis of small-scale experiments, Koolhaas [7] reported yields up to 0.77 per cent.

Physicochemical Properties.—Koolhaas and Rowaan [8] established the following properties for Java cananga oil:

Specific Gravity at 15°	0.908 to 0.925
Optical Rotation	−15° 0′ to −40° 0′
Refractive Index at 20°	1.495 to 1.506
Acid Number	0.5 to 2.0
Ester Number	15 to 35
Solubility	Soluble in 1 to 3 vol. of 95% alcohol; opalescent to turbid on further dilution
Residue on Steam Distillation	Not higher than 5%

[6] *Ber. Schimmel & Co.* (1932), 11.
[7] *Landbouw* **15** (1939), 587. Archipel Drukkerij, Buitenzorg, Java.
[8] *Indische Mercuur* **60** (1937), 504.

These figures serve in the Government Laboratories of Buitenzorg, Java, as a basis for control analysis to which every shipment of cananga oil must be submitted before an export permit is granted.

As regards the difference between the oils produced in the Bantam and in the Cheribon districts, the Bantam oils usually exhibit a slightly lower specific gravity, optical rotation, and saponification number than the Cheribon oils. These differences, in most cases not very marked, can be attributed partly to the cruder methods of distillation employed in the Bantam region.

Samples of genuine oils distilled under the author's supervision during a visit to the cananga oil producing regions in Java had these properties:

	Oils from Bantam		*Oils from Cheribon*	
	I	II	I	II
Specific Gravity at 15°....	0.915	0.913	0.921	0.923
Optical Rotation.........	−18° 45′	−16° 37′	−29° 58′	−26° 55′
Refractive Index at 20°...	1.5001	1.4999	1.5018	1.5030
Saponification Number....	14.9	15.2	29.9	29.9
Solubility at 20°..........	Not clearly soluble in 90% alcohol, up to 10 vol.	Soluble in 0.5 vol. of 95% alcohol; cloudy in more	Not clearly soluble in 90% alcohol, up to 10 vol.	Soluble in 0.5 vol. of 95% alcohol; cloudy in more

Shipments of Java cananga oil imported and examined by Fritzsche Brothers, Inc., New York, had properties varying within the following limits:

Specific Gravity at 15°/15°.....	0.910 to 0.921
Optical Rotation..............	−15° 53′ to −31° 50′
Refractive Index at 20°........	1.4961 to 1.5019
Saponification Number.........	9.8 to 41.6
Solubility....................	Not clearly soluble in 90% alcohol up to 10 vol.

The odor of cananga oil can be described as strong and flowery, reminiscent of ylang ylang, but far less pleasant, the delightful top notes of the latter being lacking. Cananga oil possesses a somewhat harsh, "leathery" note; this, however, may be quite desirable in the scenting of soaps.

Rectified Cananga Oil.—Most European and American essential oil houses offer a rectified cananga oil, rectification being carried out either with live steam or by dry distillation *in vacuo*. The process entails a loss of up to 25 per cent. The rectified oils usually exhibit a lower specific gravity and saponification number than the original (native distilled) oils.

The chief advantage of the rectified oils consists in their lighter color and a somewhat better solubility. As disadvantage may be mentioned the fact that the rectified oils generally possess less fixation value than the original oils, a part of the high boiling constituents being eliminated in the course of rectification.

Rectified cananga oils prepared by Fritzsche Brothers, Inc., New York, had properties varying within the following limits:

Specific Gravity at 15°/15°	0.903 to 0.908
Optical Rotation	−12° 30′ to −31° 30′
Refractive Index at 20°	1.4960 to 1.4989
Saponification Number	4.7 to 24.6
Solubility	Not clearly soluble in 90% alcohol up to 10 vol. Occasionally some oils are soluble in 9.5 to 10 vol. of 90% alcohol

Adulteration.—In the past, oil of cananga was frequently adulterated by the natives with fatty oils, such as coconut, castor, and sesame, which latter does not congeal readily in the freezing test for the presence of fatty oils. About twenty years ago, mineral oils (kerosene, lubricating oils, etc.) made their appearance as adulterants. To put an end to the practice of adulteration and thus protect the reputation of the oil on the world market, the Government of Java enacted legislation by which every drum of oil has to be examined in the Government Laboratories of Buitenzorg, before a certificate of purity and an export license are granted. The contents of every drum must meet the physicochemical properties established by Koolhaas and Rowaan (see above); otherwise, the lot will be condemned. The presence of fatty oils is checked by freezing, and by determination of the percentage of residue obtained on steam distillation. Normal oils exhibit a distillation residue ranging from 2.5 to 4 per cent. The acid number of this residue varies between 0.5 and 6.0, its ester number between 35 and 100. Oils with a distillation residue higher than 5 per cent are rejected. A test for the presence of mineral oils consists in oxidation with fuming sulfuric acid (see "Oleum Test," Vol. I of this work, p. 332, and also the monograph on "Citronella Oil," Vol. IV of this work, p. 108). Simmons [9] observed that the flash point of a pure cananga oil was 93° C., that of two samples adulterated with petroleum 82° to 83°.

Chemical Composition.—It is generally assumed that most of the compounds occurring in oil of ylang ylang are present also in oil of cananga, but in different proportions. The difference between the two oils seems to be one of quantitative rather than of qualitative composition. Oil of

[9] *Perfumery Essential Oil Record* **25** (1934), 167.

cananga contains more sesquiterpenes and sesquiterpene alcohols, but much less esters than oil of ylang ylang. In the case of the latter oil, a substantial part of the esters present consists of the lower boiling esters, because ylang ylang oils (particularly the so-called "Extra" and "First Quality" oils) are actually the lower boiling fractions of the total oil obtained during steam distillation of the flowers. Differences in the quantitative composition of the two oils must be attributed also to the excessively long distillation of the cananga flowers in primitive stills, which causes considerable hydrolysis of the esters originally present in the oil.

Since it is a high-priced oil, ylang ylang oil has been thoroughly investigated, and numerous compounds have been identified (see the monograph on "Oil of Ylang Ylang," "Chemical Composition"). As has been said, they occur probably also in oil of cananga, but in different proportions. In view of the importance of ylang ylang oil, cananga oil has been somewhat neglected by the essential oil chemists, and literature cites surprisingly few investigations. Elze [10] found that the oil contains about 0.2 per cent of *nerol* (diphenylurethane m. 50°–50.5°), and about 0.3 per cent of *farnesol* which he isolated by means of the acid phthalic ester. As regards the sesquiterpenes, Reychler [11] noted, years ago, that cananga oil is particularly rich in sesquiterpenes, chiefly *cadinene* (identified by its dihydrochloride m. 117°).

Total Production.—Prior to the occupation of Java by Japanese forces during World War II, annual exports of cananga oil from Java were as follows: [12]

Kilograms		*Kilograms*	
1928	18,905	1935	17,048
1929	16,049	1936	18,902
1930	10,055	1937	22,788
1931	9,267	1938	15,611
1932	11,289	1939	14,825
1933	11,965	1940	13,289
1934	15,687		

During the war, of course, conditions in Java were utterly confused. After the war the civil strife raging in the interior of the island made it impossible for the government to control production, and substantial quantities of oil were probably smuggled out of Java, mostly to Singapore, without examination by the Government Laboratories. Lately, however, conditions have improved substantially, with the government gradually regaining control over exports: the oil is shipped in galvanized iron drums

[10] *Chem. Ztg.* **34** (1910), 857.
[11] *Bull. soc. chim.* [3], **11** (1894), 407, 756, 1045; **13** (1895), 140.
[12] Private communication from Dr. J. A. Nijholt, Buitenzorg, Java.

holding 50 to 100 kg. of oil. Principal shipping port has always been Batavia (Tandjong Priok) and to a much lesser extent, Cheribon.

In 1947 only 2 metric tons of cananga oil were exported from Java; in 1948, about 3 tons; in 1949, 7.388 tons. (The latest figure is taken from official statistics. However, it is quite possible that in reality larger quantities were exported because of a lively smuggling trade.[13])

Use.—Oil of cananga can best be described as a lower-grade ylang ylang oil. It is generally used to replace the latter where price plays an important role. The odor of cananga oil is harsher than that of ylang ylang, but usually more lasting. The oil serves to great advantage in the scenting of soaps and in all kinds of technical preparations, where the higher priced ylang ylang oil cannot be employed.

SUGGESTED ADDITIONAL LITERATURE

H. C. Reed, "Cananga of the Pacific Islands," *Perfumery Essential Oil Record* **27** (1936), 211.

OIL OF YLANG YLANG

Essence d'Ylang Ylang   *Aceite Esencial Ylang Ylang*   *Ylang Ylangöl*

Oleum Anonae

History and Geographical Distribution.—As was pointed out in the preceding monograph on "Oil of Cananga," the tree *Cananga odorata* Hook. f. et Thomson (*Canangium odoratum* Baill.) occurs in two forms, viz., *macrophylla* and *genuina*. On steam distillation, the flowers of the former yield oil of cananga, the latter oil of ylang ylang, one of the most important essential oils. It is highly appreciated in perfumery because of its delightfully sweet and strong odor. Reminiscent of jasmine, oil of ylang ylang constitutes almost a complete perfume in itself.

Cananga odorata (fam. *Anonaceae*) is probably a native of the Moluccas, or of the Philippines, where it was once extensively cultivated, particularly near Manila, for the production of ylang ylang oil. The Manila oil was for a long time considered the finest of all types of ylang ylang oil, and before 1900 held practically a world monopoly. Since then, however, the center of the industry has shifted to the French possessions in the Indian Ocean.

[13] *Ibid.*

The introduction of ylang ylang into this area dates back to about 1770 when a French expedition, under Captain d'Etchevery, cruised through these waters collecting spice and other plants on Ceram and neighboring islands; these plants, including ylang ylang, were brought to Réunion, a French island in the Indian Ocean, about 400 miles east of Madagascar. For a long time the colonists of the island paid little attention to ylang ylang, although soil and climatic conditions on Réunion are well adapted to its growth. It was only toward the end of the nineteenth century that planters on Réunion came to recognize the importance of the tree and the prosperity which would result from the production of this valuable oil. From 1892 on, and particularly in 1905/1906, large plantings were started on the northeast and northwest coasts, and a high-grade oil was produced. In 1909 there were about 200,000 ylang ylang trees on Réunion.

At the turn of the century the plant was introduced into Madagascar for intensive cultivation, particularly on the small island of Nossi-Bé (located off the northwestern tip of Madagascar) where altitude, climate, and soil conditions are most favorable to its growth. This development must be attributed largely to the activities of the Rev. F. Raimbault who, during many years of untiring work, started many plantations and distilleries on Nossi-Bé, originating also the famous brand of oil "Pères Missionaires." Expansion of the industry continued, and a few years later the ylang ylang tree was introduced also into the Comoro Islands, northwest of Madagascar, between Madagascar and the east coast of Africa. On these tropical islands, particularly on Anjouan and Mayotte, conditions are ideal for the successful cultivation of ylang ylang, and large quantities of oil have been produced.

In the past 50 years the center of the ylang ylang oil industry has thus shifted from Manila first to Réunion Island, then to Madagascar (particularly Nossi-Bé) and the Comoro Islands. Manila has lost its former importance and now produces only negligible quantities of oil, amounting to less than 1 per cent of the world's production.

Prior to World War II, the world's annual output of ylang ylang oil averaged about 40 metric tons, 20 to 23 tons of which came from Nossi-Bé, 2 to 3 tons from the adjacent areas of Madagascar, 12 to 14 tons from the Comoro Islands, and 1 to 1.5 tons from Réunion Island.

In 1950 there were eight steam distilleries and twenty smaller direct fire distilleries in Nossi-Bé, four steam distilleries and four direct fire distilleries in Madagascar, about ten medium-sized steam distilleries and a number of direct fire distilleries on the Comoro Islands, and two small steam distilleries on Réunion Island. The steam distilleries in and near Manila, formerly noted for their efficiency, have disappeared, a casualty of World War I, when most of the ylang ylang plantations were eliminated to make room for housing projects. In 1939 there was just one small distillery left, even this on part-time operation. (Cf. the section "Ylang Ylang Oil from the Philippine Islands.")

Botany.—According to Loher,[1] the word ylang ylang comes from the vernacular "Alang-ilang," by which name this tree is known in most of the provinces of the Philippines. The term implies something "hanging" or "fluttering," and describes the drooping condition of the flowers which tremble in the slightest wind. Another explanation of the origin of the name, often noted in literature, has it that ylang ylang, in Malayan, means "the flower of flowers." [2]

The ylang ylang tree (*Cananga odorata* Hook. f. et Thomson, forma *genuina*) thrives in a moist, tropical climate, at sea level, near the seacoast, and in rich volcanic or fertile sandy soil. If left alone, it grows to considerable height (15 to 20 m.), but to facilitate collection of the flowers the trees are kept low (2 to 3 m.) by topping, and by bending the branches downward. The flowers are grouped in bunches of 2 to 20 and more. They possess 3 little sepals, 6 petals from 4 to 8 cm. long, narrow with pointed ends more or less curved backward, with peduncles from 3 to 5 cm. long.[3] At the moment of blossoming the flowers are green and thickly covered with hairs, which give them a greenish-white appearance. The flower matures quickly and then assumes a yellow color. When fully developed (about twenty days after blossoming) the flower has a deep yellow color. It is at this period that the flower contains a maximum of oil, and that the quality of the oil is at its highest. At this point the flowers should be harvested (see also below).

The fruit consists of a fleshy, pear-shaped pod, about 4 cm. long, with a soft pulp containing up to eleven seeds, which, when ripe, exhibit a dark maroon color.

On the islands of the Indian Ocean there seem to exist two varieties or forms of *Cananga odorata,* viz., the true ylang ylang tree, locally called "la bonne," undoubtedly *Cananga odorata* forma *genuina,* and another form distinguishable by its larger leaves and flowers. This latter is probably *Cananga odorata* forma *macrophylla,* the cananga tree exploited in Java for the production of cananga oil. The leaves of the second form, on being crushed in the hand, smell, not of ylang ylang but slightly of pepper. This

---

[1] Gildemeister and Hoffmann, "Die Ätherischen Öle," 3d Ed., Vol. II, 579, footnote 2.

[2] This, however, appears quite unlikely because in the Malayan language the letter "y" is not used at the beginning of a word. Ylang ylang is thus not a Malayan term. There is a word "ilang" in Indonesian (Malayan), but it means "to disappear," which meaning has no significance in relation to the flower in question. (Private communication from Dr. J. A. Nijholt, Buitenzorg, Java.)

[3] *Parfums France* **8** (1930), 350. Cf. P. Advisse Desruisseaux, "L'Ylang Ylang," Paris (1911), 15.

odor, however, fades rapidly, and is replaced by a faint ylang ylang scent. On steam distillation the flowers of this form give a low yield of oil; moreover, the oil is of poor quality, and possesses a lower specific gravity than true ylang ylang oil. Planters on the islands of the Indian Ocean are making efforts to exterminate this form and to limit cultivation to the true ylang ylang.

As far back as 1828 Blume [4] noted that the flowers of the *wild Cananga odorata* tree are almost odorless. The relatively high content of volatile oil in the flowers of the *cultivated* ylang ylang tree can be explained, perhaps, by the fact that the plant has been under extensive cultivation for generations, and that in the course of this "domestication" it has undergone marked physiological (if not morphological) changes. This may also be one of the reasons why the flowers of the cananga tree, which in Java grows semiwild, give a lower yield of oil, and an oil of poorer quality than do those of the cultivated true ylang ylang tree.

I. Ylang Ylang Oil from Nossi-Bé, Madagascar and the Comoro Islands

Planting and Cultivating.—Propagation of the ylang ylang tree is usually by seed, rarely by cuttings. For the purpose the seed is planted in a seed bed in the month of March. During the following June or July the young plants are placed into small bamboo pots. They must be protected from sun and wind and watered frequently. After sufficient care in the nursery, the young trees, which have developed several leaves, are planted out in the fields, into holes spaced at 6 m., in rows 6 m. apart. This is done in January, at the beginning of the rainy season. The soil should be well drained, loose, fertile, and deep (with no rocky subsoil), because the ylang ylang tree develops a deep tap root. Best results are obtained on argillaceous-sandy soil, or on well-drained soil of volcanic decomposition, on the slopes of hills. Sandy-alluvial soil is less desirable; laterite should be avoided. If necessary, the soil has to be enriched with cow dung. Frequent weeding is necessary to foster free development of the young trees. During the first year they should be shaded by means of an interspersed crop such as maize, manioc or "ambrevade" (*Cajanus indicus*). After one and a half to two years the first flowers appear. At the beginning of the third year the trees have reached a height of 2 to 3 m., and are then topped, which has the effect of directing growth toward the lateral branches.[5] Periodical pruning of the small shoots prevents the tree from growing too

[4] "Flora Javae," Brussels (1828–9).
[5] For details see *U. S. Dept. Agr., Fed. Expt. Sta. Puerto Rico, Rept.* 1939 (1940), 36; *ibid.* 1940 (1941), 33.

bushy. Keeping the tree at a height of only 2 to 3 m. greatly facilitates harvesting of the flowers. In the third year of growth a small number of flowers can be gathered; the fourth year produces a normal crop. If properly attended and sheltered from the wind a plantation lasts twenty-five years or more; under favorable conditions an age of even fifty years may be reached.

The ylang ylang tree is not easily susceptible to diseases and suffers very little from pests or parasites.

Flower Harvest.—The ylang ylang tree bears flowers throughout the year; harvesting, therefore, continues all year around, and hardly a day passes during which the large distilleries are idle for want of flower material. There are, however, three harvesting seasons which can be clearly distinguished:

1. The principal harvest, from April to June, right after the rainy season.

2. A moderate harvest during the dry season, from the end of September to November. This coincides with the spring season on the islands of the Indian Ocean. The flowers are drier and contain more essential oil than during the rainy season. The quality of the oil (specific gravity, etc.) is also higher.

3. The season of heavy rains (January to March). At this time the flowers are heavy with moisture, and weigh more than they do during other periods. Yield of oil, therefore, is subnormal, and the quality none too good.

These harvesting periods, however, are by no means clearly defined; they fluctuate from year to year with the arrival of the rains.

Apparently ylang ylang flowers contain more essential oil during the night (particularly just before daybreak) than during the day: a visitor driving through the countryside at night, or at dawn, will be surprised by the sweet fragrance which pervades groves, fields, and villages. Obviously, the flowers cannot be gathered in the dark; the best time to pick them is therefore in the early morning, from just after sunrise up to about 9 or 10 o'clock, at the latest.

The flowers are picked by crews of women and girls, who collect them in baskets which they carry on their heads. Only the fully developed, yellow flowers should be gathered, because the green (underdeveloped) flowers contain less essential oil than the yellow ones. Moreover, their oil is of poor quality, possessing a low specific gravity and exhibiting a "green," flat odor, the result, chiefly, of the absence of volatile esters and ethers, for which the oil from fully matured ylang ylang flowers is so highly appreciated. Great care must also be exercised not to crush the flowers during the picking. Damaged flowers readily fade, turn black, and cause fermentation of the sound material in the same basket. This point is of particular importance in the rainy season, when the flowers are usually wet and tend to ferment during transport to the distilleries.

Since the harvesters are paid by the weight of the material they bring to the distillery, they are in general tempted to pick the flowers rapidly and indiscriminately, and to collect green flowers along with fully developed, yellow ones. Careful distillers, therefore, check every basket on arrival, and on finding green or crushed flowers, insist on the woman's emptying the basket and selecting only yellow and undamaged flowers, for which alone she is then paid. If this procedure is strictly followed, the harvesters soon learn to be careful, and the quality of the oil will benefit greatly. Green and yellow flowers cannot always be clearly distinguished, but there is a simple test that eliminates any doubt as to shade of color: on the inner base fully matured flowers have two small reddish spots, caused probably by the presence of traces of indole.

Not all distilleries exercise sufficient care in the control of the incoming flower material to be able to produce a high-grade oil; indeed many of the small-scale distillers, particularly natives and East Indians, are quite careless in this respect, with the result that they encounter difficulties in producing "Extra" qualities of oil (see below). Even among the larger distilleries there are only a few who insist upon impeccable flower material. No wonder, then, that the quality of the oils produced by some of the larger enterprises markedly excels that supplied by many of the small distillers.

The flowers should be distilled immediately after arrival from the groves, since otherwise they start to fade and ferment, particularly when left in the baskets. Fermentation manifests itself by development of heat in the flower mass. Careful producers spread the flowers on concrete floors in a thin layer, and distill them as soon as possible after arrival. Here, too, the large distilleries with sufficient floor space and still capacity have the advantage over the small producers.

Yield of Flowers.—The yield of flowers per tree and year depends upon the location of the plantation, its age, and the care exercised in its cultivation, as well as upon the climate, soil, and altitude. The yield per year and tree varies from 5 to 20 kg. of flowers, the average being 10 kg. Since the end of World War II the yield of flowers on Nossi-Bé has been declining sharply, the result of neglect of the plantations during the war, when trees were no longer pruned and reached considerable height, at the expense of flower growth.

Distillation.—Distillation of the ylang ylang flowers, as carried out on the islands of the Indian Ocean and in Manila, is actually a fractionation by steam. The reader should keep this clearly in mind, in order to under-

stand the following pages, which will deal with the physicochemical properties of the various fractions (grades or qualities) of oil obtained in the
course of distillation. In the case of the ylang ylang flowers it so happens
that the first fractions of oil carried over by steam contain the most aromatic and valuable constituents of the oil (esters and ethers), whereas the
later fractions consist chiefly of sesquiterpenes which have little odor value.
A gradual lowering of the quality of the oil thus takes place from the first
to the last fraction, in the process of distillation. On the French islands in
the Indian Ocean (and to some extent also in the Philippines) the common
practice is to call the first fraction "Extra," and the following ones
"Première," "Seconde," and "Troisième," or in English, "Extra," "First,"
"Second," and "Third." There are, however, no fixed rules as to the degree
of fractionation, and every producer follows his own method, which depends
upon the brand he has established on the market, or sometimes upon the
demand on the part of a customer. Some distillers cut their fractions
according to specific gravity, the majority according to time (hours of
distillation). In the former case the value of the fractions stands in direct
relation to their specific gravities, whereas in the latter case the value of a
fraction is inverse to the time of distillation. In other words, specific gravity
is highest in the top fractions and diminishes as distillation proceeds. It
would also be possible to cut the various fractions according to their ester
numbers, because the first fractions consist chiefly of esters, whereas the last
fractions contain only very little ester. However, assay of the ester number
of each fraction is more complicated than determination of the specific
gravity, and entirely out of question in the small distilleries. The simplest,
although not the most exact, method consists in fractionation according to
hours of distillation. It is the process applied by most of the larger, and
nearly all the smaller, distilleries.

Distillation of ylang ylang flowers is carried out either in small, directly-
fired stills (water distillation), or in larger stills heated with steam (a
modified form of water *and* steam distillation). The first type of distillation accounts for the bulk of ylang ylang oil (about two-thirds). Only the
large distilleries are equipped with steam stills, which, incidentally, yield
the best grades of oil.

Distillation in Directly-fired Stills.—The directly-fired stills are usually of
the "Deroy" type, holding about 500 liters. The width of a retort should
be equal to, in fact preferably greater than, its height (cf. Vol. I of this
work, p. 148). The steam-connecting pipe (column or gooseneck) should
be sufficiently tall to permit provision for automatic return of the distillation waters (cohobation); a long gooseneck, moreover, allows for sharper
separation of the various fractions than a short one. Retort, column, and
condenser are made of copper, well tinned on the inside. Aluminum has

also been found to give good results, particularly in regard to the color of the oil. The retort should be provided with a false bottom (grid) to prevent the flower charge from contacting the hot, directly-fired bottom. (There are many stills—owned by natives, East Indians and Chinese—which do not contain a false bottom; oils produced in such stills often exhibit a slightly harsh, "burnt" odor.)

Stills of this general type can be found throughout the producing regions, either single, in pairs, or, in the larger distilleries, arranged in battery form.

The fire beneath each retort is kindled with wood, which the planters obtain mostly from their own property. Frequently the firewood originates from mangrove trees ("palétuviers"), which grow most abundantly along the shores of the islands. Mangrove wood, however, contains much tannic matter, and it is claimed that this wood, when employed for direct fire distillation, imparts a peculiar, smoky odor to the oil. There may be some truth in this: a strong smoky odor may often be noted in food prepared above a mangrove fire.

The charge of flowers in a directly-fired still should not be too large; if it is, distillation will be ineffective. For example, a retort of 500-liter capacity should not be charged with more than 50 kg. of flowers. Water is poured into the retort to about two-thirds of its volume, and then heated almost to the boiling point. At this stage the flowers are placed into the retort (above the false bottom), and heating is continued. The large quantity of water causes the flowers to move freely, and to yield their oil readily to the passing steam bubbles, and at the same time prevents the flowers from "packing" and adhering to the hot walls of the retort. Where the water supply is scarce, as on some of the Comoro Islands, the directly-fired stills have to be opened repeatedly during distillation, and the still contents stirred, to prevent "burning" of the flowers against the walls—an altogether inefficient and undesirable procedure.

The advantage of first heating the water in the retort almost to boiling point and then placing the flowers into the hot water has been learned by experience. It has been found that, with this procedure, actual distillation starts more quickly, and a better quality of oil (with a higher percentage of "Extra" and "First" fractions) is obtained. If, on the other hand, the flowers are charged into cold water, considerable time will elapse before the liquid reaches the boiling point. Meanwhile, the delicate flowers undergo partial fermentation, and some of the esters may be hydrolyzed, with the result that the distillate may be subnormal in regard to "Extra" and "First" qualities. Moreover, the prolonged action of hot water upon the flowers appears to cause decomposition of those still unknown complex compounds that yield the sesquiterpenes composing the higher fractions of the oil. In short, the distillate will contain a high percentage of "Second" and

"Third" fractions, and a correspondingly lower quantity of "Extra" and "First" fractions. Some distillers seem unable to obtain any "Extra" quality at all, probably because they do not adhere to the above rule.

Distillation should proceed quickly, yet smoothly, and at a uniform rate. The fire must be continuously watched, and "coups de feu" (boiling over) guarded against. An even flow of the distillate is of fundamental importance; uniform fractionation cannot be achieved without it. As has been pointed out, most distillers (particularly the smaller ones) cut their fractions according to hours. Such a practice, however, is out of the question when the condensate does not run uniformly, and the rate of distillation varies. Violations of this rule are frequent among smaller producers who let native foremen supervise distillation.

Distillation of one batch lasts up to 20 hr. or more, and during the night the native worker charged with attendance of the fire and control of the distillation may fall asleep. The fire may go down, and the rate of distillation slow up. Awakening, the worker quickly fills the oven to capacity, stirs up the fire, and the distillate again runs very freely. By such abuse of the process of "cutting according to hours," native workers obtain fractions lacking in uniformity.

Ample supply of cooling water in the condenser is another important condition for successful distillation. The distillate must be kept very cool, especially at the beginning when the more volatile constituents distill over. Wherever the water supply is scarce, as on some of the Comoros, distillation must be carried out at a much slower pace, although this is not desirable in general.

All direct fire stills are constructed so as to permit automatic return of the distillation waters into the still; in other words, they are automatically *cohobated* during the distillation of the flowers.

Distillation of one charge in directly-fired stills lasts up to 22 hr. It starts usually at 10 A.M. and is completed early the following morning, when the last runs of the "Third" quality are collected. This leaves just enough time to clean the stills before the new flower supply arrives, and distillation of the new batch starts again. Small quantities of the last runs of the "Third" quality have a tendency to settle in pockets and dents of old condenser coils, and if not removed, distill over with the "Extra" fraction of the new batch, lowering its specific gravity. Moreover, the inside walls of old condensers are often covered with a soft, sponge-like residue—consisting of partly resinified compounds—which tends to absorb oil, viz., the last runs of the "Third" quality. For these reasons, retorts as well as condensers should be cleaned, at regular intervals, by blowing steam through them. This is easy in steam stills, but more difficult and time-consuming in directly-fired stills.

From his own observations in the field, the author can state that most white producers, even the small ones, try to adhere to the rules by which a high-grade of ylang ylang oil can be obtained. This is not always ·so with the numerous native, East Indian, and Chinese distillers, and with those white producers who let native labor do all the work, without strict supervision.

The rules to be observed may be summarized as follows:

1. Only fully matured, yellow flowers should be harvested, and early in the morning.

2. The flowers must not be damaged during picking.

3. The flowers must be quickly transported to the distillery, and distilled immediately upon arrival.

4. The water in the directly-fired stills should be heated almost to boiling point before the flowers are charged into the retort.

5. Distillation must proceed smoothly, at a uniform and lively rate, and under careful supervision.

6. The various fractions must be cut sharply.

7. Condensation must be efficient.

8. The condensers—in fact all parts of the still—must be carefully cleaned at regular intervals.

Unless these precautions are strictly observed, it will be difficult to obtain an "Extra" quality at all. This explains why many native distillers produce only "First," "Second," and "Third" qualities, and seem simply unable to obtain an "Extra" quality.

Having no facilities for export abroad and lacking experience as shippers, small producers must sell their oil to local dealers and brokers, who, in turn, supply the exporting houses according to the orders the latter receive from abroad. On the local markets the oil lots are evaluated chiefly according to their specific gravities, a gravity of 0.900 serving as base, and a slight premium being paid for each unit above 0.900. (The temperature at which the gravity is taken varies from 26° to 28°, the average on the islands being 27° C.) Because of this practice, small operators see no reason to pay special attention to the production of oils with a very fine odor, and content themselves with supplying lots that meet the general requirements in regard to specific gravity. They are not even interested in cutting the various fractions very sharply, because by mixing different fractions and selling the blends according to gravity the producers obtain, in the end, the same price. For the same reason, many small operators do not bother to produce "Extra" qualities, but rather force distillation to the limit, so as to obtain a larger quantity of "Second" and "Third" qualities.

Because of their lower price, there is a greater demand for "Second" and "Third" qualities than for "Extra" and "First."

(It should be mentioned at this point that distillation of ylang ylang flowers, like that of cananga flowers in Java, can also be carried out without any fractionation whatsoever. When this is done, the *total* or "complete" oil will be obtained. Distillation in this case is usually stopped after 15 hr. A "complete" oil can also be prepared by first fractionating in the usual way, and subsequently mixing the various fractions after distillation. Today there is comparatively little demand for "complete" oils; the lots offered on the market under this label are either truly "complete" oils [obtained without fractionation], or mixtures, chiefly of "Third" and "Second" qualities, with very little, if any, "First," and no "Extra" at all.)

During his stay on the islands, the author had occasion to study distillation (fractionation) of ylang ylang from all angles, and in a number of distilleries, small and large—French, native and Indian-owned. The following data apply only to oils produced in directly-fired stills; the steam-distilled oils will be discussed separately and later. It should be mentioned that specific gravity and yield of oil are generally higher during the dry season than during the rainy season. They also depend upon the soil, and the altitude at which a plantation is located.

Several small operators in Nossi-Bé, the principal ylang ylang producing region in the world, regulate distillation (fractionation) according to the time schedule given below. (Note that the figures under the column headed "Length of Distillation" indicate the hours required for distilling *each* particular fraction alone, *after any previous fractions have been separated.* The total at the bottom of the column "Length of Distillation" shows the number of hours necessary to complete distillation of *all* fractions.)

Length of Distillation (in hr.)	Fraction (Quality)	Quantity of Fraction (in g.)	Specific Gravity at 27°
1½.................	"Extra"	200	~0.960
2 to 2½............	"First"	200	~0.940
3½.................	"Second"	200	~0.920
6...................	"Third"	400	~0.910
13 to 13½		1,000	

By more prolonged distillation, a larger quantity of "Third," or a "Fourth" quality (specific gravity ~0.890) could be obtained, but such fractions have little odor value; moreover production becomes uneconomical because of high fuel consumption.

Another scheme of fractionation by hours, practiced by a few producers on Nossi-Bé is this:

Length of Distillation (in hr.)	Fraction (Quality)	Percentage of Fractions (%)	Specific Gravity at 27°
½	"Extra"	5	~0.950 to 0.970
½	"First"	5	~0.930 to 0.950
6	"Second"	40	~0.915 to 0.930
8	"Third"	50	~0.910 to 0.915
15		100	

A small producer in Nossi-Bé, whom the author visited, fractionated according to the following simple scheme:

Length of Distillation (in hr.)	Fraction (Quality)	Specific Gravity at 27°
6	"First"	~0.940
6	"Second"	~0.920
8	"Third"	~0.910
20		

. This distiller was not interested in producing an "Extra" quality.

Incidentally, the specific gravities noted in this case are those by which "First," "Second," and "Third" qualities are generally defined on the local market of Nossi-Bé.

A small, but progressive, French producer in Nossi-Bé obtained these fractions from 3,280 kg. of flowers:

Fraction (Quality)	Quantity of Fraction (in kg.)	Length of Distillation (in hr.)	Specific Gravity at 27°
"Extra"	15	1	0.950 and higher
"First"	15	3	0.940 and higher
"Second"	20	5	0.920 and higher
"Third"	30	9	~0.912
	80	18	

The average small-scale distillers obtain the following fractions from 100 kg. of flowers:

Fraction (Quality)	Quantity of Fraction (in g.)	Length of Distillation (in hr.)	Specific Gravity at 27°
"Extra" and "First"	700 to 800	~4	0.930 to 0.950
"Second"	400 to 500	~6	0.915 to 0.930
"Third"	700 to 800	8 to 12	0.905 to 0.915
	1,800 to 2,100	18 to 22	

Some operators stop distillation earlier than others because production of the "Third" quality is hardly remunerative. During the dry season, distillation of "Extra" and "First" qualities requires more time than during the rainy season, because the flowers then contain a greater amount of esters and ethers.

One kilogram of ylang ylang oil produced in directly-fired stills consists approximately of the following:

	Grams
"Extra" and "First" Quality........	400
"Second" Quality..................	250
"Third" Quality...................	350
	1,000

That the yield of oil (and of the various fractions) fluctuates daily, depending upon the weather, can be seen from the records of a small-scale, French producer in Nossi-Bé.

Date	Quantity of Flowers (in kg.)	Fraction (Quality)	Length of Distillation (in hr.)	Quantity of Fraction (in g.)
Oct. 4.........	60	"Extra"	1	200
		"First"	3	200
		"Second"	5	230
		"Third"	9	845
			18	1,475
Oct. 5.........	62	"Extra"	1	205
		"First"	3	200
		"Second"	5	240
		"Third"	9	920
			18	1,565
Oct. 6.........	60	"Extra"	1	200
		"First"	3	190
		"Second"	5	205
		"Third"	9	840
			18	1,435
Oct. 7.........	65	"Extra"	1	235
		"First"	3	120
		"Second"	5	240
		"Third"	9	935
			18	1,530

Date	Quantity of Flowers (in kg.)	Fraction (Quality)	Length of Distillation (in hr.)	Quantity of Fraction (in g.)
Oct. 8.........	49	"Extra"	1	150
		"First"	3	115
		"Second"	5	240
		"Third"	9	630
			18	1,135
Oct. 9.........	64	"Extra"	1	240
		"First"	3	180
		"Second"	5	295
		"Third"	9	925
			18	1,640

Distillation in Steam Stills.—Since steam distillation requires more complicated and expensive apparatus than direct fire distillation, only the bigger establishments are equipped for it. There are eight steam distilleries on Nossi-Bé, four in the adjoining section of Madagascar proper, and several on the Comoro Islands. Together these produce about one-third of the total output of ylang ylang oil. In these establishments the stills are arranged in battery form, the live steam being supplied by a powerful steam generator. The individual retorts are usually slightly wider than high, with a capacity ranging from 500 to 1,000 liters. Each retort is provided with a false bottom (grid), an indirect steam coil or steam jacket for indirect heating, and a perforated steam coil for distillation with direct steam.

The process actually applied is a combination of direct steam distillation, and water *and* steam distillation. After the retort and condenser have been thoroughly cleaned by blowing live steam through the system, water is pumped into the retort up to about 1 in. above the false bottom (grid). By means of the steam jacket or indirect steam coil the water is then heated to about 70° C, and the flowers are quickly charged into the retort above the grid. The flowers are not entirely covered by water (as in the case of direct fire distillation), only the lower part of the charge being immersed. Distillation is started slowly by carefully injecting live steam through the perforated steam coil. The volatile oil distills over easily, and the top fractions, which consist chiefly of esters and ethers, are collected in the oil separator. After a while the direct steam is shut off, distillation then being continued by heating with indirect steam. Later, direct steam may again be resorted to, and this sort of alternation may be continued. There are no fixed rules, and each distiller follows his own method, which may be influenced by the condition of the flower material, the season, and other factors.

Each retort is provided with a gooseneck sufficiently tall (1 to 2 m.) to permit (a) provision for automatic return of the distillation waters (cohobation), and (b) adequate separation of the various fractions (qualities). Some distillers do not cohobate the condensed water, but rather let the condensate flow through a series of oil separators until the water is clear. It appears advisable, however, to cohobate at least the "Extra" and "First" runs, because these oils possess a high specific gravity, the condensate easily turns "milky," and does not readily separate in the separators. In the case of the "Second" and "Third" runs, on the other hand, cohobation is not so necessary, because their specific gravities are substantially lower than that of water, and they readily separate from the distillation water.

The most efficient method of separating the various fractions in the course of steam distillation (as in the case of direct fire distillation) is to cut the fractions according to specific gravity. In this procedure the hours required for the distillation of each fraction, and the quantity of oil obtained, vary with the seasons. The method offers the advantage that the various fractions may be drawn directly from the oil separators, without any subsequent blending and adjustment.

The following table illustrates fractionation according to specific gravity, as carried out in one of the leading steam distilleries of Nossi-Bé (from 100 kg. of flowers):

Specific Gravity at 27°	Fraction (Quality)	Length of Distillation (in hr.)	Quantity of Fraction (in g.)
~0.955	"Extra"	~3	~400
~0.942	"First"	~3	~400
~0.922	"Second"	4 to 6	300 to 400
0.910 to 0.912	"Third"	6 to 8	800 to 1,400
		16 to 20	2,000 to 2,500

There are only a few steam distilleries where the fractions are cut according to their specific gravities. In most steam distilleries, as in practically all the distilleries which work with directly-fired stills, fractionation is carried out according to length of time (hours). To render this latter method more effective and reliable and to prevent irregular flow of the condensate, the steam pressure should be kept constant by means of automatic reducing valves.

Below are two typical time schedules of fractionation by hours of distillation, as carried out in two steam distilleries (in Nossi-Bé, and Madagascar proper).

In the first example, 50 kg. of flowers yielded:

Length of Distillation (in min.)	Fraction (Quality)	Quantity of Fraction (in g.)
10 to 15	"Extra"	~80
20	"Surfine"	~70
45	"First"	~150
120	"Second"	~250
Balance	"Third"	~450
22 hr.		1,000

The other distillery visited by the author was equipped with steam stills of 1,000 liters capacity. Distillation was not "pushed"; therefore the yield of "Third" quality was only small:

Length of Distillation (in hr.)	Fraction (Quality)	Yield of Fraction	Specific Gravity at 25°
1½ to 2	"Surfine"	50%	0.555 to 0.970 (usually ~0.960)
1	"First"	16%	~0.945
1	"Second"	16%	~0.930 and higher
1	"Third"	16%	~0.920
Not longer than 6 hr.		~100%	

In the Comoro Islands several steam distilleries operate according to the time schedule below (flower charge 100 kg.):

Length of Distillation (in hr.)	Fraction (Quality)	Quantity of Fraction (in g.)	Specific Gravity at 27°
4 to 5	"Extra"	~800	0.950 to 0.965
4	"First"	~500	0.930 to 0.950
4	"Second"	~500	0.915 to 0.930
2 to 5	"Third"	~200	0.910 to 0.915
(Depending upon quality of flowers)			
14 to 18		~2,000	

Another example of steam distillation in the Comoro Islands (flower charge also 100 kg. of flowers):

Fraction (Quality)	Quantity of Fraction (in g.)	Specific Gravity at 27°
"Extra"......	Up to 1,000	0.950 to 0.970
"First".......	~500	0.930 to 0.950
"Second" ⎱	600 to 700	~0.925
"Third" ⎰		0.910 to 0.915

Up to 2,200

In Mayotte (Comoro Islands) 60 kg. of flowers, on steam distillation, yielded:

Length of Distillation (in hr.)	Fraction (Quality)	Quantity of Fraction (in g.)	Specific Gravity at 27°
1½.......	"Extra" ⎱	~450	~0.960
1½.......	"First" ⎰		
4........	"Second" ⎱	~550	0.921 to 0.922
8........	"Third" ⎰		

1,000

Steam Distillation vs. Direct Fire Distillation.—Weighing the respective advantages and disadvantages of steam distillation and direct fire distillation, we find the balance heavily in favor of steam distillation, at least in the larger distilleries.

Distillation (fractionation) in steam stills has the following in its favor:

1. The oil obtained has a more delicate perfume, free from "burnt" off-odors.

2. Because of the more regular and uniform flow of the distillate, and the sharper fractionation, standardization of the various fractions is possible.

3. There being less influence of boiling water upon the flower charge, and more rapid distillation:

(a) less hydrolysis of the esters occurs; therefore, the fractions, and particularly the "Extra" and "First" fractions, possess a higher ester content, and

(b) less decomposition of certain complex compounds (still unknown) takes place, with correspondingly less formation of sesquiterpenes, the chief constituents of the last fractions. In other words, compared with direct fire-distilled oils, the steam-distilled oils contain a higher percentage of "Extra" and "First" fractions, and a lower percentage of "Second" and "Third" fractions.

On the other hand, directly-fired stills offer these advantages:

1. Lower initial investment.
2. Simpler operation, in general.
3. Lower cost of fuel, because individual stills are fired according to the quantity of flower material arriving at the distillery.

A word should be added, at this point, about the production of "Extra," "First," "Second," and "Third" qualities during the dry and rainy seasons. From October to November (the dry period) flowers contain a maximum of oil, the oil, moreover, being of best quality. At that time, steam stills give a very large yield of "Extra" quality, little "First," and a normal amount of "Second" and "Third" qualities. On the other hand, directly-fired stills, during the dry season, yield "Extra" and "First" qualities in the usual proportion. During the rainy season the flowers are heavy with moisture and give a low yield of poor quality oil. At this time, steam stills are nevertheless able to produce "Extra" and "First" qualities (in about equal amount), whereas directly-fired stills often yield no "Extra" quality at all, producing only "First," "Second," and "Third" qualities (in normal amounts).

Yield of Oil.—Like the quality, the yield of oil depends upon many factors—primarily the season, and the length of distillation. Under normal conditions (i.e., with 18 to 22 hr. of distillation) the yield of oil ranges from 2 to 2.25 per cent. If distillation is prolonged—as is often done in the case of direct fire distillation—additional quantities of oil may be obtained, but these last runs consist chiefly of sesquiterpenes, have little odor value, and hardly pay for the fuel consumed. No producer extends distillation beyond 36 hr.

Quality of Oil.—In the foregoing we have discussed, in detail, the many factors which influence the quality of ylang ylang oil, viz., climate, soil, altitude, weather, condition of the flowers, and method of distillation. Fractionation in the course of distillation, though imperfect, results in a concentration of the most aromatic and valuable constituents in the first fractions.

It is often asserted, particularly in the older literature, that the Manila oils possess by far the best odor, with the Réunion, Nossi-Bé, and Comoro oils following in that order. This, however, no longer holds true. True Manila oils are not readily available, their production having fallen almost to the vanishing point. Moreover, lots exported from the Philippine Islands may consist of bulkings, containing so-called Albáy oils, which are produced by natives under very primitive conditions. In other words, Manila oils are no longer an important factor in the ylang ylang oil industry. A similar, although less marked, development has taken place on Réunion Island, which at present produces only small quantities of oil (but of very good quality).

In addition, the quality of ylang ylang oil depends not so much upon geographical origin as upon the condition of the flower material and the method of distillation. In this respect the quality of the oils produced in the modern steam distilleries of Nossi-Bé surpasses that of all other oils. Oils from the Comoro Islands are reputedly somewhat lower in quality than the leading Nossi-Bé brands, but here, too, much depends upon the producer. The author has had occasion to examine some Comoro oils that compared favorably with the best Nossi-Bé oils. It should be mentioned here that the large distilleries in Nossi-Bé do not market their output through local brokers, but export it directly through their principals or sales organization in France. In other words, the leading brands are usually entirely privately distilled and contain no admixed native-distilled oils. Those top grades are therefore of strictly controlled and uniform quality.

Present Situation.—Much of what was said above, particularly in regard to fractionation and quality of the oils, applies to the years prior to World War II. More recently, however, the picture has changed considerably— at least temporarily. Therefore it would seem important also to describe the situation of Madagascar's ylang ylang industry at the present time. Unable to export ylang ylang oil during the war, large-scale producers had to neglect their vast plantations, which demand a great deal of field management and much labor. Trees, no longer pruned regularly, have now grown to considerable height—at the expense of flower growth. Older trees, being no longer replaced by young ones, have now reached the end of their productivity. As a result, the yield of flowers has declined alarmingly on Nossi-Bé during the last few years, in 1950 reaching about one-half of the prewar figures. This particularly affected the large distilleries, which before the war used to supply the trade with their well-known brands of "Extra," "First," "Second," and "Third" qualities. At the time of this writing (1951) it is almost impossible to obtain substantial lots of really good "Extra" quality ylang ylang oil. At present most of the oil coming from Madagascar and adjacent islands (Nossi-Bé and the Comoros) is supplied by small operators—chiefly Indians and Chinese—who now produce only "First" and "Second," very rarely "Third," and almost never "Extra," qualities. Oils now being offered on the market as of "Extra" quality are actually poor "Firsts" at the best, if not mediocre "Seconds." In other words, the quality of Madagascar ylang ylang oil has undergone a general decline. This situation, of course, may prove to be only temporary. The present very high prices will undoubtedly induce large-scale producers to start on a long-range program of replantation, but it takes about five years for a newly planted tree to reach full productivity.

Physicochemical Properties.—In the course of the distillation of ylang ylang flowers, the distillate is separated into various fractions; as distillation proceeds the specific gravity and the ester content of these fractions decrease, whereas the optical (laevo-) rotation and the refractive index increase. The "Extra" (top) fraction exhibits the highest specific gravity and ester number, and the lowest optical rotation and refractive index, whereas the "Third" (last) fraction has the lowest specific gravity and ester number, and the highest optical (laevo-) rotation and refractive index. The quality of the various fractions can, therefore, be determined by means of their physicochemical properties. The "Extra" and "First" fractions possess the strongest and finest odor, because they contain the highest percentage of esters, ethers, and phenols. The last fractions consist chiefly of sesquiterpenes, and therefore have only little odor value.

It is difficult to establish definite limits for the physicochemical properties of each fraction, because in fractionating almost every producer follows his own method. Moreover, many of the lots reaching overseas markets consist of bulkings, made up by the producers, local dealers, and exporters in accordance with their own standards or with demands from abroad.

The following will describe the properties of oils and fractions distilled under the author's supervision while he studied production of ylang ylang oil on the islands. The properties reported below thus represent those of *genuine* fractions, but not those of bulkings.

The first example is that of oils obtained by direct fire distillation in a small but well equipped distillery on the island of Nossi-Bé:

Nossi-Bé (Direct Fire Distillation)

Fraction	Specific Gravity at 15°	Optical Rotation	Refractive Index at 20°	Saponification Number	Solubility at 20°
"Extra"........	0.955	−27° 23′	1.5020	126.0	Soluble in 0.5 vol. of 90% alcohol; turbid in 2 vol. and more
"First"........	0.953	−44° 50′	1.5070	119.0	Soluble in 0.5 vol. of 90% alcohol; turbid to cloudy in 1 vol. and more
"Second"......	0.949	−55° 5′	1.5112	96.6	Not clearly soluble in 90% alcohol up to 10 vol.

Oils distilled in Nossi-Bé in a steam still had these properties:

Nossi-Bé (Steam Distillation)

Fraction	Specific Gravity at 15°	Optical. Rotation	Refractive Index at 20°	Saponification Number	Solubility at 20°
"Extra"	0.960	−38° 15′	1.5061	128.8	Soluble in 0.5 vol. of 90% alcohol; cloudy in 1.5 vol. and more
"First"	0.946	−56° 12′	1.5105	89.6	Not clearly soluble in 10 vol. of 90% alcohol
"Second"	0.938	−61° 10′	1.5100	81.2	Not clearly soluble in 10 vol. of 90% alcohol
"Third"	0.923	−51° 22′	1.5095	51.8	Not clearly soluble in 10 vol. of 90% alcohol

The following two groups of oils were produced in one of the largest steam distilleries of Nossi-Bé, under most carefully controlled conditions, and from the best flower material. Note the high ester numbers and specific gravities of the first fractions! The odor of these oils was excellent; they represent the highest grade of oil. It should be mentioned that the so-called "Super Extra" fraction is not commercially available; it was produced on

Nossi-Bé (Steam Distillation)

Fraction	Specific Gravity at 15°	Optical Rotation	Refractive Index at 20°	Saponification Number	Solubility at 20°
"Super Extra"	0.987	−20° 15′	1.5009	198.8	Soluble in 0.5 vol. of 90% alcohol; turbid to cloudy in 2 to 2.5 vol. and more
"Extra"	0.977	−25° 52′	1.5005	184.8	Soluble in 0.5 vol. of 90% alcohol; turbid in 1.5 vol. and more
"First"	0.971	−35° 0′	1.5045	154.0	Soluble in 0.5 vol. of 90% alcohol; cloudy in 1.5 vol. and more
"Second"	0.935	−54° 10′	1.5050	85.4	Not clearly soluble in 10 vol. of 90% alcohol
"Third"	0.922	−55° 28′	1.5099	54.6	Not clearly soluble in 10 vol. of 90% alcohol

Nossi-Bé (Steam Distillation)

Fraction	Specific Gravity at 15°	Optical Rotation	Refractive Index at 20°	Saponification Number	Solubility at 20°
"Super Extra".	0.976	−26° 35'	1.5005	186.2	Soluble in 0.5 vol. of 90% alcohol; opalescent in 2.5 vol. and more
"Extra".......	0.968	−35° 51'	1.5030	158.7	Soluble in 0.5 vol. of 90% alcohol; cloudy in 1.5 vol. and more
"First"........	0.957	−36° 56'	1.5028	130.7	Soluble in 0.5 vol. of 90% alcohol; cloudy in 1.5 vol. and more
"Second"......	0.942	−57° 18'	1.5101	84.0	Not clearly soluble in 90% alcohol up to 10 vol.
"Third".......	0.930	−58° 0'	1.5119	58.8	Not clearly soluble in 90% alcohol up to 10 vol.

the occasion solely for the purpose of demonstrating to the author that it is actually of extraordinary quality.

Oils obtained by steam distillation in a large distillery on Madagascar proper (adjacent to Nossi-Bé) were also of very good quality. The relatively high specific gravity and ester number of the "Third" fraction prove that distillation was not "pushed" too far. In fact, the entire distillation lasted only 6 hr.

Madagascar (Steam Distillation)

Fraction	Specific Gravity at 15°	Optical Rotation	Refractive Index at 20°	Saponification Number	Solubility at 20°
"Extra".......	0.967	−34° 45'	1.5011	160.5	Soluble in 0.5 vol. of 90% alcohol; turbid in 1.5 vol. and more
"First"........	0.954	−49° 54'	1.5048	119.5	Soluble in 0.5 vol. of 90% alcohol; cloudy with more
"Second"......	0.944	−76° 0'	1.5147	105.5	Not clearly soluble in 90% alcohol up to 10 vol.
"Third".......	0.940	−76° 50'	1.5141	76.5	Not clearly soluble in 90% alcohol up to 10 vol.

That oils of excellent quality are produced also on the Comoro Islands can be seen from the next example. The oils were obtained by steam distillation. Note the high specific gravity of the "Extra" fraction. (High-grade Comoro oils are noted for this characteristic!)

Comoro Islands (Anjouan) (Steam Distillation)

Fraction	Specific Gravity at 15°	Optical Rotation	Refractive Index at 20°	Saponi-fication Number	Solubility at 20°
"Extra".......	0.986	−23° 44′	1.5051	154.0	Soluble in 0.5 vol. of 90% alcohol; cloudy in 1.5 vol. and more
"First"........	0.963	−33° 0′	1.5030	136.3	Soluble in 0.5 and 1 vol. of 90% alcohol; cloudy with more
"Second"......	0.934	−52° 20′	1.5094	74.7	Not clearly soluble in 90% alcohol up to 10 vol.
"Third".......	0.915	−54° 18′	1.5085	45.7	Not clearly soluble in 90% alcohol up to 10 vol.

Another group of oils from the Comoro Islands, also produced in a modern steam distillery, had these properties:

Comoro Islands (Anjouan) (Steam Distillation)

Fraction	Specific Gravity at 15°	Optical Rotation	Refractive Index at 20°	Saponi-fication Number	Solubility at 20°
"Extra".......	0.976	−31° 45′	1.5041	172.2	Soluble in 0.5 vol. of 90% alcohol; turbid to cloudy in 1.5 vol. and more
"Second"......	0.942	−61° 10′	1.5102	95.2	Not clearly soluble in 10 vol. of 90% alcohol
"Third".......	0.922	−59° 10′	1.5095	56.0	Not clearly soluble in 10 vol. of 90% alcohol

And a last sample:

Comoro Islands (*Mayotte*) (Steam Distillation)

Fraction	Specific Gravity at 15°	Optical Rotation	Refractive Index at 20°	Saponi- fication Number	Solubility at 20°
"Extra".......	0.960	−49° 12′	1.5065	142.8	Soluble in 0.5 vol. of 90% alcohol; cloudy in 1.5 vol. and more
"Second"......	0.948	−56° 10′	1.5101	102.2	Not clearly soluble in 10 vol. of 90% alcohol

Trabaud [6] cited these figures for "Extra," "First," "Second" and "Third" fractions, and for the "Total" (complete) oil:

	"Extra"	"First"	"Second"	"Third"	"Total"
Specific Gravity at 15°.....	0.9686	0.9556	0.9396	0.9191	0.9406
Optical Rotation..........	−31° 52′	−33° 4′	−54° 36′	−61° 44′	−48° 22′
Acid Number.............	3.9	3.9	3.9	3.0	2.25
Ester Content, Calculated as Acetate of $C_{10}H_{18}O$...	55.5%	50.3%	37.3%	21.0%	37.4%
Solubility in:					
90% Alcohol...........	1:0.5	1:0.5 then insoluble	insoluble	insoluble	insoluble
95% Alcohol...........	1:0.5	1:0.5 then insoluble	1:0.5	1:1	1:0.5

Chiris [7] reported the following properties for the various fractions of ylang ylang oil:

	"Extra"	"First"	"Second"	"Third"
Specific Gravity at 15°	0.942 to 0.946; exceptionally up to 0.970	0.932 to 0.950	0.922 to 0.931	0.910 to 0.920
Optical Rotation..	−35° 0′ to −45° 0′	−38° 0′ to −50° 0′	−46° 0′ to −58° 0′	−30° 0′ to −40° 0′
Refractive Index at 20°	1.498 to 1.508	1.508 to 1.512	∼1.510	1.506 to 1.510
Acid Number.....	Up to 2.4	Up to 1.6	Up to 1.6	Up to 1.6
Ester Content, Calculated as Acetate of $C_{10}H_{18}O$	42 to 54%; exceptionally up to 65%	28 to 38%	20 to 28%	12 to 18%
Free Alcohol Content, Calculated as $C_{10}H_{18}O$	8.5 to 14%; exceptionally up to 15.5%	7 to 11%	4 to 8.5%	3.5 to 6%

[6] *Perfumery Essential Oil Record* **28** (1937), 406.
[7] *Parfums France* **8** (1930), 350.

(It should be noted that these limits were established in 1930 and therefore apply to oils produced prior to that date. Today the picture has changed considerably, the oils now reaching the market exhibiting somewhat different properties.)

The solubility of ylang ylang oil, in general, is poor. This statement applies to all qualities (fractions). Usually from 0.5 to 3 vol. of 95 per cent alcohol are required to give a clear solution, which often becomes cloudy on addition of more alcohol.

The color of the oils ranges from light amber to yellow-brown. The "Extra" quality exhibits the deepest color, probably because of the presence of phenolic substances. This type of oil, incidentally, is the most sensitive to light, and if exposed to it easily turns brown, losing at the same time its originally delicate and strong odor. In general, ylang ylang oils should be stored in well-sealed containers, protected from light and air. If correctly stored, the properties undergo little change, except, perhaps, the specific gravity and refractive index, which may increase, and the optical rotation and solubility, which may decrease. When exposed to light and air, the changes are marked and take place rapidly: the specific gravity increases (even above 1.0); the refractive index also increases, while the optical rotation decreases rapidly. The solubility may improve a little. The ester number shows no change, but the ester number after acetylation increases substantially and abnormally. Aging of the oil affects particularly the sesquiterpenes, transforming them into oxygenated and acetylizable resinous compounds. The color of the oil, in general, becomes darker, and the odor flatter.

Total Production of Ylang Ylang Oil.—In the years immediately preceding World War II, the total yearly production of ylang ylang oil in the world averaged about 40 metric tons. According to Muller,[8] a leading expert in the industry, it was distributed approximately as follows:

	Metric Tons
Nossi-Bé	20 to 23
Comoro Islands	12 to 14
Madagascar Proper	2 to 3
Réunion Island	1 to 1.5

Production on the Philippine Islands, which in 1913 had been more than 2 metric tons, was only 0.5 to 1 ton per year from 1935 to 1939.

For several years after 1945, production of ylang ylang oil on Nossi-Bé declined by some 50 per cent, the result chiefly of a marked diminution in the inflorescence of the tree. (See above section, "Present Situation.")

---

[8] Private communication from Mr. Charles Muller, Ambanja, Madagascar.

Because of this decline Nossi-Bé in 1947 exported only 9,125 kg. of ylang ylang oil, and only 3,083 kg. in 1948. In 1949, however, exports increased again to an estimated 23,000 or 24,000 kg.; exports in 1950 were 24,000 or 25,000 kg. (Official figures are difficult to obtain, and it seems quite possible that part of the oil exported from Nossi-Bé originated in the Comoros, and was transshipped via Nossi-Bé.)

The Comoro Islands, which were expected to supply about 4,000 kg. of oil in 1951, were hit by a terrible cyclone on January 4 of that year. More than 50 people were killed and so much damage was done to the plantations that no oil could be produced in 1951. According to official reports it will require at least three years for normal conditions to be restored.

On Réunion Island production of the oil is also diminishing rapidly and may soon reach the vanishing point. Here, however, the causes are chiefly of an economic nature, the sugar industry absorbing all available labor at much higher wages than the ylang ylang growers can afford to pay.

Adulteration.—Most of the compounds occurring in the "Extra" and "First" fractions as chief constituents are readily available in synthetic form: benzyl acetate, methyl benzoate, p-cresol methyl ether, for example. Adulteration of ylang ylang oil with these low-priced synthetic aromatics, therefore, is frequent. As a matter of fact, addition of such synthetics to a "Second" quality oil would change its physicochemical properties (specific gravity, optical rotation and ester content) in such a way that the "Second" quality would approach the characteristics of a genuine "First" quality—except for the odor. Fortunately these synthetics cannot be employed on the islands, since imports of any of them would immediately become known to the customs officials, who would prohibit their use for the purpose intended. Moreover, small producers are not familiar with these products and would not know how to employ them.

The only form of adulteration practiced on the island—actually it should be called sophistication rather than adulteration—consists in the admixture of old or poor lots to fresh and good lots. As was pointed out above, the specific gravity of an improperly stored oil increases on standing, while the optical rotation decreases. In other words, with the passage of time the specific gravity and optical rotation of an old "Third" quality approach those of a freshly distilled "Second" quality oil. Or, a poorly distilled oil may be bulked with a properly distilled lot; this would affect the properties very little, but would result in a marked deterioration of the odor of the good lot. In selecting ylang ylang oil, the buyer, therefore, should pay considerable attention to the odor, which in properly distilled oils is ethereal, fruity, flowery and sweet. An oil carefully produced in directly-fired stills may exhibit a much finer odor than a lot carelessly distilled in steam stills. The best protection against adulteration is to buy only oils marketed by well-known large distilleries and under guaranteed labels.

Terpeneless Ylang Ylang Oil.—In the foregoing we have seen that the value of an ylang ylang fraction depends chiefly upon its content of oxygenated compounds. The sesquiterpenic portion makes no more than a weak contribution to the typical ylang ylang odor; on the other hand, the sesquiterpenes are responsible for the insolubility of the oil in alcohol. Even "Extra" qualities contain about 35 per cent of sesquiterpenes. To offer ylang ylang oils more soluble than the natural ones, sesquiterpeneless oils have been prepared (cf. Vol. I of this work, p. 218). These oils consist chiefly of the aromatic compounds (alcohols, esters, phenols, phenol ethers, etc.); in other words, the sesquiterpeneless oils combine the advantages of highest odor concentration and good solubility in alcohol.

Trabaud[9] prepared a sesquiterpeneless ("deterpenated") oil from an "Extra" quality, by an original process, in the cold, and reported these properties:

Specific Gravity at 15°.............	0.995
Optical Rotation..................	−23° 32′
Ester Content, Calculated as Acetate of $C_{10}H_{18}O$.....................	66%
Solubility.......................	Soluble in 25 vol. of 70% alcohol; in 1.5 vol. of 75% alcohol; in 0.7 vol. of 80% alcohol

II. Ylang Ylang Oil from Réunion Island

As was mentioned in the beginning of this monograph, the ylang ylang oil industry of Réunion Island had its start at the turn of the century; it reached its peak after World War I, particularly between 1920 and 1928, when several modern distilleries produced substantial quantities of a very high-grade oil. Extensive plantations existed at that time in the region of "La Plaine" and "Rivière-des-Galets." However, when the world-wide depression set in, prices of the oil fell to such low levels that these plantations had gradually to be abandoned, and by 1936 nothing was left of the once beautiful and well-kept groves which had formerly supplied the distilleries with flower material. In 1939 only about 12,000 ylang ylang trees remained on Réunion Island; these were owned by two distilleries, one in Piton (St. Paul) and the other in Grand-Pourpier (St. Paul). Extension to Réunion of the new social laws enacted in France just prior to World War II increased the cost of labor on the island to such an extent that the ylang ylang industry could no longer compete with that of Nossi-Bé, Mada-

9 *Perfumery Essential Oil Record* **28** (1937), 406.

(Left) Production of ylang ylang oil in Nossi-Bé (Madagascar). Inside view of a large distillery. The still tops are lifted, permitting the charging of the flowers into the stills. (Right) A modern ylang ylang still in Nossi-Bé. The still top can be lifted by a hoist and the still can be tilted for discharging of the exhausted flowers. Note the oil separator beneath the condenser on the left, and the U-shaped pipe conducting the distillation water back into the still for auto-cohobation during distillation of the flowers. (Bottom) Close-up of an oil separator. Note the clear supernatant layer of oil and the milky distillation water beneath, which has to be returned to the still for automatic cohobation. *Photos Fritzsche Brothers, Inc., New York.*

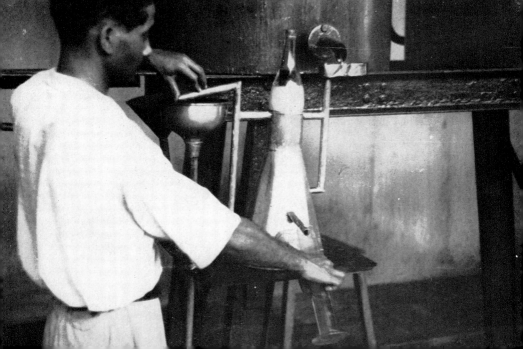

(*Top*) Production of ylang ylang oil in Manila, Philippine Islands. Harvesting of the flower material. Because of the great height of the trees, most of the flowers cannot be hand-picked but must be torn off the branches with long poles. (*Bottom*) Ylang ylang, Manila. Discharge of the spent flower material from a still. *Photos Fritzsche Brothers, Inc., New York*

gascar itself, and the Comoro Islands. After the end of the war conditions became even more serious, and at present the number of ylang ylang trees on Réunion Island has been reduced to about 10,000. Moreover, only a part of the harvest is gathered. This condition is simply the result of the high cost and shortage of labor—which latter has been absorbed by the all-powerful sugar industry of the island. In order to procure the necessary labor, ylang ylang growers would have to pay such high wages that the cost of the oil would be exorbitant. If this situation continues the ylang ylang trees on Réunion Island will undoubtedly be abandoned, or even destroyed, which would mean the end of this once flourishing industry. At present the production of ylang ylang oil on Réunion Island no longer offers any economic incentive, and it may well be that the industry will follow the way of the Réunion coffee and cocoa industries—which have practically reached the vanishing point.

Harvest and Distillation.—The flower harvest lasts from the end of October to June. Only fully developed flowers are collected; gathering starts in the early morning. The flowers are transported to the two distilleries and distilled immediately upon arrival, distillation lasting from about 11 A.M. to 3 P.M. Meanwhile, the women collect a second batch of flowers which is distilled right after the first batch, viz., from 4 to 8 P.M. Distillation is carried out in a number of modern steam stills by a modified process of water *and* steam distillation. The stills are equipped with a grid (false bottom); beneath the grid is a perforated steam coil through which live steam is injected. An automatic steam-reducing valve keeps the steam pressure uniform throughout the operation and permits sharp and uniform cutting of the fractions when distilling and fractionating according to a time schedule. One hundred and fifty liters of water are pumped into the still and heated almost to the boiling point. At the proper moment 100 kg. of flowers are quickly charged into the still and distillation is carried out as rapidly as possible. This prevents hydrolysis of the esters and assures a high yield of "Extra" and "First" fractions.

One hundred kilograms of flowers yield:

Length Distillation (in hr.)	Fraction (Quality)	Quantity of Fraction (in g.)
3	"Extra" and "First"	~650
1 to 2	"Second"	~250

The time allowed for the distillation of the "Second" fraction is determined by the working hours remaining on one particular day, the steam pressure left in the steam generator, and other factors.

The two distilleries on Réunion Island consider the "Third" quality an inferior product, not worth the high cost of labor and fuel prevailing on the island. Réunion producers believe the "Third" quality to be a product which should be left to Nossi-Bé, Madagascar proper, and particularly to the Comoro Islands, where costs of production are much lower. On special demand, Réunion producers may nevertheless occasionally supply a "Third" quality; this, however, is the exception rather than the rule.

Physicochemical Properties.—Genuine ylang ylang oils from Réunion Island examined in the laboratories of Fritzsche Brothers, Inc., New York, shortly before and after World War II had properties varying within the limits given in Table 11.1.

TABLE 11.1

Fraction	Specific Gravity at 15°	Optical Rotation	Refractive Index at 20°	Saponification Number	Solubility
"Extra H"	0.959 to 0.961	−37° 55′ to −44° 45′	1.5007 to 1.5022	149.3 to 153.1	Soluble in 0.5 vol. of 90% alcohol, turbid with more
"Extra E"	0.952 to 0.954	−49° 55′ to −52° 22′	1.5045 to 1.5054	128.8 to 133.5	Soluble in 0.5 vol. of 90% alcohol, turbid with more
"Second"..	0.912 to 0.913	−64° 20′ to −65° 0′	1.5058 to 1.5060	59.7 to 61.6	Not clearly soluble in 90% alcohol up to 10 vol.

The properties noted above apply to oils produced during the last fifteen years. According to Garnier,[10] the properties of Réunion ylang ylang oils produced formerly were somewhat different, the optical rotation of the "Extra" qualities ranging from −40° 5′ to −50° 5′, and the saponification number usually being about 175.

An outstanding characteristic of the Réunion ylang ylang oil used to be its remarkably fine, delicate, and very flowery top note, and the absence of any "burnt" cananga off-odor in the "Extra" and "First" fractions. Recently, however, it has become increasingly difficult to produce the former very high-quality oil. Owing to the labor shortage mentioned above, it is not always possible to harvest the flowers at precisely the right moment— an all-important factor in the quality of ylang ylang oil.

[10] Private communication from Mr. Robert Garnier, Paris.

III. Ylang Ylang Oil from the Philippine Islands

As was explained above, the ylang ylang oil industry actually originated in the Philippines; Manila oils achieved a high reputation before they were gradually forced off the world markets by cheaper oils from Réunion Island, Madagascar (Nossi-Bé), and the Comoro Islands. The displacement of the Manila oils began toward the end of the last century (1892) and progressed from year to year. In 1914 approximately 2,500 kg. of oil were exported from the Philippinés, in 1928 about 560 kg., and in 1948 only 5 kg.

Oil of ylang ylang thus joins the number of other products from the Philippines, including indigo and coffee, which have ceased to be produced on these islands. However, so high was the reputation of the Manila ylang ylang oils that even today perfumers inquire about them, not being aware that they are now produced in very limited quantities. It is principally for this reason that the development of the Philippine ylang ylang industry will be reviewed briefly here. The following account is based primarily upon the author's observations in the course of a visit to the islands in 1939.[11]

The industry had its modest beginning when, during the 'sixties of the last century, Albertus Schwenger, a sailor, was stranded on the islands. Inspired by the delightful fragrance of the ylang ylang flowers, versatile Albertus conceived the idea of rigging up a still on a wheelcart, with which he journeyed through the countryside, distilling small quantities of ylang ylang flowers. First to produce ylang ylang oil on a commercial scale was F. Steck (1858 to 1880), owner of the pharmacy which later became known as Botica Boie.[12] Steck had his nephew, Paul Sartorius, come over from Germany, to take charge of the production of the oil. After his uncle's death, Sartorius introduced his oil under the label "Ylang Ylang Oil Sartorius." This became one of the most famous brands on the market. Other drug houses in Manila (most of them German) followed, and at the beginning of the twentieth century the Manila ylang ylang oil was represented by the well-known labels "Sartorius," "Witte," "Siegert," and "Dr. Jaehrling." Extensive ylang ylang groves were planted in the vicinity of Manila, the natives were taught to grow ylang ylang trees in their patch gardens, and large distilleries, equipped with the most modern stills from Germany, were erected in Manila.

The bulk of the flower material came from the small properties of the natives, each comprising from 10 to 20 trees. In season, the flowers were harvested every morning and sold to the distillers in Manila through field brokers. Due to the primitive trucking facilities prevailing at that time,

[11] Cf. Guenther, *Am. Perfumer* **41** (July 1940), 34.
[12] Cf. "1830 to 1930 Centennial Memorial of Botica Boie."

transport to the distilleries was often delayed. As a result, the flowers were frequently picked in the afternoon, kept overnight in sacks submerged in water, and transported to Manila in the early morning. Thousands of kilograms of flowers were distilled daily during the height of the season.

The entry of the United States into World War I brought about a fundamental change in the production of the Manila ylang ylang oil. The German firms owning the distilleries encountered difficulties in exporting their product. Practically all of these firms passed ultimately into the hands of the Alien Property Custodian. Production of the oil came to a standstill, and the natives who formerly had taken good care of their trees began to neglect them, since they could find no buyers for the flowers. Moreover, the war boom caused a great influx of native labor into Manila, and many small ylang ylang groves and gardens in Manila were destroyed, to make room for huts which could be rented to the newly arrived war-workers at high prices. Between 25,000 and 30,000 ylang ylang trees were cut down—a heavy blow to the industry.

After the end of World War I the exporters tried to revive the industry. They soon found themselves forced to dismantle their distilleries, however, because they could not obtain sufficient flower material. At the same time competition on the part of Réunion Island, Madagascar (Nossi-Bé), and the Comoro Islands made itself increasingly felt.

When the author visited the Philippines in 1939 there was only one grove of ylang ylang left near Manila. This was in Novaliches and contained about 6,000 trees. These had grown to great height, which made gathering of the flowers quite difficult. Only the flowers within reach of the harvesters could be selected by hand, whereas the flowers growing on the higher branches had to be torn down with long bamboo poles, specially constructed. Careful selection of the flower material, a prerequisite for the production of a high-grade oil, was therefore impossible. The torn-off flowers were simply collected from the ground, gathered in baskets and carted to the distillery in the grove. There were two principal flowering harvests, one from the middle of February to May, and the other from the middle of June to October.

The distillery near the grove, which had been erected before World War I, was equipped with five copper stills of excellent workmanship. Each still contained a grid on which the flowers were charged. Steam was injected beneath the grid, the process being a steam *and* water distillation. The distillation water of the first batch was used again for the next flower batch; in other words, the distillation waters were cohobated.

The yield of the first fraction varied from 0.3 to 0.5 per cent. About 350 kg. of flowers were usually required to produce 1 kg. of oil.

Two samples of genuine Manila ylang ylang oils, distilled under the author's supervision in 1939, had the following properties:

	First Quality	*Second Quality*
Specific Gravity at 15°....	0.961	0.957
Optical Rotation.........	−26° 33′	−56° 50′
Refractive Index at 20°...	1.4932	1.5110
Acid Number............	1.1	1.9
Saponification Number....	160.9	111.4
Solubility at 20°.........	Soluble in 0.5 vol. of 90% alcohol; opalescent in more	Soluble in 0.5 vol. of 90% alcohol; opalescent to cloudy in more

World War II effected a further decline in the production of Manila ylang ylang oil. According to Umbreit,[13] the number of the trees on the above-mentioned plantation in Novaliches has diminished to about 2,000, and distillation has not been resumed since the liberation of Manila. In the meantime wages and living expenses in Manila have increased to such an extent that today the cost of collecting the flower material for the purpose of distillation is almost prohibitive. In the last few years the flowers actually harvested on the plantation have been sold as fresh flowers, for the making of leis, and have brought very good prices. Production of oil of ylang ylang under such conditions is obviously uneconomical. Moreover, the plantation in question has now been incorporated into Quezon City. The new capitol and other government buildings are being erected nearby, and the value of the land is now much too high for a plantation. Probably the land will soon be sold for real estate, and this will be the end of the Manila ylang ylang industry.

It should be mentioned here that Manila ylang ylang oil is not the only type produced on the Philippine Islands. In the provinces of Albáy, Bohol, Camarines, and Mindoro there are numerous ylang ylang trees (probably originally planted, but long since escaped from cultivation) which now grow neglected in a semiwild and wild state. Natives harvest the flowers of these trees by cutting entire branches and collecting all flowers without regard to their condition. This method of harvesting has a detrimental effect upon the trees as they grow taller and taller. Harvesting plainly becomes increasingly difficult, if not impossible.

The stills in these provinces are of primitive construction, with the result that the oils (known in the trade as Albáy oils) are usually of poor quality and not uniform. Occasionally a small lot of fair quality may reach Manila, from where it is then shipped under the label of Manila oil. Between the date of the liberation and 1949 about 350 kg. of Albáy oils were exported.

[13] Private communication from Mr. F. C. Umbreit, Manila, October 1949.

In 1948, however, no oil was produced because the natives in Albáy demanded too high prices for the flower material.

CHEMICAL COMPOSITION OF YLANG YLANG OIL

The chemical composition of ylang ylang oil is complex; an "Extra" quality, for example, contains more than 30 constituents belonging to some ten classes distinguished by different functional groups. We owe our present knowledge of the chemistry of the oil chiefly to the exhaustive work of Glichitch and Naves [14] who, in 1932, published their investigations of genuine oils produced in the Comoro Islands. Prior to that date our knowledge of the subject had been based upon work carried out upon oils from the Philippine Islands (Manila oils).

The presence of the following compounds had been observed in Manila ylang ylang oils by Gal,[15] Flückiger,[16] Reychler,[17] Schimmel & Co.,[18] Darzens,[19] and Bacon: [20]

d-α-Pinene. (Schimmel & Co.)

l-Linaloöl and Geraniol. These two terpenic alcohols occur in the oil partly free, partly in esterified form (Reychler).

Benzyl Alcohol. (Schimmel & Co.)

Creosol. (Schimmel & Co.)

Eugenol. (Flückiger)

Isoeugenol. (Schimmel & Co.)

p-Cresol. (Darzens)

p-Cresol Methyl Ether. (Reychler)

Eugenol Methyl Ether. (Schimmel & Co.)

Safrole. (Bacon)

Isosafrole. (Bacon)

p-Cresyl Acetate. (Darzens)

Benzyl Acetate. (Schimmel & Co.)

[14] *Parfums France* **10** (1932), 7, 36.
[15] *Compt. rend.* **76** (1873), 1482.
[16] *Arch. Pharm.* **218** (1881), 24.
[17] *Bull. soc. chim.* [3], **11** (1894), 407, 576, 1045; **13** (1895), 140.
[18] *Ber. Schimmel & Co.*, April (1900), 48; October (1901), 57, 58; April (1902), 64; April (1903), 79.
[19] *Bull. soc. chim.* [3], **27** (1902), 83.
[20] *Philippine J. Sci.* **3** (1908), A, 65.

Methyl Benzoate. (Schimmel & Co.)

Methyl Salicylate. (Schimmel & Co.)

Benzyl Benzoate. (Schimmel & Co.)

Methyl Anthranilate. (Schimmel & Co.)

Formic Acid. (Bacon)

Acetic Acid. (Reychler, Schimmel & Co.)

Valeric Acid(?). (Bacon)

Benzoic Acid. (Gal, Schimmel & Co.)

Salicylic Acid. (Schimmel & Co.)

Cadinene. The sesquiterpene gave a dihydrochloride m. 117° (Reychler).

A Sesquiterpene Hydrate. Colorless and odorless needles m. 180° (Schimmel & Co.)

The compounds contributing most to the odor of the oil are the esters, phenols, and phenol ethers.

Glichitch and Naves, whose work is cited above, succeeded not only in confirming the presence, in Comoro oils, of all compounds previously found in Manila oils, but also in identifying quite a number of other substances. Moreover, the investigation of these two authors furnished some insight into the *quantitative* composition of an "Extra" fraction of ylang ylang oil from the Comoro Islands:

Aldehydes and Ketones (0.1% to 0.2%)	Acetone, Furfural, Benzaldehyde, Compounds giving a semicarbazone m. 209°–212°;
Basic Substances (0.1%)	Methyl Anthranilate, Bases with a nicotine-like odor;
Terpenes (0.3% to 0.6%)	*d*-α-Pinene;
Phenols and Phenol Ethers (3%)	*p*-Cresol, *p*-Cresol Methyl Ether, A Phenol m. 116°–118°, Eugenol, Isoeugenol, Methyl Salicylate, Benzyl Salicylate(?), Higher Phenols;

Alcohols and Esters
(52% to 64%)

Free and Combined

Methyl Benzoate,

$\Bigl[$ *l*-Linaloöl,
α-Terpineol,
Benzyl Alcohol,
Phenyl Ethyl Alcohol,
Geraniol,
Nerol,
Farnesol,
Nerolidol,
l-Cadinol,
A monocyclic sesquiterpene alcohol $C_{15}H_{26}O$, b_5 142°–144°, d_{20} 0.9150, α_D −5° 8′, n_D^{20} 1.4937, hydrochloride m. 81° (which may be identical with bisabolene trihydrochloride),
A solid sesquiterpene alcohol, crystallized in large prisms m. 138°;

Sesquiterpenes
(33% to 38%)

A slightly laevorotatory, bicyclic sesquiterpene $C_{15}H_{24}$, b_{12} 114°–116°, d_{20} 0.9026, α_D −2° 18′, n_D^{20} 1.4966; it yielded no crystallized hydrochloride or nitrosate;

A strongly laevorotatory, bicyclic sesquiterpene, b_{12} 116°–118°, d_{20} 0.9057, α_D −79° 21′, n_D^{20} 1.5052; it yielded no crystallized derivative;

d-Caryophyllene, b_{12} 120°–121°, d_{20} 0.905, α_D +6° 17′, n_D^{20} 1.5005; nitrosate m. 152°–153°;

An aliphatic sesquiterpene $C_{15}H_{24}$, b_9 136°–138°, d_{15} 0.8502, α_D +1° 15′, n_D^{15} 1.5425; it yielded no crystallized derivatives and may be related to sesquicitronellene;

A laevorotatory sesquiterpene (fraction), b_{12} 132°–134°, d_{20} 0.890, α_D −110° 48′, n_D^{20} 1.5068; hydrochloride m. 118°–119° (identical with cadinene dihydrochloride);

A dextrorotatory sesquiterpene (fraction), b_9 134°, d_{20} 0.8822, α_D +11° 15′, n_D^{20} 1.5195; hydrochloride m. 118°–118.5° (identical with cadinene dihydrochloride).

As regards the last two sesquiterpenes, Glichitch and Naves arrived at the conclusion that they are probably sesquiterpenes which, under the influence of hydrochloric acid, became isomerized into cadinene— as is the case with copaene, for example.

The esters occurring in oil of ylang ylang are formates, acetates, valerates, esters of acids C_5, C_6, C_8 and C_{10}, and benzoates of the above listed alcohols. Benzyl acetate and benzyl benzoate are particularly important.

The percentages in which constituents occur in ylang ylang oil vary substantially with the quality (fraction) of the oil. Those noted above apply to the "Extra" quality. Naves and Glichitch [21] were able to determine the proportions of constituents in the other qualities of oil by dividing the compounds into three groups (A, B, and C), according to their boiling range:

Boiling Range at 10 mm. pr.	"Extra" (%)	"First" (%)	"Second" (%)	"Third" (%)
A. b_{10} 115°............	37–42	24–27	5– 7	2.5– 5
B. b_{10} 115° to 155°.....	33–38	33–38	63–67	77 –82
C. b_{10} 155° to 185°.....	12–16	12–18	18–20	10 –13

It may be postulated that this distribution of constituents in the several fractions of the oil depends not only upon their volatility, but also upon the nature of the cleavage which takes place when the complex compounds containing these constituents are broken down. Basing their calculation upon the total quantity of benzyl benzoate/sesquiterpenes present, Glichitch and Naves noted that the percentage of benzyl benzoate shows the following average values for the different qualities (fractions) of ylang ylang oil:

	Per Cent
"Extra"........	40
"First"........	34
"Second"......	32
"Third"........	19

Benzyl benzoate is less volatile and hence more difficult to carry over with steam than the sesquiterpenes. In the opinion of Glichitch and Naves, the increasing proportions of the latter in the fractions collected at the latest stage of distillation therefore indicate that the sesquiterpenes are actually obtained (freed) by the splitting of certain as yet unidentified complex compounds (precursors).

Further support for this postulate may be found in the fact that the *distilled* oils of ylang ylang contain a fairly high percentage of sesquiterpenes, whereas ylang ylang oils *extracted* from the flowers by means of petroleum ether are almost entirely lacking in these hydrocarbons. Since the *total* oil of ylang ylang (*distilled* from the flowers with a yield of about 2 per cent) contains, on the average, 60 to 65 per cent of sesquiterpenes, the result is a yield, from the flowers, of about 0.7 to 0.8 per cent of oxygenated substances. However, *extraction* of the flowers with petroleum ether results in a *total yield* of less than 1 per cent, and practically no sesquiterpenes are present. Therefore, it seems highly probable that the sesquiterpenes contained in the distilled oil are formed in the course of distillation of the flowers, from complex compounds (precursors) insoluble in petroleum ether,

[21] *Parfums France* **10** (1932), 40.

and nonvolatile with steam. At the time of their writing, Glichitch and Naves could not advance any theory as to the nature of these complex compounds.

[In this connection it should be mentioned that more recently Naves [22] has investigated a concentrated extract of clove buds (benzene as solvent) for the presence of caryophyllene, and found that the extract contained not caryophyllene, but epoxydihydrocaryophyllene (the caryophyllene oxide of Treibs [23]). Steam-distilling the clove buds after they had been extracted with benzene, Naves obtained a volatile oil, composed chiefly of caryophyllene. The latter, therefore, is not a natural, biological constituent of the clove buds, but must have originated under the influence of boiling water —cf. Vol. II of the present work, p. 716.]

That the sesquiterpenes present in distilled ylang ylang oils are actually formed in the course of distillation was finally proved in 1937 by Trabaud,[24] who steam-distilled flowers previously extracted with petroleum ether. Trabaud obtained an oil with these properties:

> Specific Gravity at 15°....... 0.9025
> Optical Rotation............ −59° 0′
> Ester Content.............. 18%

The high laevorotation of this oil clearly indicates the presence of sesquiterpenes. The odor of the oil compared with that of "Third" qualities.

In the course of their work on oil of ylang ylang, Glichitch and Naves isolated a relatively large number of compounds, but they could not identify many others that exist in the oil in minute quantities only. Among them are low-boiling aldehydes and ketones, basic substances accompanying methyl anthranilate, high-boiling phenols, esters of higher aliphatic acids, and lactones. These trace substances play a most important role in the odor of the oil, but their elucidation will require further and painstaking examination.

Concrete and Absolute of Ylang Ylang

The first experimental extractions of ylang ylang flowers with volatile solvents (cf. Vol. I of this work, pp. 198 ff.) were carried out in Manila by Bacon,[25] who, using petroleum ether as solvent, obtained yields of concrete

---

[22] *Helv. Chim. Acta* **31** (1948), 378.

[23] *Chem. Ber.* **80**, No. 1 (1947), 56.

[24] *Perfumery Essential Oil Record* **28** (1937), 406. According to private information gathered by the author on Réunion Island, similar experiments were made some years ago by Garnier and Défaud.

[25] *Philippine J. Sci.* **4** (1909), A, 127.

ranging from 0.7 per cent to 1.0 per cent. Bacon reported these properties for a concrete of ylang ylang:

> Specific Gravity at 30°/4°........... 0.940
> Refractive Index at 30°............. 1.4920
> Ester Number..................... 135
> Ester Number after Acetylation...... 208

In 1911 Charles Garnier, of Paris, installed an extraction plant on Réunion Island and started the commercial production of concrete and absolute of ylang ylang. These, because of their remarkably true-to-nature odor, were readily absorbed by the perfume industry in Paris.[26] In 1920 Garnier's plant was transferred to Piton, near St. Paul, on Réunion Island. Other producers entered the field, erecting two extraction plants on Nossi-Bé (Madagascar), and one in Bambao (Comoro Islands). Prior to World War II several hundred kilograms of concrete were produced yearly in these plants.

According to Naves and Mazuyer,[27] the yield of concrete varies between 0.80 and 0.95 per cent. The concrete gives from 75 to 80 per cent of alcohol-soluble absolute; on steam distillation the concrete yields from 51 to 63 per cent of a volatile oil.

Chiris[28] reported the following properties for two concretes from the Comoro Islands (A and B), and for the two corresponding absolutes (C and D):

	A	B	C	D
Specific Gravity at 15°..............	1.0317	1.024	1.0436	1.0369
Optical Rotation....................	...	...	−7° 0′	−5° 15′
Refractive Index at 20°.............	...	1.5200	1.5255	1.5227
Acid Number......................	17.68	10.08	9.8	9.1
Ester Number.....................	148.4	177.45	189	200
Ester Number after Acetylation *....	201.95	244.2	...	289.6
Phenol Content...................	4%	10%	...	12.5%
Solubility........................	Partly soluble in 95% alcohol		Soluble with turbidity in 1.2 vol. and more of 80% alcohol	

* Determined on products that had been freed of phenols, and calculated upon the original product.

Naves, Sabetay and Palfray[29] described the properties of four volatile oils (I, II, III, and IV) obtained by distillation of ylang ylang concretes:

[26] Private information.
[27] "Les Parfums Naturels," Paris, Gaulthier-Villars (1939), 291.
[28] *Parfums France* **8** (1930), 359.
[29] *Perfumery Essential Oil Record* **28** (1937), 336.

	I	II	III	IV
Specific Gravity at 15°/15°	1.017	1.022	1.019	1.026
Optical Rotation	−6° 30′	−5° 54′	−6° 4′	−6° 16′
Refractive Index at 20°	1.5112	1.5008	1.5106	1.5143
Acid Number	2.8	3.2	1.9	2.4
Ester Number	216.4	226.2	231.4	212.5
Phenol Content	21.5%	17.0%	19.6%	24.0%

The chemical composition of the extracted ylang ylang oils (concretes and absolutes) has not yet been elucidated. It can be assumed that they contain most of the oxygenated substances occurring in the distilled oil (cf. above section on "Chemical Composition"). The sesquiterpenes which, by cleavage of complex compounds, are formed during distillation of the flowers, are not present in the extracted ylang ylang oil.

On the other hand, the extracted oils undoubtedly contain a number of high boiling components not present in the distilled oil, and which contribute greatly to their high fixation value. Being nonvolatile with steam, these high boiling substances are not carried over during distillation of the flowers, and therefore are lacking in the distilled oils.

In general, the odor of the extracted oils is sweeter, warmer and more "velvety" than that of the distilled oils. The perfume of the extracted oils reproduces that of the living flowers to a remarkable degree. An absolute of ylang ylang is almost a perfume in itself. No wonder, then, that it has found great favor among expert perfumers, and is used in some of their finest creations.

In connection with what has been said above, several experiments carried out by Muller,[30] in Nossi-Bé, and not yet published in literature, may be of interest:

1. Extraction of ylang ylang flowers with petroleum ether yields from 0.9 to 1 per cent of concrete. The specific gravity of this petroleum ether concrete slightly exceeds 1.0 at 15°. The concrete yields from 75 to 80 per cent of alcohol-soluble absolute.

2. Extraction of ylang ylang flowers with benzene yields from 2.5 to 3 per cent of concrete, with a specific gravity of about 0.967 at 20°. The yield of concrete by extraction with benzene is thus much higher than that obtained by extraction with petroleum ether. The benzene concrete yields 85 per cent of alcohol-soluble absolute, the latter containing 49 per cent of esters.

[30] Private communication from Mr. Charles Muller, Ambanja, Madagascar. Mr. Muller has been connected with the ylang ylang industry of Nossi-Bé and Madagascar for many years and is intimately acquainted with all technical problems pertaining to the preparation of ylang ylang products.

3. Steam distillation for 10 or 12 hr. of the flowers remaining after extraction with petroleum ether yields 1.2 per cent of a volatile oil (d_{32} 0.9025) with a weak odor.

4. Ylang ylang flowers remaining after extraction with petroleum ether were extracted with benzene. This procedure yielded 2 per cent of a product (d_{15} 0.930) with a sesquiterpene-like odor.

These experiments were carried out just prior to the outbreak of World War II; the products obtained could not be shipped to France for chemical study. However, from the results of the experiments, it can be assumed that ylang ylang flowers contain certain complex compounds insoluble in petroleum ether but soluble in benzene. In the course of the usual hydrodistillation of the flowers, these complex substances are decomposed, yielding sesquiterpenes and other compounds.

In the commercial preparation of concrete of ylang ylang with petroleum ether as solvent, two types of concrete are obtained:

(a) The straight concrete of ylang ylang to which nothing is added.

(b) The concrete of ylang ylang *to which is added* the volatile oil obtained by steam distillation of the residual flowers after extraction with petroleum ether. In the ordinary distillation of these residual flowers with live steam for a few hours, the yield of volatile oil (d_{15} 0.930) amounts to about 0.2 per cent. A concrete of ylang ylang to which this volatile oil has been added will have a specific gravity slightly below 1.0 (at 15°), and an ester content lower than that of the unmixed concrete. The end product, therefore, consists of about 80 per cent of straight concrete and 20 per cent of added volatile oil (from the residual flowers).

Use of Ylang Ylang Oil and Absolute

Oil of ylang ylang is one of the most important perfume raw materials. It must, however, be employed and dosed with great discretion. In the hands of an experienced perfumer the oil produces remarkable effects, imparting floral top notes to an otherwise dull and flat composition. Ylang ylang oil blends particularly well with jasmine, lilac, gardenia, lily of the valley, and similar scents. Some of the masterpieces of French perfumery owe their delightful tonalities in part to a skillful combination of bergamot, rose, ylang ylang, and vanilla. The employment of too much or of a poor grade of oil, however, may easily spoil a blend, producing a cloying or faded note, respectively.

In high-grade perfumes only "Extra" qualities of ylang ylang oil should be used. "First" and "Second" grades lend themselves for use in cosmetics;

"Third" qualities, backed up with synthetic aromatics, are well suited for the scenting of soaps.

Absolute of ylang ylang, almost a perfume in itself, possesses particularly warm, velvety, and lasting notes. Much more than the distilled oil, it exhibits spicy (eugenol) and sweet (vanilla) tonalities, blending into floral as well as oriental compositions.

SUGGESTED ADDITIONAL LITERATURE

H. C. Reed, "Cananga of the Pacific Islands," *Perfumery Essential Oil Record* **27** (1936), 211.

CHAPTER XII

ESSENTIAL OILS OF THE PLANT FAMILY *OLEACEAE*

CONCRETE AND ABSOLUTE OF JASMINE

Next to the rose, the jasmine supplies the most important and indispensable natural flower oil employed in modern perfumery. The fragrant white blossoms of the jasmine plant (probably native to the southern foothills of the Himalayas), since time immemorial have been used in India for ceremonial purposes and for the scenting of ointments. The plant was brought to North Africa and Spain, apparently by the conquering Moors, for the name jasmine is of Arabic origin (*Yasmin* or *Ysmyn*). In Southern France there are two varieties of *Jasminum* (fam. *Oleaceae*), viz., *J. officinale* L. (Type), which grows wild in the higher altitudes of the Maritime Alps, and *J. o.* var. *grandiflorum* L., the so-called Spanish jasmine, which is cultivated for the extraction of its natural flower oil. First introduced from Spain, the variety *grandiflorum* has been extensively grown in the Grasse region of Southern France for about two hundred years; in fact the cultivation of jasmine for perfumery has given rise to that unique and charming natural flower oil industry, for which the Grasse region has become world famous. The jasmine fields are located in the lowlands between the Maritime Alps and the Mediterranean Sea, or on the southern slopes of these mountains, which protect the flower fields from cold north winds. Jasmine plantations extend from Vence (Alpes Maritimes) in the east to Seillans (Var) in the west, centers of cultivation being Mouans-Sartoux and Mougins between Cannes and Grasse, and Pégomas in the lovely and fertile Siagne Valley.

Gradually increasing in the course of many years, total production of jasmine flowers in Southern France reached its maximum in 1927, with about 1,500 metric tons. Since then it has been declining steadily for a number of reasons, among them high cost and shortage of labor in France, and rising value of land, which can be used to better advantage by growing vegetables for the tourist trade on the Côte d'Azur. The demand for homesites in the area is growing, and many former jasmine fields have been divided up to provide room for small homes or fashionable villas. In addition, World War II brought about a serious shortage of fertilizers; cultivation of food crops became more important than that of flowers. Consequently, most of the older jasmine plantations that had passed the peak of productivity several years previously were discontinued and have not been renewed. In 1939, Southern France produced only about 700 metric tons of jasmine flowers; in 1947, the total acreage of jasmine in the Grasse region was about 200 hectares, yielding approximately 600 metric tons of flowers. In 1949 the output rose to 800 tons.[1] However, for the reasons explained above, jasmine production in the Grasse region will probably never regain the monopolistic position which it held years ago.

This course of events is not at all surprising; it was foreseen a long time ago by enterprising pioneers in the natural flower oil industry, who started planting jasmine in other Mediterranean countries, where conditions of climate, soil, and particularly of labor, are much more favorable than in the vicinity of the fashionable French Riviera.

In 1912, for example, the late Charles Garnier began experimenting with jasmine near Cairo in Egypt, which country has since produced up to 100 metric tons of flowers per year. Jasmine plantings were also laid out in Syria and Algeria. Sicily and Calabria (Italy), the most serious competitors of Grasse, now produce yearly approximately 500 metric tons of jasmine flowers, yielding about 1,500 kg. of concrete (55 per cent from Sicily, and 45 per cent from Calabria). Large plantations have also been started recently in Morocco (near Khémisset, on the road between Meknès and Rabat). Once they attain full production, and if no unforeseen event intervenes, these fields will supply flowers at prices with which the growers in Grasse cannot easily compete. However, as far as quality is concerned, that of the jasmine concretes and absolutes from the Grasse region is still supreme.

CULTIVATION, HARVEST AND FLOWER YIELD

(a) **Southern France.**—With the exception of soils that contain too much clay or gravel, any type of land may be used, provided it is irrigable, permeable, and well exposed to the sun. The summers must be warm, the winters mild. Prior to planting, the soil should be thoroughly cleaned of any old roots; if this is not done the jasmine plants will be attacked by fungi and root diseases, and the new planting may soon perish. In Southern France the starting of a jasmine plantation entails considerable expense, especially if the planting is laid out on the slope of a hill, since this will have to be terraced for irrigation. A jasmine field lasts from ten to fifteen years; on light, well-aired and well-drained soil the life-span may be much longer. After jasmine, crop rotation must be practiced, and years must pass before jasmine can again be grown on the same field.

In Southern France jasmine is planted by grafting *Jasminum grandiflorum*, the Spanish type of jasmine, on stocks of the wild growing *J. officinale*, which is a hardier plant, more resistant to frost and root diseases

---

[1] *Soap, Perfumery & Cosmetics* **23** (1950), 912.

than the former. *Jasminum officinale* stocks are planted in March, 10 cm. apart, in rows 80 cm. apart. Grafting with *J. grandiflorum* is done in the second or third year. Every year the bushes are trimmed to keep them low and to facilitate harvesting of the flowers. One hectare usually comprises 100,000 to 150,000 stocks. In the first year one hectare produces from 1,200 to 2,000 kg. of flowers, in the second and in the following years from 3,000 to 4,000 kg. As the plantation becomes older, stocks die out in increasing number.

According to Arnaud,[2] the jasmine plant is very sensitive to frost and must be hilled up every fall to prevent damage by frost. In addition, the plant is easily attacked by root rot; therefore, the soil should be plowed deeply, prior to planting, and any detritus must be carefully removed and burned. The rows should be spaced 1 m. to 1.5 m. apart, which means about 50,000 plants per hectare.

Depending upon the weather, the flowering season commences about July 20 and lasts until the beginning of November. During the first two weeks the yield of flowers is low; with normal weather it reaches its maximum between August 10 and October 20. The flowers collected in August are heavier (8,000 to 10,000 blossoms per kg.) than those gathered in October (14,000 blossoms per kg.). Moreover, the August and September flowers contain substantially more perfume than the October flowers. Obviously, weather conditions also play an important role, warm weather and ample sunshine producing a much larger flower crop, and flowers with more perfume, than cool or rainy weather.

The flowers have to be picked early in the morning, because they contain a maximum of perfume at daybreak. After ten o'clock the perfume seems to disappear. Flowers gathered at noon or in the afternoon yield much less flower oil than those collected very early in the day (cf. below). The flowers have to be picked very carefully. A skilled woman or girl can collect up to 3 kg. of flowers in 6 hr.

(b) Italy.—The jasmine industry of Italy has lately become a serious competitor to that of the Grasse region. Jasmine is now cultivated on a large scale in Calabria and Sicily. According to La Face,[3] plantings are scattered throughout the Province of Reggio (particularly near Reggio, Melito, Brancaleone, and Rocella Jonica); in Sicily jasmine fields are located in the Provinces of Messina and Syracuse (near Syracuse and Avola). Because of the mild winters in Sicily and Calabria it is not necessary to graft the jasmine as in Southern France, and *Jasminum grandiflorum* can be planted directly, by means of cuttings. These are prepared

---

[2] *Ind. parfum.* **4** (1949), 367.
[3] Private communication from Dr. F. La Face, Reggio Calabria.

in March and placed directly in the field. Or cuttings are grown in a nursery for a sufficiently long time to develop a strong root system. The young plants are then set out in the field during March of the following year. Prior to planting, the ground must be well prepared by hoeing to a depth of about 80 cm. For fertilization, organic fertilizers enriched with chemicals are commonly employed. In the second year after planting, the bushes have to be trimmed, by pruning the branches, which grow 30 to 40 cm. above ground. This must be repeated every year during the lifetime of a planting. Frequency of irrigation is determined by the nature of the soil; usually the fields are irrigated once every week, from June to October. Cultivation consists of one hoeing in March (after pruning), two or three hoeings in summer, and one hoeing generally at the end of the harvest.

The flower harvest commences toward the end of June, and continues to the end of October. According to La Face,[4] the flowers, on extraction with petroleum ether, give a normal yield and quality of concrete throughout this period—which is quite remarkable. The yield in flowers, from the third year on, averages 4,500 kg. per hectare; in many cases yields up to 5,000 and even 5,500 kg. have been reported.

The same author reported the following percentage yield of flowers in the various months of the harvest:

	Per Cent
July	25.60
August	28.58
September	23.33
October	16.33
November	6.16
	100.00

(c) **Egypt.**—In Egypt, where jasmine was first planted for perfumery purposes in 1912, the plantations are located near Cairo. As in Sicily and Calabria, no grafting is necessary, the mildness of the winters excluding any damage by frost. The Nile provides ample irrigation through a well organized system of canals and ditches. Propagation is effected by layering, which produces vigorous plants. The shrubs are planted 1 m. apart, in rows 2 m. apart, one hectare containing about 5,000 plants. The individual bushes are permitted to grow much larger than in Southern France. Nevertheless they are trimmed every year, so that the flowers stay within reach of the children who do the harvesting. Because of the heat prevailing during the months of harvest, the picked flowers must be processed immediately after arrival in the extraction plant.

[4] *Boll. ufficiale staz. sper. ind. essenze deriv. agrumi, Reggio Calabria,* **18** April/June (1948), 3.

(d) **Morocco.**—The first attempts to grow jasmine in Morocco for the extraction of its perfume on a modern, industrial scale were made in 1941 by an experienced grower and manufacturer of natural flower oils in the Grasse region of Southern France.[5] Numerous initial difficulties and obstacles had to be overcome, however, and much work had to be accomplished before—in 1950—the plantings began to produce.

In Southern France jasmine flowers are cultivated on small fields and patches, a large number of plants being set out per hectare, and most of the work being carried out by individual farmers—with the old-fashioned mattock.

In Morocco these time-honored practices have been discarded; the above-mentioned producer has followed the modern methods employed by the grape growers of Algeria, who are among the most efficient agriculturists in their field. The primary task was the choice of a suitable site, ideal from the point of view of soil, climate, labor, and irrigation. It was necessary to find a location where jasmine would not only grow well, but also yield a maximum of oil. The next problem was selection of plants giving a high yield of flowers with a strong and fine perfume; the carefully selected slips were planted in nurseries. With powerful tractors the soil was then dug to a considerable depth, and the plants set out as are grape vines in Algeria, i.e., with 2.5 m. between rows and 1 m. between plants. This arrangement permitted planting of only 4,000 cuttings per hectare. After two years, the plantations were fitted out (like vineyards producing the better quality of table grapes) with metal supports and three lines of metal wire. Under these conditions the plants may attain a height of 1.8 or even 2 m.

As regards cultural care, the methods used in the Algerian vineyards were employed on the new jasmine fields of Morocco—i.e., frequent dressings with large quantities of fertilizers and manure, almost weekly tillings, and two major cultivations per year. The work is done not by hand, but with caterpillar tractors adapted to the purpose (as for work in vineyards). During the flowering period irrigation is practiced every week. By careful plant selection it has been possible to escape diseases.

In Southern France the harvest begins at the end of July, and usually ends at the beginning of November, when the first frosts deprive the flowers of most of their perfume. In Morocco, on the other hand, the flowers can be collected as early as June, and as late as the middle of December. Up to 6,000 kg. of flowers may thus be harvested per hectare, while a field is at its height of development.

---

[5] The author is greatly obliged to his friend and close collaborator for many years, Mr. Pierre Chauvet, Seillans (Var), France, for much of the information contained in this section.

As in France, collection of the flowers begins at dawn. In Morocco, the work is done by hundreds of native boys and girls, who arrive at the plantation before daybreak and enliven the harvest with laughter and song; their remuneration means substantial pocket money for them, and a welcome side-income for their parents.

The collected flowers are taken to the centrally located extraction plant *every hour* and processed at once. This is a most important factor, which has contributed greatly to the high quality of the new Moroccan jasmine flower oil produced near Khémisset, on the highway between Rabat and Meknés. (In Southern France, jasmine cultivation is practiced on small fields scattered over a comparatively extended area; frequently, hours lapse between collection of the flowers and their arrival at the various extraction plants.) It might be expected that the hot climate of Morocco would result in products inferior in odor to those obtained in the cooler climate of Southern France. However, two factors have obviated any such development: the careful selection of the planting material, and the processing of the flowers within a very short time after picking.

METHODS OF EXTRACTION

Jasmine represents the classic example of those flowers which, even after having been detached from the plant, continue developing and emitting their natural perfume, until they fade and deteriorate. The chief problem of oil extraction thus resolves itself into that of capturing the full amount of perfume which the flower contains at the moment of picking, plus the quantity emitted after picking. The solution was reached many years ago by the introduction of the process of *enfleurage* (cf. Vol. I of the present work, pp. 189–196). Formerly, *enfleurage* was the only method by which the natural flower oil was isolated from jasmine flowers in the Grasse region. The procedure, however, is a most delicate one, requiring a great deal of hand labor, which today renders the final extracts very costly, despite the relatively high yield of natural flower oil. For this reason, *enfleurage* has now been largely abandoned, in favor of extraction with volatile solvents, particularly petroleum ether (cf. Vol. I of this work, p. 200). Today only about 15 per cent of the total jasmine harvest in the Grasse region is treated by *enfleurage*, 85 per cent being submitted to solvent extraction. The latter is a more modern process, yielding concrete and absolute flower oils, which represent the perfume of the living flower in a truly remarkable way.

As regards the yield of flower oil obtained by *enfleurage* on the one hand, and extraction with solvents on the other, it is obviously not permissible to base figures merely upon the concentrated extracts (absolute of *enfleurage*, and concrete or absolute of extraction) because these concentrates retain large amounts of inert fats and waxes, which are not part of the actual odoriferous principle, i.e., the volatile oil. The latter can be isolated from the concentrated extracts by distillation with superheated steam *in vacuo*, or by co-distillation with ethylene glycol *in vacuo* (cf. Vol. I of this work, pp. 213, 215). It is not produced commercially, but only in the laboratory, in order to determine the content of volatile oil in a concentrated floral extract, or to examine the purity of a concrete or absolute.

When comparing the yield of jasmine flower oil obtained by the process of *enfleurage* with that of one obtained by solvent extraction, it is necessary to use the *volatile* oils as bases of calculation. Numerous investigations along these lines have been carried out in the course of the last fifty years, particularly by Passy,[6] Hesse and his collaborators,[7] Erdmann,[8] Jeancard and Satie,[9] von Soden,[10] Viard,[11] Établissements Chiris,[12] and Naves.[13] As a result of these investigations it is now known that *enfleurage* yields from 2 to 2½ times as much perfume as extraction with solvents.

Several authorities have attributed the higher yield of natural flower oil obtained by *enfleurage* to the presence of glycosides in the plant, which by the action of diastases are split in the course of *enfleurage*, freeing the odorous aglycones. According to Naves,[14] however, any such theory has no basis in fact; and all attempts to explain the continued formation of perfume in the jasmine blossom, even after picking, have ended in complete failure.

Concrete and Absolute of Extraction

Yield.—According to Naves and Mazuyer,[15] 1,000 kg. of jasmine flowers on extraction with petroleum ether, yield from 2.8 to 3.4 kg., in exceptional cases from 2.4 to 3.8 kg. of concrete. Girard [16] noted yields varying between 0.28 and 0.30 per cent. These figures agree well with those observed by the present author during his activities in the flower oil industry of Southern France. La Face [17] recently reported that in Sicily and Calabria

[6] *Compt. rend.* **124** (1897), 783. *Bull. soc. chim.* [3], **17** (1897), 519.

[7] *Ber.* **32** (1899), 565, 765, 2611; **33** (1900), 1585; **34** (1901), 293, 2916, 2928; **36** (1903), 1465; **37** (1904), 1457. *Chemistry Industry* **25** (1902), 1.

[8] *Ber.* **34** (1901), 2281.

[9] *Bull. soc. chim.* [3], **23** (1900), 555.

[10] *J. prakt. Chem.* [2], **69** (1904), 267.

[11] *Parfums France* **4** (1926), 35.

[12] *Ibid.* **6** (1928), 229.

[13] *Ibid.* **8** (1930), 296. Cf. Naves and Mazuyer, "Les Parfums Naturels," Paris (1939), 130, 223, 229.

[14] *Perfumery Essential Oil Record* **39** (1948), 214. Cf. Langlais and Bollinger, *Ind. parfum.* **2** (1947), 78, 123.

[15] "Les Parfums Naturels," Paris (1939), 225.

[16] *Ind. parfum.* **2** (1947), 184.

[17] Private communication from Dr. F. La Face, Reggio Calabria, November (1949).

the yield of concrete is seldom less than 0.3 per cent; during the most vigorous period of flower development it may be as high as 0.33 per cent.

In general, the yield of concrete depends upon several factors, among them the weather, the month of the harvest, and the hours at which the flowers have been picked. According to Rovesti,[18] the yield reaches its maximum in the second half of August and at the beginning of September. Flowers picked from 6 to 10 A.M. yield twice as much flower oil as those collected in the afternoon, or one and one-half as much as flowers gathered at daybreak (the perfume of the latter, however, is finer).

As regards the months of harvest, La Face [19] observed the following yields of concrete:

	Per Cent
July	0.3273
August	0.3322
September	0.3200
October	0.3180
November	0.3000

On treatment with high-proof alcohol in the usual way, concrete of jasmine from the Grasse region yields from 45 to 53 per cent of alcohol-soluble absolute (Naves and Mazuyer). Concretes prepared at the beginning of the flowering season give 53 to 55 per cent, while those prepared at the end of the season yield only 42 to 44 per cent of absolute (Girard). According to La Face,[20] concretes from Sicily and Calabria usually give from 50 to 54 per cent of absolute, but those prepared from October flowers only 45 to 48 per cent. Experiments carried out by La Face [21] showed the following yields of absolutes from concretes that had been obtained during the various months of the flower harvest:

	Per Cent
July	52.30
August	54.80
September	53.60
October	48.10
November	30.66

It should be mentioned here that all the properties and yields reported above refer to concretes and absolutes derived by extraction of the flowers with petroleum ether. Some manufacturers occasionally use benzene for this purpose, but the concretes as well as absolutes obtained by extraction

---

[18] *Rivista ital. essenze profumi* **10** (1928), 185.

[19] *Boll. ufficiale staz. sper. ind. essenze deriv. agrumi, Reggio Calabria,* **18** April/June (1948), 3.

[20] Private communication of Dr. F. La Face, Reggio Calabria, November (1949).

[21] *Boll. ufficiale staz. sper. ind. essenze deriv. agrumi, Reggio Calabria,* **18** April/June (1948), 3.

of the jasmine flowers with benzene are very dark colored, solid or viscous masses, with a "cooked" off-odor, reminiscent of marmalades. They have not been well received in the perfume industry and will, therefore, not be discussed here.

Physicochemical Properties.

(a) *Concrete.*—Concrete of jasmine is a reddish-brown, waxy mass, only partly soluble in 95 per cent alcohol, with a characteristic jasmine flower odor.

According to Naves and Mazuyer [22] (I) and Girard [23] (II), concretes of jasmine from the Grasse region have properties varying within these limits:

	I	II
Congealing Point	47° to 51°	...
Melting Point	49° to 52°	47° to 52°
Specific Gravity at 60°/60°	...	0.886 to 0.8987
Specific Optical Rotation (in 10% alcoholic solution)	About +5° to +12°	...
Refractive Index	...	1.4640 to 1.4658
Acid Number	9.8 to 12.6	12.6 to 15.4
Ester Number	68 to 105	...

La Face [24] reported on the properties of concretes from Sicily and Calabria obtained in the various months of the flowering season:

	Melting Point	Acid Number	Saponification Number
July	51°	11.65	111.18
August	50.2°	11.12	112.60
September	50°	9.80	118.14
October	51.5°	11.98	108.10
November	53°	8.91	77.36

The same author [25] also determined the properties of concretes prepared from flowers that had been harvested during various hours of the day.

Hours of Harvest	Properties of the Concrete				Yield of Absolute (%)	Volatile Oil Content (%)
	Odor	Melting Point	Acid No.	Saponif. No.		
5 to 7 A.M.	Flowery	49.8°	13.35	146.56	54	15.64
7 to 9 A.M.	Flowery	50.5°	12.43	122.97	53	10.66
5 to 7 P.M.	Herbaceous	51°	15.93	116.15	51.3	9.84
7 to 9 P.M.	Mediocre	50°	11.85	116.45	53	13.90

[22] "Les Parfums Naturels," Paris (1939), 227.
[23] *Ind. parfum.* 2 (1947), 184.
[24] *Boll. ufficiale staz. sper. ind. essenze deriv. agrumi, Reggio Calabria,* 18 April/June (1948), 9.
[25] *Ibid.*

(b) *Absolute.*—Freshly prepared absolute of jasmine is a viscous, clear, yellow-brown liquid, soluble in 95 per cent alcohol, and possessing a beautiful odor, characteristic of the live flowers. On aging, the absolute darkens and usually becomes turbid, gradually separating a grayish deposit; the odor assumes a deeper tonality and the color changes to red.

Girard [26] (I) and Naves and Mazuyer [27] (II) reported these properties for absolute of jasmine:

	I	II
Specific Gravity	d_4^{20} 0.9290 to 0.9550	d_{15} 0.962 (minimum)
Optical Rotation	$+2°$ 23' to $+4°$ 95'	...
Refractive Index at 20°	1.4822 to 1.4935	1.4860 to 1.4920
Acid Number	4.2 to 17.2	25 to 30 (average)
Ester Number	96.4 to 147.6	124 to 194
Benzyl Benzoate Content	4.5 to 6.5%	...
Indole Content	0.08 to 0.20%	...
Methyl Anthranilate Content	0.15 to 0.35%	...

Rovesti [28] examined absolutes of jasmine from Liguria (Italy) and noted these properties:

Specific Gravity at 15°	0.931 to 0.970
Optical Rotation	$+0°$ 30' to $+2°$ 48'
Refractive Index	1.4807 to 1.525
Ester Content, Calculated as Benzyl Acetate	41% to 65%
Indole Content	0.3% to 0.7%
Methyl Anthranilate Content	0.2% to 0.8%

La Face [29] determined the properties of absolutes prepared in Sicily and Calabria from flowers harvested in the various months of the season:

	Specific Gravity	Acid Number	Ester Content, Calculated as Benzyl Acetate (%)
July	0.9637	14.60	38.29
August	0.9679	21.66	39.58
September	0.9654	17.36	38.85
October	0.9425	15.81	30.87
November	0.9170	16.81	18.09

The same author also studied the properties of absolutes from flowers that had been picked during different hours of the day.

[26] *Ind. parfum.* **2** (1947), 184.
[27] "Les Parfums Naturels," Paris (1939), 227.
[28] *Rivista ital. essenze profumi* **10** (1928), 185.
[29] *Boll. ufficiale staz. sper. ind. essenze deriv. agrumi, Reggio Calabria,* **18** April/June (1948), 7.

Hour of Harvest	Specific Gravity	Optical Rotation	Acid No.	Ester Content, Calculated as Benzyl Acetate (%)	Ester Content, Calculated as Benzyl Acetate of the Volatile Fraction (%)
				Properties of the Absolute	
5 to 7 A.M.	0.9758	+0° 18′	14.26	40.85	69.75
7 to 9 A.M.	0.9651	+2° 18′	15.11	37.26	59.00
5 to 7 P.M.	0.9542	+3° 7′	15.60	31.88	56.20
7 to 9 P.M.	0.9542	+2° 0′	15.10	31.94	57.50

Naves and Grampoloff [30] determined the properties of 18 lots of Sicilian and Calabrian jasmine absolutes, which they used in their research on the chemical composition of these absolutes (see below). Column (I) lists the wider limits, column (II) the more usual values:

	I	II
Specific Gravity at 20°/4°...	0.9290 to 0.9550	0.935 to 0.948
Optical Rotation..........	+2° 14′ to +4° 57′	+3° 6′ to +3° 54′
Refractive Index at 20°.....	1.4822 to 1.4935	1.4850 to 1.4912
Acid Number.............	4.2 to 17.2	11 to 15
Ester Number............	96.4 to 147.6	115 to 142
Indole Content...........	0.08 to 0.20%	...
Methyl Anthranilate Content	0.15 to 0.35%	...

Naves and Grampoloff reported that in the *concretes*, the indole content varied from 0.06 to 0.14 per cent; it was highest at the beginning of the flowering season.

(c) *Volatile Oil.*—On direct steam or water distillation the jasmine flowers give such a low yield of oil that this process cannot be employed industrially.

As has already been mentioned, the volatile oils of jasmine referred to in literature are derived by distillation of the concretes or absolutes and serve chiefly for assaying the quality of a concrete or absolute. For this reason they will be described here.

Von Soden [31] examined the distillates from jasmine concrete prepared industrially in Grasse during the first half of the flower harvest (I) and during the second half of the harvest (II):

	I	II
Yield of oil from 100 kg. of flowers....	0.0777%	0.0718%
Specific Gravity at 15°.............	0.9955	0.967
Optical Rotation.................	−1° 0′	Illegible
Acid Number....................	2.5	3.5
Ester Number...................	190.0	161.5

[30] *Helv. Chim. Acta* **25** (1942), 1500.
[31] *J. prakt. Chem.* [2], **69** (1904), 268.

Walbaum and Rosenthal [32] described the properties of steam-volatile oils derived from three concretes of jasmine:

	I	II	III
Yield in per cent	15.4	14.0	13.5
Acid Number	2.7	1.4	2.4
Ester Number	196.4	208.0	213.5

Naves, Sabetay and Palfray [33] reported the following properties of steam-volatile oils obtained from concretes of French origin (Grasse). The yield of volatile oil varied from 10 to 19 per cent.

	I	II	III	IV	V
Specific Gravity at 15°	0.996	0.989	1.004	0.988	0.998
Optical Rotation	+3° 30'	+2° 34'	+3° 16'	−0° 40'	+0° 12'
Refractive Index at 20°	1.4971	1.4982	1.4956	1.4949	1.4938
Acid Number	2.24	2.80	3.10	3.50	2.70
Ester Number	193.0	204.2	216.1	194.1	224.2

Girard [34] noted properties of volatile oils derived from absolutes by distillation with superheated steam at 88°–92° and 35–40 mm. pr.:

Specific Gravity at 20°/4°	0.966 to 1.0106
Refractive Index at 20°	1.4920 to 1.5041
Optical Rotation	−2° 64' to +3° 18'
Acid Number	0.1 to 6.7
Ester Number	165 to 227
Indole Content	0.10 to 0.31%
Methyl Anthranilate Content	0.22 to 0.40%

Studying the properties of volatile oils derived by co-distillation with ethylene glycol at reduced pressure from Sicilian and Calabrian concretes and absolutes that had been obtained at various months of the harvest, La Face [35] arrived at these results:

	From Concrete		From Absolute	
		Ester Content, Calculated		Ester Content, Calculated
	Distillate	as Benzyl Acetate of the	Distillate	as Benzyl Acetate of the
Month	(%)	Distillate (%)	(%)	Distillate (%)
July	13.14	59.30	27.47	58.75
August	14.45	57.80	28.87	58.40
September	14.47	59.21	31.00	58.77
October	10.90	53.0	25.00	50.80
November	5.24	39.0	18.09	40.80

[32] *Ber. Schimmel & Co., Jubiläums-Ausgabe* (1929), 200.
[33] *Perfumery Essential Oil Record* **28** (1937), 336.
[34] *Ind. parfum.* **2** (1947), 184.
[35] *Boll. ufficiale staz. sper. ind. essenze deriv. agrumi, Reggio Calabria,* **18** April/June (1948), 3.

Distilling Sicilian and Calabrian concretes of jasmine with superheated steam at 88°–92° and 35–40 mm. pr., Naves and Grampoloff [36] obtained from 10.6 to 23.5 per cent (in most cases from 16 to 21 per cent) of volatile oil, which they used for their research on the chemical composition of jasmine flower oil (see below). Column (I) lists the wider limits, column (II) the more usual values:

	I	II
Specific Gravity at 20°/4°	0.966 to 1.0106	0.968 to 0.986
Optical Rotation	−2° 38′ to +3° 11′	Usually dextro-rotatory
Refractive Index at 20°	1.4920 to 1.5041	1.4950 to 1.5012
Acid Number	0.1 to 6.7	2.8 to 4.4
Ester Number	165 to 227	184 to 210
Indole Content	0.10 to 0.31%	...
Methyl Anthranilate Content	0.22 to 0.40%	...

The distillates from concretes obtained during the first part of the harvest exhibited the highest specific gravity, refractive index, and ester content, and were generally dextrorotatory.

Conclusions.—The production of high-grade concretes and absolutes of jasmine is by no means an easy matter. It requires a great deal of experience and many factors must be considered. To summarize, the quality of the finished flower oils depends upon the following:

1. Geographical location. Flowers from higher altitudes seem to contain a finer perfume than those from the lower altitudes and plains.

2. Weather conditions. During warm and sunny weather the flowers give a better yield and quality of oil than during cloudy and particularly rainy weather.

3. The season. During the height of the season (second half of August to middle or end of October) yield and quality of the flower oil is higher than at the beginning and at the end of the season.

4. Time of harvest. Flowers picked in the morning up to 10 A.M. are of far better quality (yield and perfume) than those collected at noon or in the afternoon. This is particularly important in hot countries—e.g., Egypt, Algeria and Morocco.

5. Speed of flower transport. No time should be lost in transporting the harvested flowers to the extraction plant, where they must be worked up immediately after arrival. This can only be done if the extractors have a sufficiently large capacity to process all the flowers right after delivery, even during the height of the season. Many omissions are committed in this respect, and often the flowers are kept in heaps on the floor of the

[36] *Helv. Chim. Acta* **25** (1942), 1500.

extraction plant for long hours. During this waiting period they exhale a great deal of their perfume (see above), which is then simply lost. In addition, they start to fade and wither, yielding then flower oils of poor quality and low in indole content (see below). In the hot climate of North Africa, it is most important to speed up delivery of the flower material and to process them without any loss of time. The flat note, which characterizes some of the concretes and absolutes from North Africa, is perhaps a result of neglect in this respect. Concretes and absolutes from cooler sections of the Grasse region in Southern France are renowned for their richer and usually much more flowery perfume.

6. Process of extraction. That utmost care must be exercised in the actual extraction process is self-understood. Temperatures during the concentration of the extracts should be kept as low as possible. The solvent must be of highest purity and, on evaporation, leave not the slightest trace of a kerosene odor (cf. Vol. I of this work, pp. 200 ff.).

Absolute of Enfleurage

The process of *enfleurage* was developed in Southern France during the eighteenth century and has never been employed outside of the Grasse region on a commercial scale. As was mentioned above, today only about 15 per cent of jasmine flowers harvested in the vicinity of Grasse are processed by *enfleurage,* the balance being extracted with volatile solvents. This decline has resulted from the gradual increase in the cost of labor—*enfleurage* requires a great deal of skilled hand labor—and from the shortage of fat, which was most acute during, and for a time after, World War.II. In the not too far distant future, *enfleurage* will probably be abandoned altogether, as was the case some years ago with *maceration* of roses and orange blossoms (for details see Vol. I of this work, pp. 189, 198).

Absolute of jasmine *enfleurage* is a viscous oil of dark reddish-brown color, and soluble in 95 per cent alcohol. On aging, the color changes to a dark red, the absolute often separates a deposit, and loses its original solubility in 95 per cent alcohol. The odor is reminiscent of the live flowers but exhibits a decided fatty off-note, the latter becoming more pronounced with age. When evaporated on testing paper, the absolute displays a slight odor of vanillin, caused by the presence of gum benzoin, traces of which are added to the fatty *corps* for protection against rancidity.

No purpose would be served in reporting physicochemical properties of absolute of *enfleurage*, because they depend entirely upon the degree of concentration of the alcoholic washings (*extraits de pommade*), and upon the purification of the absolute thus obtained (freezing and removal of fat, etc.). Every manufacturer uses his own methods in this respect and no universal standards for physicochemical properties have been established. More uniform are the volatile oils derived in the laboratory by steam distillation, at reduced pressure, of absolute of *enfleurage,* or by co-distillation with ethylene glycol at reduced pressure (see above).

Naves, Sabetay and Palfray [37] obtained from 15.5 to 26.1 per cent of volatile oils, by submitting absolute of *enfleurage* to distillation with superheated steam at reduced pressure. Seven steam volatile oils exhibited properties varying within these limits:

Specific Gravity at 15°.......	0.993 to 1.047
Optical Rotation............	+2° 10' to +3° 40'
Refractive Index at 20°......	1.4944 to 1.5015
Acid Number..............	2.2 to 7.5
Ester Number..............	234.0 to 268.8

More than fifty years ago, Hesse and Müller [38] steam-distilled the odoriferous extracts of a jasmine *pommade* and obtained volatile oils containing from 69.1 to 73 per cent of esters (calculated as benzyl acetate), 2.5 per cent of indole, and 0.5 per cent of methyl anthranilate. The yield of steam-volatile oil, calculated upon the weight of the flowers, was 0.17 per cent.

Absolute of Chassis

After 24 to 48 hr. in contact with the *corps,* the flowers are removed from the *chassis (défleurage).* However, they still contain substantial quantities of perfume that have not been absorbed by the fat. On extraction with petroleum ether, these partly exhausted flowers yield a second-grade flower oil, known in the trade as absolute of *chassis* (cf. Vol. I of this work, p. 197). It is a viscous, light brown liquid, soluble in 95 per cent alcohol, exhibiting a quite strong and lasting odor which, however, differs from that of the live flowers. This may partly be the result of the low indole content of the absolute. On aging, a fatty note becomes noticeable. The quantities of absolute of *chassis* offered every year on the market are small and will be further reduced as the process of *enfleurage* is less and less employed.

As in the case of absolute of *enfleurage,* little purpose is served in reporting physicochemical properties for absolute of *chassis.* However, those of the volatile oil obtained in the laboratory by steam distillation of the latter offer more interest. Naves and Mazuyer [39] examined such a volatile oil and found:

[37] *Perfumery Essential Oil Record* **28** (1937), 337.
[38] *Ber.* **32** (1899), 565, 768.
[39] "Les Parfums Naturels," Paris (1939), 231.

Specific Gravity at 15°............ 0.962
Optical Rotation................. +2° 42′
Refractive Index at 20°........... 1.4902
Acid Number.................... 4.9
Ester Content, Calculated as Benzyl
  Acetate...................... 47.8%

CHEMICAL COMPOSITION OF JASMINE FLOWER OIL

According to Naves,[40] the chemical composition of the jasmine flower oil obtained by solvent extraction closely resembles that obtained by *enfleurage,* provided that in the process of recovery *all* the odorous substances formed by the flowers have been collected. It is, therefore, permissible to discuss the chemistry of the extraction oil along with that of the *enfleurage* oils.

The first investigation of the oil was carried out by Verley,[41] who isolated a substance of the empirical molecular formula $C_9H_{10}O_2$; this he thought to be phenylglycol methylene acetal and named it "Jasmal." Shortly afterward, however, Hesse and Müller [42] proved that the substance $C_9H_{10}O_2$ was not phenylglycol methylene acetal, but benzyl acetate, the chief constituent of jasmine flower oil (benzyl acetate and phenylglycol methylene acetal have the same empirical molecular formula $C_9H_{10}O_2$). The early work of Hesse and Müller, and later investigations by Hesse [43] alone, demonstrated that the jasmine flower oil obtained by *enfleurage* is composed approximately of the following:

	Per Cent
Benzyl Acetate	65.0
d-Linaloöl	15.5
Linalyl Acetate	7.5
Benzyl Alcohol	6.0
Jasmone	3.0
Indole	2.5
Methyl Anthranilate	0.5
Phenols and Bases with a Narcotic Odor	Traces

Since the publication of Hesse's fundamental research, our knowledge of the chemical composition of jasmine flower oil has been largely increased, owing to the efforts of several other investigators—among them Erdmann,[44]

[40] *Perfumery Essential Oil Record* **39** (1948), 214.
[41] *Compt. rend.* **128** (1899), 314. *Bull. soc. chim.* [3], **21** (1899), **226.**
[42] *Ber.* **32** (1899), 565, 765.
[43] *Ber.* **32** (1899), 2611; **33** (1900), 1585; **34** (1901), 291, 2916, 2929; **37** (1904), 1457. *Z. angew. Chem.* **25** (1912), 363.
[44] *Ber.* **34** (1901), 2282; **35** (1902), 27.

(*Top*) Flowers of *Jasminum officinale* var. *grandiflorum*. (*Bottom*) Jasmine flower harvest in the Grasse region of Southern France. *Photos Fritzsche Brothers, Inc., New York.*

(*Top*) Jasmine flower harvest in the Grasse
gion of Southern France. *Photo Fritz*
Brothers, Inc., New York. (*Left*) An exter
jasmine plantation in Morocco. Note the v
across the field which will serve to support
bushes as they grow higher. *Photo Mr. P*
Chauvet, Seillans (Var), France. (*Bottom*)
tive children bringing the morning's harves
jasmine flowers to the centrally located ext
tion plant near Khémisset (Morocco). *P*
Mr. Pierre Chauvet, Seillans (Var), Fran

von Soden,[45] Elze,[46] Louveau,[47] Sabetay and Trabaud,[48] and particularly Naves and Grampoloff.[49] The last two authors examined volatile jasmine oils derived by steam distillation of (petroleum ether) concretes and absolutes from Sicily and Calabria, as well as from the Grasse region. They were able to confirm the presence, in their oils, of the constituents previously reported by the earlier workers. In addition, they identified a number of compounds never before observed in jasmine oil.

The literature of the last fifty years indicates that jasmine flower oil contains:

Benzyl Acetate. The chief constituent, identified more than fifty years ago by Hesse and Müller (see above).

Linalyl Acetate. Reported by the same authors.

Benzyl Benzoate. In 1942, Naves and Grampoloff [50] found that Italian, as well as French jasmine absolutes, contain from 4.5 to 6.5 per cent of benzyl benzoate, which they characterized by preparation of the *p*-nitrobenzyl *m*-nitrobenzoate m. 143°–143.5°.

Benzyl Alcohol. Identified years ago by oxidation to benzaldehyde and benzoic acid m. 121.5° (Hesse and Müller).

Geraniol. Elze [51] isolated geraniol by means of its acid phthalate, and its calcium chloride addition compound.

Nerol. Reported by the same author. Confirmed by Naves and Grampoloff, who prepared the neryl allophanate m. 84°–84.5°, and the geranyl allophanate m. 124°–124.5°.

l-α-Terpineol. First observed in jasmine oil by Elze, and later confirmed by Naves and Grampoloff (allophanate m. 133°–134°).

d- and *dl*-Linalool. First reported as *d*-linalool by Hesse and Müller. Naves and Grampoloff found that the alcohol is present in jasmine flower oil as a mixture of *d*- and *dl*-linalool (α_D +12° 14′).

An Alcohol(?). Among the alcohols which reacted with phthalic anhydride, Naves and Grampoloff noted one with an odor reminiscent of β,γ-hexenol ("Leaf Alcohol"— cf. Vol. I of this work, p. 158).

Farnesol. First isolated by Elze [52] from the fraction b_5 145–155°; presence confirmed by Naves and Grampoloff, who oxidized the alcohol to farnesal (semicarbazone m. 133°).

[45] *J. prakt. Chem.* [2], **69** (1904), 256, 267. *Ber.* **37** (1904), 1458.
[46] *Chem. Ztg.* **34** (1910), 912; **50** (1926), 782. *Riechstoff Ind.* (1926), 181.
[47] *Rev. marques parfumerie* **10** (1932), 482; *ibid.*, Special Number "Le Jasmin," Paris (1936).
[48] *Compt. rend.* **208** (1939), 1242.
[49] *Helv. Chim. Acta* **25** (1942), 1500.
[50] *Ibid.*
[51] *Chem. Ztg.* **34** (1910), 912.
[52] *Ibid.* **50** (1926), 782.

Nerolidol. Also reported in jasmine oil by the last two authors.

An Alcohol $C_{18}H_{34}O$. According to Naves and Grampoloff, absolute of jasmine contains from 26 to 37 per cent of an alcohol $C_{18}H_{34}O$, or a mixture of isomers of the same empirical molecular formula. This alcohol is sparingly volatile and plays an important role in the fixation and cohesion of the odorous principles of jasmine absolute.

Eugenol. Sabetay and Trabaud [53] reported that jasmine absolute contains 0.142 per cent of eugenol; Naves and Grampoloff found 0.25 per cent. The last two authors identified eugenol by means of its 2,4-dinitrophenyl ether m. 115°–115.5°.

p-Cresol. First reported as a constituent of jasmine oil by Elze; [54] later confirmed by Naves and Grampoloff, who prepared its 2,4-dinitrophenyl ether m. 93°.

Creosol. First observed in the fraction $b_{1.5}$ 96°–100° by Naves and Grampoloff; 2,4-dinitrophenyl ether m. 119°–120°.

Lactones(?). In the phenolic fractions, these authors also noted the presence of lactones with a powerful, lasting, and fruity odor. On treatment of the alkaline solutions with carbon dioxide, the lactones are regenerated from the salts of their hydroxy acids.

Benzaldehyde. Identified by Naves and Grampoloff through the phenylhydrazone.

Jasmone. First reported by Hesse (see above) in the jasmine oil obtained by *enfleurage*. Naves and Grampoloff observed only small quantities of jasmone in the concretes and absolutes of extraction, and arrived at the conclusion that this ketone is contained chiefly in the jasmine *enfleurage* products. They isolated jasmone by means of Girard and Sandulesco's reagent P, and identified it through the semicarbazone m. 207.5°–208°, and the 2,4-dinitrophenylhydrazone m. 121°–122.5°.

A Ketone $C_{12}H_{16}O_3$. Using reagent P of Girard and Sandulesco, Naves and Grampoloff also obtained a ketone, or ketolactone of the empirical molecular formula $C_{12}H_{16}O_3$. It had an herbaceous odor, and gave a 2,4-dinitrophenylhydrazone m. 166°–166.5°. The ketone or ketolactone was probably isomeric with the calythrone of Penfold and Simonsen.[55]

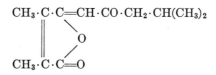

Benzoic Acid. In free form. Identified through its melting point 121.5°–122° (Naves and Grampoloff).

Methyl Anthranilate. Reported first by Hesse (see above) as a constituent of jasmine *enfleurage* oil. Naves and Grampoloff found only traces of methyl anthranilate in the concretes and absolutes of extraction which they investigated.

[53] *Compt. rend.* **208** (1939), 1242.   [55] *J. Chem. Soc.* (1940), 412.
[54] *Chem. Ztg.* **34** (1910), 912.

Indole. This important constituent was also first observed by Hesse in *enfleurage* oils, and later in the products of extraction with petroleum ether.[56] According to Naves and Grampoloff, the content of indole in jasmine extraction products should always be determined from the concrete, not the absolute. In the preparation of absolutes from concretes, the ethyl alcohol which has to be distilled off, carries over with it the greater part of the indole. Moreover, polyindoles are formed. Their presence is annoying in the study of jasmine distillates, because the indole regenerated by depolymerization of the polyindoles contaminates the greater part of the fractions distilled over.

The question of the occurrence of indole in jasmine flowers was the subject of much controversy among early workers in the field, until Cerighelli [57] by systematic experiments arrived at the following conclusion:

Indole is a natural constituent of the flowers of Spanish jasmine (*Jasminum officinale* L. var. *grandiflorum*). It appears to be present in the buds in the form of a complex compound, not yet isolated. As soon as the buds open in the morning, the indole is freed and evaporates into the atmosphere, at a rate of 0.6 to 0.8 mg. per 100 g. of flowers per hour. In the morning, 100 g. of flowers contain an average of 5 mg. of indole. Toward evening, when the flowers close, the indole disappears in the tissues of the flowers and reappears the following morning with daybreak.

Flowers picked under the usual conditions contain, per 100 g., from 3.5 to 5.0 mg. of indole at the moment of treatment (extraction); this remains in the air within the heap of flowers piled up on the floor of the extraction plant, or within the extractor, and then passes into the concrete. If the flowers are exposed freely to the atmosphere, the indole evaporates, and only 0.9 to 1.5 mg. remain in 100 g. of flowers. In a confined atmosphere 100 g. of flowers emanate from 14.5 to 19.0 mg. of indole in 24 hr., which fact explains the relatively high indole content of the *enfleurage* products.

USE OF JASMINE FLOWER OIL

Jasmine flower oil is indispensable in high-grade perfumes. In a crude way, it may be compared with butter in fine cooking. There is hardly a good perfume, of floral as well as of oriental character, which does not contain at least a small percentage of jasmine flower oil. It is, therefore, not surprising that during the years of World War II, when it was very scarce in the Western Hemisphere, American perfumers were willing to pay up to $2,000 per lb. of absolute of extraction.

As regards the use of the various types of jasmine flower oil, the absolute of extraction, although expensive, gives perhaps the best results. Some

[56] Hesse, *Ber.* **37** (1904), 1458. Cf. von Soden, *J. prakt. Chem.* [2], **69** (1904), 268.
[57] *Compt. rend.* **179** (1924), 1193.

perfumers in France prefer to purchase the concretes and from them prepare alcoholic washings which are then employed as such in handkerchief
perfumes.

The absolute of *enfleurage* has the advantage of a lower price. It is
usually a dark colored product, and has to be handled with care. Here
again, some of the older perfume houses in Europe, particularly in France,
give preference to the *extraits* (alcoholic washings) of the *pommades* and,
although this entails a great deal of labor and requires special machinery,
they prefer employing these alcoholic *extraits* in old-fashioned perfume
formulas.

Absolute of jasmine blends practically with any floral scent, lending
smoothness and elegance to perfume compositions. The same can be said
of the heavier compositions of the oriental type.

SUGGESTED ADDITIONAL LITERATURE

Georges Igolen, "L'Extraction des Parfums Naturels par les Solvants Volatils,"
Chimie & industrie **61** (1949), 466.

CONCRETE AND ABSOLUTE OF LILAC

The common lilac, *Syringa vulgaris* L. (fam. *Oleaceae*), a native of Iran
(Persia), was introduced to Europe three hundred to four hundred years
ago. It has been cultivated widely in many parts of Europe, and now
grows also semiwild. In the course of years, other species of *Syringa* have
been imported from the Orient to Europe; from the point of view of odor,
these are perhaps more interesting than *Syringa vulgaris*. Because of the
popularity of the delightful fragrance of the lilac, many efforts have been
made to capture its scent by various methods, but without success. In 1938
Igolen [1] extracted flowers of *Syringa vulgaris* with peroleum ether and obtained 0.24 to 0.36 per cent of a solid concrete with a dark green color.
Extraction with benzene yielded 0.6 per cent of a hard concrete with a
black-green color. The odor of the two concretes was quite disagreeable,
far remote from that of the living flowers.

On treatment with alcohol in the usual way the concrete obtained by
extraction with petroleum ether gave 38 per cent of an alcohol-soluble vis-

[1] *Parfums France* **16** (1938), 117.

cous absolute with a green color. The absolute contained 8.72 per cent of a steam-volatile oil with these properties:

>
> Specific Gravity at 15°....... 0.9594
> Optical Rotation at 20°...... −4° 20′
> Refractive Index at 20°...... 1.4876
> Acid Number............... 16.8
> Ester Number............... 59.92

The oil did not contain any indole.

Extracting flowers of *Syringa vulgaris* L. with petroleum ether Sabetay [2] obtained 0.377 per cent of a concrete with the following properties: m. 44°, acid number 15.8, ester number 94.7. The product had a "green" odor, reminiscent also of sweet basil.

Almost forty years ago, Kerschbaum [3] investigated the chemical composition of lilac flower oil and reported the presence of farnesol, without however giving any proof.

The natural flower oil of lilac is not produced commercially,[4] as none of the extraction methods, including steam distillation, yields products representing the odor as it is contained in the flowers. On the other hand, numerous synthetic compounds have been developed which reproduce the delightful scent of lilac quite remarkably.

[2] *Ind. parfum.* **5** (1950), 86.

[3] *Ber.* **46** (1913), 1733.

[4] Recently Sabetay [*Ind. parfum.* **5** (1950), 86] reported that a house in Grasse succeeded in developing a satisfactory concrete. According to Meunier, (*ibid.* 28) the solvent used for this purpose is butane.

CHAPTER XIII

ESSENTIAL OILS OF THE PLANT FAMILY *AMARYLLIDACEAE*

CONCRETE AND ABSOLUTE OF TUBEROSE

Introduction.—*Polyanthes tuberosa* L. (fam. *Amaryllidaceae*), a native probably of Central America or Mexico, is cultivated in the Grasse region of Southern France, and lately also in Morocco, for the extraction of its natural flower oil, which forms an important adjunct in the formulation of high-grade perfumes. For many years tuberose flower oil has been one of the most valuable and expensive of the perfumer's raw materials. Plantations are located near Pégomas, Auribeau, and Mandelieu, in the fertile and lovely valley of the Siagne River, not far from Cannes and Grasse (A.M., France). Formerly, up to 75 metric tons of flowers were harvested every year and processed by the flower oil manufacturers of the Grasse region; recently, however, the quantity has declined substantially. It was about 30 tons in 1926, only 17 tons in 1927, and 15 metric tons in 1950.[1] On the other hand, newly established plantings in Morocco (near Khémisset, between Rabat and Meknès) are being expanded. Cultivation and harvest of tuberose require a great deal of labor, which has become quite high-priced in France, whereas it is still cheap in Morocco. The cost of land in southern France has also risen, since the ground is increasingly required for the cultivation of vegetables and other food products.

Being very sensitive to winterkill the plants cannot remain in the field during the cold months. Therefore, the bulbs must be dug out every November, stored over the winter in an airy and dry place, and replanted in April. (The suckers, detached from the principal bulb and placed in a nursery, develop into plants, which bear flowers only in the third year. For this reason, the quantity of flowers cannot be increased rapidly, if sudden demand should require an increase in production.) The plant thrives in a porous alluvial soil, but in rich soil lasts seldom more than two or three years. In practice the plantings are renewed every year, with a change in location. Only rarely is a field kept in tuberose longer than one or two years.

According to Mazuyer,[2] the horticultural variety of tuberose exploited in Southern France for the extraction of its perfume is that with single flowers; the double flowered variety usually goes to the cut flower trade. The flowers on top of the long stalk are grouped in spike-shaped clusters,

---

[1] *Soap, Perfumery & Cosmetics* **23** (1950), 914.
[2] "Production et Culture de Tubéreuse," *J. parfum. savon.* **21** (1908), 195.

15 to 20 cm. long. The flowering period begins in July, reaching its maximum toward the middle of August, and lasting to the end of September, at which time a secondary blooming takes place. The flowers are collected every morning just when they start to open. They are picked by hand, flush with the stalk. Great care has to be exercised to select only those flowers which are just starting to open, because fully opened flowers would wither and fade during the process of *enfleurage* (48 hr. for each batch!— see below), and spoil the perfume of the *pommade*. On the average, 1,000 tuberose plants yield from 25 to 30 kg. of flowers per year.

METHODS OF EXTRACTION

The tuberose is one of those plants the flowers of which continue to develop their natural perfume for some time *after they have been harvested.* The problem therefore resolves itself into that of capturing the additional quantities of natural flower oil emitted by the flowers after picking, and before withering. This was accomplished many years ago by the introduction of the old-fashioned *enfleurage* process, which to a limited extent is still practiced in the Grasse region of Southern France (for details see Vol. I of this work, pp. 189 ff.). In recent years the process of *enfleurage* has been partly replaced by the modern method of extraction with petroleum ether (Vol. I, p. 200), which requires much less labor than *enfleurage,* and although giving a much smaller yield of natural flower oil, nevertheless produces concretes and absolutes of very strong odor, representative of the true perfume of the living flowers. (To avoid misunderstanding, it should be pointed out here that steam or water distillation of the tuberose flowers directly gives only a very small yield of oil; moreover, the oil thus obtained is of very poor odor. The process therefore cannot be employed. The steam-volatile oils described below are not derived directly from the flowers. They are distilled in the laboratory from the several concretes and absolutes, in order to determine their content of volatile oils, and to study the physicochemical properties and chemical composition of these oils—cf. Vol. I of this work, p. 213.)

Years ago Hesse [3] submitted two separate batches of tuberose flowers of 1,000 kg. each to *enfleurage,* and to extraction with volatile solvents, in order to compare the quantities of *actual* flower oil (volatile oil) obtained by the two processes. He determined these quantities by steam-distilling the concentrated extracts and the residual flowers (the latter for the sake of completeness). Hesse found that 1,000 kg. of tuberose flowers, on *enfleurage,* yielded 801 g. of steam-volatile oil. To this must be added 78 g. of vola-

[3] *Ber.* **36** (1903), 1459.

tile oil from the residual flowers (absolute of *chassis*—cf. Vol. I of the present work, p. 197). Extraction of 1,000 kg. flowers with petroleum ether, on the other hand, yielded only 56 g. of steam-volatile oil (at the beginning of the harvest only 36 g.). To this must be added 10 g. of oil, which Hesse obtained by the process of steam-distilling the residual flowers after the extraction with petroleum ether. Considering that this process is not employed in industrial practice, and taking into account the fact that in the process of *enfleurage* the absolute of *chassis* is actually obtained as a valuable by-product, it appears that in the case of tuberose flowers, *enfleurage* yields about fifteen times more volatile flower oil than does extraction with petroleum ether.

Enfleurage.—For the purpose of *enfleurage* only freshly picked flowers, which are still closed, should be used. These open up on the *chassis,* where they are kept for 48 hr., before being replaced by fresh flowers. One kg. of *corps* is treated with 2.5 to 3.5 kg. of flowers in the course of the harvest. About 150 kg. of flowers are required to yield 1 kg. of absolute of *enfleurage.* The residual flowers, removed from the *chassis* every 48 hr., still contain some natural flower oil and are submitted to extraction with petroleum ether, giving the so-called absolute of *chassis.* The yield of the latter varies from 1.2 to 1.5 per cent, calculated upon the flowers. The absolute of *enfleurage* and the absolute of *chassis* are usually offered to the trade in separate form, but some houses in Grasse combine the two products, marketing them under the label "absolute of *enfleurage.*"

The true (pure) absolute of *enfleurage* contains from 11.4 to 14.8 per cent of steam-volatile oil. Hesse [4] (I), and Naves, Sabetay and Palfray [5] (II) reported these properties for volatile oils obtained by steam distillation of absolutes of *enfleurage:*

	I	II
Specific Gravity at 15°.	1.009 to 1.035	...
Optical Rotation.	$-2°\,30'$	...
Refractive Index at 20°.	...	1.5352 and 1.5136
Acid Number.	32.7	10.6 and 29.3
Ester Number.	243 to 280	243 and 205.0
Methyl Anthranilate Content.	3.2 to 5.4%	...

Hesse [6] also examined a steam-volatile oil obtained from an absolute of *chassis:*

Specific Gravity at 15°.	1.043
Optical Rotation.	$-3°\,21'$
Saponification Number.	225.4
Methyl Anthranilate Content.	2.0%

[4] *Ibid.,* 1464.
[5] *Perfumery Essential Oil Record* **28** (1937), 337.
[6] *Ber.* **36** (1903), 1465.

The absolute of *enfleurage* is a brown, semisolid, alcohol-soluble liquid, the viscosity of which depends upon temperature and method of purification. It possesses a characteristic odor of tuberose flowers with a somewhat fatty "by-note." The odor of the absolute of *chassis* is less pronounced.

Solvent Extraction.—Extraction of tuberose flowers with petroleum ether yields from 0.08 to 0.11 per cent, in exceptional cases as much as 0.14 per cent, of concrete (Naves and Mazuyer [7]). In the author's experience about 1,150 kg. of tuberose flowers are required to yield 1 kg. of concrete.

Concrete of tuberose is a light to dark brown, waxy, quite hard mass, only partly soluble in high-proof alcohol. Walbaum and Rosenthal [8] (I), Naves and Sabetay [9] (II), and Girard [10] (III) reported these properties for concrete of tuberose:

	I	II	III
Congealing Point	56.9°	49°–50°	...
Melting Point	...	...	57°
Specific Gravity at 60°	...	...	0.8951
Refractive Index at 60°	...	...	1.4601
Acid Number	...	52.2	56
Ester Number	...	76.4	63
Saponification Number	117.6	...	119

The concrete contains from 3 to 6 per cent of steam-volatile oil, for which Hesse [11] (I), and Elze [12] (II) reported the following properties:

	I	II
Specific Gravity at 15°	1.007	1.003
Optical Rotation	−3° 45′	−3° 15′
Acid Number	22.0	25.0
Ester Number	224.0	230.0
Methyl Anthranilate Content	1.13%	1.4%

On treatment with alcohol in the usual way, the concrete yields from 18 to 23 per cent of alcohol-soluble absolute, a very viscous, or semiliquid brownish mass of powerful and lasting odor, truly reminiscent and characteristic of the perfume of the living flower. A genuine absolute examined by Naves and Mazuyer [13] exhibited these values:

Congealing Point	21° to 22°
Specific Gravity at 25°	0.982
Refractive Index at 25°	1.4916
Acid Number	84.6
Ester Number	138.2

[7] "Les Parfums Naturels," Paris (1939), 282.
[8] *Ber. Schimmel & Co., Jubiläums-Ausgabe* (1929), 193.
[9] Cf. "Les Parfums Naturels," Paris (1939), 283.
[10] *Ind. parfum.* 2 (1947), 217.   [12] *Riechstoff Ind.* 3 (1928), 154.
[11] *Ber.* 36 (1903), 1463.   [13] "Les Parfums Naturels," Paris (1939), 283.

Three absolutes of extraction, prepared in Liguria (Italy) by Rovesti [14] had the following properties:

Specific Gravity at 15°..........	1.1211	0.988	1.061
Optical Rotation..............	−5° 41′	−1° 7′	−3° 15′
Ester Number after Acetylation..	168	156	217

CHEMICAL COMPOSITION

The chemical composition of tuberose flower oil was investigated by Hesse,[15] Schimmel & Co.[16] (almost simultaneously but independently), and more recently by Elze,[17] who reported the presence of the following compounds:

Geraniol and Nerol. Both alcohols in free form, as acetates, and probably also as propionates (Elze). No proof, however, was furnished for the occurrence of geraniol and nerol, as propionates, in the oil.

Farnesol. Also identified by Elze.

Benzyl Alcohol. Free and in ester form (Hesse).

Methyl Benzoate. Identified by Schimmel & Co., and by Elze.

Benzyl Benzoate. Characterized by Hesse, and Elze.

Methyl Salicylate. Identified by Hesse in the flower oil obtained by *enfleurage*, but not in the product of extraction with volatile solvents. (Judging from the odor of genuine absolute of extraction, methyl salicylate is probably present also in the latter—the author.)

Methyl Anthranilate. Identified by Schimmel & Co., and by Hesse. The last named worker calculated that the flower oil of tuberose derived by *enfleurage* contains 56 times more methyl anthranilate than the product of extraction with volatile solvents.

Eugenol. Identified by Elze.

Butyric Acid and perhaps Phenylacetic Acid(?). Reported by Hesse.

About fifty years ago Verley [18] investigated a tuberose flower oil, and asserted that it contains about 10 per cent of a ketone $C_{13}H_{20}O$, to which he assigned the name "tuberone." However, none of the chemists who later attempted to isolate a ketone of this formula from natural tuberose oil succeeded in confirming Verley's claim.

[14] *Profumi ital.* **3** (1925), 243.
[15] *Ber.* **36** (1903), 1459.
[16] *Ber. Schimmel & Co.,* April (1903), 74.
[17] *Riechstoff Ind.* **3** (1928), 154.
[18] *Bull. soc. chim.* [3], **21** (1899), 307.

Use

Pure absolute of extraction of tuberose is perhaps the most expensive natural flower oil at the disposal of the modern perfumer. Therefore it can be used in perfumes of only the highest grade. In lower priced creations the absolute of *enfleurage* will give sufficiently good results. Both products are employed in scents of the heavier type, floral as well as oriental. Tuberose flower oil is an important base, particularly in gardenia perfumes.

Suggested Additional Literature

R. Arnaud, "Culture des Plantes à Parfums en France," *Ind. parfum.* 4 (1949), 335.

CONCRETE AND ABSOLUTE OF NARCISSUS

Three species of *Narcissus* (fam. *Amaryllidaceae*) are processed in the Grasse region of southern France for the extraction of their natural flower oils:

1. *Narcissus poeticus* L., which grows wild in many parts of the Départements Alpes Maritimes, Var, and Basses Alpes, particularly in the Maures Mountains and on the plateau of Caussols, near Castellane, Comps, Valensole, Riez, Moustier, Aups, etc., at altitudes of 1,000 m. and higher. The flowers are white and possess a strong perfume. The harvest takes place after that of the cultivated flowers (see below), i.e., in March, April, and even in May.

2. *Narcissus tazetta* L., which is largely cultivated, but also grows wild. The color of the flowers ranges from light to dark yellow. Plantations are located in the Grasse region, near Cogolin, Magagnosc, Opio, Valbonne, Tourette, Grasse, etc. A planting lasts from three to four years, and yields 300 to 400 kg. of flowers per hectare in the first year, and 900 to 1,200 kg. in the third year. The harvest takes place in March. According to Naves and Mazuyer,[1] on a well-cultivated field one harvester can collect 0.4 to 1.0 kg. of flowers per hour. Only the flowers are gathered and used for extraction.

[1] "Les Parfums Naturels," Paris (1939), 237.

3. *Narcissus jonquilla* L., the common jonquil. The concrete and absolute derived from this species are described in the next monograph.

Recently from 30 to 60 metric tons of *Narcissus poeticus* and *Narcissus tazetta* have been processed yearly in the extraction plants of the Grasse region. Of the two species the former is by far the more important.

Steam distillation of the flowers gives such a poor yield of oil that other means have to be resorted to in order to isolate the natural perfume. Years ago the method employed for this purpose was maceration with hot fat (see Vol. I of this work, p. 198), but now the flowers are extracted almost exclusively with petroleum ether (see Vol. I, p. 200).

The yield of concrete ranges from 0.21 to 0.45 per cent, in most cases from 0.25 to 0.28 per cent for *Narcissus tazetta*. The yield of concrete from *Narcissus poeticus* is lower, varying between 0.20 and 0.26 per cent (Naves and Mazuyer). The concrete gives from 27 to 32 per cent of alcohol-soluble absolute. The concrete contains 2.2 to 3.5 per cent of steam-volatile oil (the latter, however, is not a product of commerce).

Physicochemical Properties.—Concrete of narcissus is a waxy mass of yellow-green color, which turns lighter on aging. The odor is quite characteristic of the live flowers. Naves and Mazuyer [2] (I), and Igolen [3] (II) reported these properties for two concretes of narcissus:

	I	II
Melting Point	~50°	55° to 56°
Acid Number	24	22.4
Ester Number	60	52.5

The absolute is a viscous, alcohol-soluble liquid, of green-brown color. The above-named workers reported the following values for two absolutes of narcissus:

	I	II
Specific Gravity at 15°	0.960	0.9728
Refractive Index at 20°	1.4884	1.4928
Acid Number	38	39.2
Ester Number	88.6	87

Naves, Sabetay and Palfray [4] submitted a concrete of narcissus to distillation with superheated steam *in vacuo* (see Vol. I of this work, p. 215) and obtained a volatile oil with these properties:

Specific Gravity at 15°	0.994
Optical Rotation	+0° 40'

[2] *Ibid.*
[3] *Compt. rend.* **214** (1942), 234.
[4] *Perfumery Essential Oil Record* **28** (1937), 336.

Refractive Index at 20°.............. 1.4988
Acid Number....................... 5.6
Ester Number..................... 184.0

More recently, Igolen [5] applied Sabetay's method of codistillation *in vacuo* with ethylene glycol to a concrete of narcissus (see Vol. I of this work, pp. 215 and 216). The volatile oil amounted to 11.5 per cent of the concrete and exhibited the following values:

Specific Gravity at 15°.............. 0.9714
Optical Rotation at 20°.............. −5° 48′
Refractive Index at 20°.............. 1.5050
Acid Number...................... 11.2
Ester Number..................... 78.14
Ester Number after Acetylation....... 171.14
Methoxy Content.................. 3.7%

An investigation of this volatile oil revealed that it contained these compounds as principal constituents:

Eugenol.

Benzyl Alcohol.

Cinnamyl Alcohol.

Benzaldehyde.

Benzoic Acid, free and in ester form.

The compounds which impart to the oil its characteristic odor have not yet been identified, however.

Use.—Concrete and absolute of narcissus have lately been gaining in importance. The absolute is used in high-grade perfumes of the French type, to which it imparts exquisite, strong and "heavy" tonalities, difficult to identify. Absolute of narcissus blends particularly well with jasmine.

SUGGESTED ADDITIONAL LITERATURE

R. Arnaud, "Culture des Plantes à Parfums en France," *Ind. parfum.* **4** (1949), 368.

[5] *Compt. rend.* **214** (1942), 234.

CONCRETE AND ABSOLUTE OF JONQUIL

Narcissus jonquilla L. (fam. *Amaryllidaceae*), the common jonquil, bears three to eight flowers of golden yellow color. Their perfume is peculiar, strong, but more delicate than that of *Narcissus poeticus* L. or *N. tazetta* L. For the extraction of its perfume, the jonquil is cultivated in the Grasse region of Southern France, near Peyménade, Callian, Montauroux, Magagnosc, Opio, Valbonne, Auribeau, Le Tignet, Tanneron and Grasse.

The cultivation of jonquil requires considerable care; a field cannot be kept in jonquil longer than three or four years. The harvest of the flowers takes place in April. Years ago as much as 60 metric tons of jonquil flowers were processed annually in the extraction plants of Grasse, but more recently the quantity has fallen off to about 5 tons only.

Because of the delicate odor exhaled by the flowers, *enfleurage* appears to be the best method of capturing the natural perfume. However, jonquils bloom for only one month, a period insufficiently long to saturate the *corps* on the *chassis*. (See Vol. I of this work, p. 189.) For this reason, the process of maceration with hot fat (Vol. I, p. 198) at 50° to 70° is often applied. Lately this process has been replaced almost entirely by the modern method of extraction with petroleum ether (Vol. I, p. 200). In the author's experience, 450 to 500 kg. of flowers are required to yield 1 kg. of concrete which, in turn, gives 450 to 500 g. of alcohol-soluble absolute. Naves and Mazuyer [1] reported yields of concrete ranging from 0.25 to 0.51, in most cases 0.35 to 0.45 per cent. The concrete gives 40 to 55 per cent of absolute. The concrete contains from 3 to 7 per cent of a steam-volatile oil which, however, is not a commercial product.

Physicochemical Properties.—Concrete of jonquil (by extraction with petroleum ether) is a waxy mass of dark brown color. Naves and Mazuyer [2] (I), and Girard [3] (II) reported these properties for two concretes:

	I	II
Melting Point	48° to 52°	...
Specific Gravity at 60°	...	0.9313
Refractive Index	...	1.4717
Acid Number	30 to 44.2	43.4
Ester Number	88.6 to 106.8	98

Absolute of jonquil is a viscous, dark brown liquid, soluble in alcohol. Two absolutes described by the same authors exhibited the following values:

[1] "Les Parfums Naturels," Paris (1939), 235.   [3] *Ind. parfum.* **2** (1947), 188.
[2] *Ibid.*

	I	II
Specific Gravity at 15°.	0.992	0.995
Refractive Index at 20°.	1.5044	1.506
Acid Number.	42.4	39.2
Ester Number.	132.6	126.7

Naves, Sabetay and Palfray[4] (III and IV), and von Soden[5] (V) submitted three concretes of jonquil to steam distillation and obtained volatile oils with these properties:

	III	IV	V
Specific Gravity at 15°.	1.047	1.038	1.064
Optical Rotation.	−2° 12′	−1° 37′	$[\alpha]_D$ −2° 45′
Refractive Index at 20°.	1.5175	1.5120	. . .
Acid Number.	2.4	5.6	0
Ester Number.	216.0	196.0	250.0

Examining the steam-volatile oil (V), von Soden reported that it contained:

Methyl Benzoate.

Benzyl Benzoate.

Cinnamic Acid.   In ester form (including methyl cinnamate).

Linaloöl.

Methyl Anthranilate.

Indole.

Von Soden arrived at his conclusion by simply fractionating the volatile oil, noting the boiling points, and submitting the various fractions to olfactory tests.   The indole was separated by means of its picrate.

Shortly afterward, Elze[6] identified *jasmone* in jonquil flower oil, by preparation of its semicarbazone m. 201°–202°.

Use.—Absolute of jonquil is a valuable adjunct in high-grade perfumes of the French type.   It imparts "heavy" tonalities to floral as well as oriental scents.

SUGGESTED ADDITIONAL LITERATURE

R. Arnaud, "Culture des Plantes à Parfums en France," *Ind. parfum.* 4 (1949), 368.

[4] *Perfumery Essential Oil Record* **28** (1937), 336. Cf. Vol. I of the present work, p. 215.
[5] *J. prakt. Chem.* **110** (1925), 277.
[6] *Riechstoff Ind.* (1926), 181.

CHAPTER XIV

ESSENTIAL OILS OF THE PLANT FAMILY *RUBIACEAE*

CONCRETE AND ABSOLUTE OF GARDENIA

Many species of gardenia are grown in various parts of the world, particularly in India, China, and other countries of the Far East, among them *Gardenia florida* L. (fam. *Rubiaceae*), *G. grandiflora* Lour., and *G. citriodora* L., the flowers of which exhale a strong and heavy perfume. The different species of gardenia seem to contain essential oils of different composition, some of them reminiscent of jasmine, others of orange blossoms. The species with white flowers appear to have the most agreeable odor.

Because of the popularity of gardenia perfume, numerous attempts have been made in the course of years to extract the natural flower oil from gardenias, but most efforts to arrive at commercial production have been unsuccessful.

In 1912, Garnier [1] started gardenia plantations on Réunion Island. Little is known about the taxonomy of the species propagated by Garnier. The bushes were planted 1 by 1.5 m. apart, 1 hectare containing about 7,000 plants. Propagation was effected by means of cuttings. One bush produced from 100 to 260 g. of flowers per year. The harvest took place in November and December. According to information gathered by the author during a visit to Réunion Island, 3,000 to 4,000 kg. of flowers on extraction with petroleum ether yielded 1 kg. of concrete which, in turn, gave 0.5 kg. of alcohol soluble absolute. Naves and Mazuyer [2] report that 2 to 2.6 metric tons of flowers produce 1 kg. of concrete which, in turn, give 0.5 kg. of absolute. Limited quantities of this type of absolute were produced prior to the outbreak of World War II on Réunion Island and were used by some of the leading perfume houses in Paris. Small quantities found their way into the United States.

Another attempt to produce gardenia flower oil was made in California between 1936 and 1939, when a leading gardenia grower near Los Angeles tried to dispose of his surplus flowers by submitting them to a modified process of *enfleurage*. The natural flower oil thus obtained resembled the perfume of the living flowers, but did not find ready acceptance in the perfume industry of the United States. The venture in Los Angeles was, therefore, abandoned. About 50 lb. of gardenia flower oil were produced altogether in Los Angeles.

[1] Conversation with the late Mr. Charles Garnier, Paris.
[2] "Les Parfums Naturels," Paris (1939), 212.

Chemical Composition.—Almost fifty years ago Parone [3] submitted fresh gardenia flowers to maceration with paraffin oil and obtained 0.0704 per cent of a yellow oil with these properties:

$$B_{12-15}\dots\dots\dots\dots\dots\dots\dots\dots\dots\dots\dots 84°–150°$$
$$\text{Specific Gravity at } 20.5°\dots\dots\dots\dots 1.009$$
$$\text{Specific Optical Rotation at } 20°\dots\dots +2° 56'$$

According to Parone, natural gardenia flower oil contains the following compounds:

Benzyl Acetate.

Styrolyl Acetate (Methylphenylcarbinyl Acetate). $C_6H_5 \cdot CH \cdot (OCOCH_3) \cdot CH_3$.

Linaloöl.

Linalyl Acetate.

Terpineol.

Methyl Anthranilate.

The chief constituent of natural gardenia oil is benzyl acetate, but the odor of the oil is influenced particularly by styrolyl acetate.

Use.—Very little, if any, gardenia flower oil is produced at present, and this only on Réunion Island. The natural oil is, therefore, scarcely used in our industry. The decline in production of natural gardenia oil must be attributed to the fact that today it is possible to reproduce the perfume of the living flowers by skillful blends of certain synthetic aromatics, essential oils, and natural flower oils, among them benzyl acetate, styrolyl acetate, linaloöl, terpineol, oil of ylang ylang, and absolute of tuberose.

CONCRETE AND ABSOLUTE OF KARO-KAROUNDÉ

Leptactina senegambica Hook. f. (fam. *Rubiaceae*), locally known as "Karo-Karoundé," is a shrub 0.5 to 2.5 m. in height. It grows abundantly in the highlands of French Guinea, particularly in the Fonta-Djalon, at altitudes ranging from 1,000 to 1,400 m. The plant bears large white, strongly scented blossoms. The flowering period commences at the end

[3] *Boll. chim. farm.* **41** (1902), 489.

of October, as soon as the rains cease, and continues throughout the dry season.

On extraction with petroleum ether, the freshly picked flowers yield 0.133 per cent of concrete flower oil. (Compared with the yield obtained from such flowers as rose and jasmine, this is a very low figure.) On extraction with 95 per cent alcohol in the usual way, the concrete gives 60 to 65 per cent of alcohol-soluble absolute. Small quantities of concrete and absolute have been produced in Labé (French Guinea) since about 1936.

Concrete of karo-karoundé is a semisolid mass of orange-red color; it melts at 44°. The absolute, a viscous liquid, also possesses an orange-red color, which darkens with the passage of time.

Steam-distilling concrete of karo-karoundé *in vacuo*, Sabetay, Palfray and Trabaud [1] obtained 15.1 per cent of a volatile oil with these properties:

Specific Gravity at 15°/15°....................	0.9944
Optical Rotation...............................	−20° 0′
Refractive Index at 20°........................	1.5150
Acid Number..................................	17.3
Ester Number.................................	38.1
Ester Number after Acetylation.................	138
Ester Number after Cold Formylation...........	174.5
Methoxy Content (Zeisel Method)...............	3.1%
Total Phenol Content (Phenols and Acids).......	32.0%

Fractionating the volatile oil, Sabetay, Palfray and Trabaud,[2] identified *isoeugenol* by means of its acetyl derivative, m. 79°. One fraction gave a characteristic *indole* reaction. Most interesting is the presence of *cyanides* in the oil, of which about 38 per cent (calculated as benzyl cyanide) are present.

Absolute of karo-karoundé possesses an odor reminiscent of jasmine, tuberose, and orange blossoms, with by-notes of ylang ylang and carnation. The basic tone of the odor is fruity (peach-apricot) and vanilla-like.

Small quantities of the absolute are used in high-grade perfumes of the French type, to lend originality and distinction to the composition.

SUGGESTED ADDITIONAL LITERATURE

R. Paris and A. Bouquet (Fac. pharm., Paris), "Karo-Karundé (*Leptactinia senegambica*), an African *Rubiaceae*," *Ann. pharm. franç.* **41** (1946), 233. *Chem. Abstracts* **41** (1947), 7057.

[1] *Compt. rend.* **207** (1938), 540. Cf. *Perfumery Essential Oil Record* **29** (1938), 344.
[2] *Ibid.*

CHAPTER XV

ESSENTIAL OILS OF THE PLANT FAMILY *MAGNOLIACEAE*

OIL OF STAR ANISE

Essence de Badiane   *Aceite Esencial Anis Estrellado*   *Sternanisöl*
                Oleum Anisi Stellati

A. Oil of Star Anise Fruit

Introduction.—Oil of star anise, distilled from the fruit of *Illicium verum* Hook. f., fam. *Magnoliaceae*, must not be confused with oil of anise, derived from the seed (fruit) of *Pimpinella anisum*, fam. *Umbelliferae* (see Vol. IV of this work, p. 563). Both oils contain anethole as principal constituent, but the odor and flavor of the latter oil are finer than those of the former. Anise oil is produced chiefly in Central Europe, star anise oil in China and adjoining sections of French Indo-China. Under normal conditions anise oil fetches much higher prices than star anise oil.

Botany.—For a long time there has been considerable confusion in regard to the taxonomy of the tree which produces the true star anise fruit. There are several species of the genus *Illicium* growing in various parts of East Asia, and yielding fruits of similar appearance. Linnaeus [1] first named the plant *Badanifera anisata,* but shortly afterward changed the name to *Illicium anisatum* L.[2] It was only in 1886 that the younger Hooker [3] identified the tree yielding the official star anise fruit and designated it *Illicium verum* Hook. f.

There is another species of *Illicium,* cultivated in temple groves in Japan, described by several botanists in the seventeenth and eighteenth century, which was named *Illicium japonicum* Sieb. by von Siebold in 1825. In 1837 von Siebold changed the name to *Illicium religiosum* Sieb.[4] The fruits of this so-called Japanese star anise tree are poisonous, and contain an essential oil with a disagreeable odor, quite different from that of the true star anise. (See the following monograph "Oil of Star Anise, Japanese.")

The official star anise (*Illicium verum* Hook. f.) is a stately evergreen, attaining a height of 45 ft. Its trunk seldom grows to more than 10 in. in diameter. In its white bark the star anise tree resembles the birch; in

[1] "Materia medica e regno vegetabili," Stockholm, Vol. 1 (1750), 180.

[2] "Species plantarum," Stockholm (1753), 664.

[3] *Botanical Magazine,* July (1888). "Watt's Dictionary of India," Calcutta, Vol. IV (1890), 330. Holmes, *Am. J. Pharm.* **60** (1888), 503.

[4] Cf. J. J. Rein, "Japan," Vol. II (1886), 160, 307. Cf. Gildemeister and Hoffmann, "Die Ätherischen Öle," 3d Ed., Vol. II, 564.

its silhouette, the poplar. Seen from a distance, dense stands of star anise look like groups of coniferous trees.

The fruit has a star-like shape and exhibits a characteristic anise odor; hence the name star anise. It consists usually of eight boat-shaped follicles, or carpels, arranged around a central axis. The carpels taper to an almost straight beak; in the fresh state they are green, in the dried state reddish-brown and slightly wrinkled. The enclosed seed is ovoid, smooth, and, when developed, brown.

Most of the fresh (green) fruit is used by the natives for distillation of the essential oil; a small quantity is dried and exported.

Geographical Distribution.—*Illicium verum* Hook. f., a native of tropical and subtropical East Asia, grows in the planted and semiwild state in a relatively small section of East Asia. The principal producing regions of the oil lie in southeastern China (Province of Kwangsi), and in the adjacent parts of French Indo-China (State of Tonkin).

Centers of production and trading in the Chinese Province of Kwangsi are Nanning and Long-Tcheou. In the years before World War II, Nanning supplied almost 50 per cent of the 400 metric tons of oil produced annually in the world. Long-Tcheou furnished about 30 per cent.

Center of production and trading in the neighboring French Indo-Chinese state of Tonkin is Langson; about 20 per cent of the total output of star anise oil comes from Tonkin.

The oils produced in Tonkin and in Kwangsi along the border are usually of higher quality than the oil from the interior of Kwangsi.

The star anise tree was introduced to Indo-China by the Chinese, long before the conquest of Tonkin by the French. Of these early Chinese plantings only traces are left; the trees exploited at present were planted by the Thos and the Nungs, who now inhabit the star anise producing regions. The present age of these trees varies from fifty to sixty years. In view of the fact that *Illicium verum* reaches an age of one hundred years and more, trees now being exploited are comparatively young and still bear ample fruit.

The star anise industry of China and Indo-China is almost entirely in the hands of natives. The trees grow semiwild or planted near villages and isolated huts; the stills are of primitive, although clever, construction. Production of the oil is a typical oriental village industry. Prior to the outbreak of World War II, there were in Tonkin about 850 hectares covered with star anise; of these, about 800 hectares were located in the district of Langson. No new plantings have been started by the natives in the course of the last thirty or forty years. (Statistics on the star anise acreage in Kwangsi are not available.) During the author's visit to Langson, in

1938–39, a French firm developed about 100 hectares of star anise trees, which were expected to come into full production in 1940. It was the only European-owned and managed plantation, but it is more than doubtful whether this venture survived the Japanese occupation, the insurrection of the natives, and the colonial warfare after the end of World War II.

In general, the cultivation of the star anise tree has been limited to a relatively confined area of East Asia. Agricultural stations in other parts of the world have repeatedly attempted to grow the star anise tree, but most of these efforts have failed to yield a commercially worth-while crop of fruit. It would, therefore, appear that ecological factors prevailing in Kwangsi and Tonkin are particularly favorable to the growth of *Illicium verum*.

Planting and Cultivating.—Seed is collected from vigorous older trees, known for their high yield of fruit. It is removed from the fruit with a knife; only fully matured seed—easily recognized by its brown color—is retained. One kilogram of fruit yields about 1,000 seeds suitable for planting. After stratification the seeds are planted, 3 to 4 cm. apart, in a well-protected bed. Since the seed loses its germinating power rapidly, it has to be planted within three days after the harvest. Seedbeds are started usually in October or November. After the young plants have developed the fourth leaf, they are transferred to a nursery, and planted about 25 cm. apart. Here the young trees remain for three years, or until they are sufficiently strong to be planted out, 5 to 6 m. apart.

In Tonkin, most crops are planted late in the fall or during the first months of the year. In the case of star anise, however, the better practice would be to plant the seed in June or July, because it has been observed that premature plantings frequently suffer from the dryness prevailing in April and May.

During their growth, the trees require no special care, except perhaps some weeding in July and August, which is done merely to facilitate gathering of the fruit from the ground, and to exterminate all straw-like herbs and plants liable to burn and spread bush fires. Thorough plowing and some mulching in the fall provide the trees with sufficient moisture for the dry season. On a modern plantation, every tree, at the beginning of summer, should receive about 15 lb. of stable dung and 100 lb. of ammonium sulfate.

The color of the flowers ranges from white to red. The trees flower throughout the year, but most abundantly at certain seasons. Thus, fruit matures all year round, but in normal years the August to October harvest accounts for about 80 per cent of the total yearly harvest.

Development of Flowers and Fruit.—According to Drouet,[5] the star anise flowers in an unusual way. Blossoms appear from March to the end of April, but are sterile, and develop no fruit. A second flowering period begins in July or August, and lasts for two or three weeks; the flowers, larger than those of the other periods, develop into fruit prematurely, from November to January. The natives call this "the late harvest." The third flowering season starts immediately after the second, in fact, partly dovetailing with it. The flowers at this time are relatively small, and develop into fruit very late, usually in August, September, or October of the following year. This is the so-called "big harvest."

At the beginning of the fruiting period the embryonic fruits developed from the second and third flowering periods of the year are filiform and do not differ from one another. It is only in October that it becomes possible to differentiate between fruit which will ripen, prematurely, from November to January, of the same year, and fruit which will ripen from August to October of the following year, when the tree is again in full bloom.

Aside from the two harvests described above there is a less important one in March-April. However, because of the strong winds and sudden changes in temperature prevailing during this latter period, much fruit falls off the trees prematurely, after reaching only $\frac{1}{10}$ or $\frac{1}{5}$ of the normal size. This fruit is especially rich in essential oil.

Harvest of the Fruit.—To prevent breaking of the branches and damage to the trees, it has been suggested that double ladders be employed, or that only the fallen fruit be collected. Neither of these methods, however, is practical. Double ladders cannot be used on trees of such height as the star anise, whose lowest branches sometimes are 7 ft. above the ground. Nor is it advisable to collect the fallen fruit, because the star anise fruit contains a maximum of essential oil before complete maturity and should be gathered at this stage of development, in order to give a high yield of oil. Overripe, naturally dropping fruit can be used only for export in the dried state.

In actual practice, the natives, usually the village children, simply climb into the trees and, using long poles with a little hook attached to the end, detach the fruit from the branches, or shake the branches until the fruit falls to the ground. The branches are sufficiently strong to support the harvesters; moreover, the fruit is usually not so far out of reach that the harvesters cannot easily detach it.

Yield of Fruit.—A star anise planting comes into production when about ten years old. The first harvest is small, amounting to only 0.5 to 1 kg. of fruit per tree. In the fifteenth year a tree yields approximately 20 kg., and

---

[5] Conversation with Mr. L. Drouet in Langson, Tonkin, French Indo-China. Cf. Eberhardt, *Bull. agence gén. colonies Paris* No. 222 (1927), 447.

the planting becomes profitable. Full production sets in after the twentieth year, with a yield of about 30 kg. of fruit per tree. The average life span of a tree is 80 to 100 years or more. The productivity of a star anise tree, like that of most nontrimmed fruit trees, varies considerably from one year to another. A tree yielding 30 kg. of fruit in one year may give only 1 kg. the following year.

According to Drouet,[6] the productivity of a star anise tree may be estimated as follows:

A tree thirteen to twenty-five years old may yield from 0 to 5 kg. of fresh fruit in poor years,. from 5 to 10 kg. in normal years. A tree older than twenty-five years may yield from 0 to 10 kg. in poor years, from 10 to 20 kg. in normal years, and from 20 to 40 kg. in very good years. All these figures apply to fresh fruit.

Other estimates are those of Eberhardt[7] who reported that a worth-while harvest cannot be expected before the tenth year; from the tenth to the twentieth year a tree yields 30 to 35 kg. of fruit (which yield about 600 g. of oil). From the twentieth year on a tree produces, on the average, 40 to 45 kg. of fruit.

Drying of the Fruit.—On drying in flat baskets, by exposure to the sun for about 10 days, 100 kg. of fresh (green) fruit yield from 25 to 30 kg. of dried fruit. To preserve it better, the Chinese frequently boil the fruit for a few minutes before drying; however, it then loses part of the essential oil. The dried fruit is exported as such and represents the well-known, brown star anise of the drug market.

Distillation.—The fruit is distilled mostly in the fresh (green) state. If too much fruit has accumulated at a distillation post, it may be kept for about ten days, or even longer, provided it is spread out in a thin layer and frequently turned over, to prevent fermentation. Occasionally the demand for dried fruit for export slackens and prices then fall to very low levels. Whenever this happens, the natives use their accumulated stocks of dried fruit for distillation, hoping to obtain a better return from the essential oil.

Production of the oil is entirely in native hands. The stills used for the purpose resemble those employed for the distillation of cassia oil in China (cf. Vol. IV of this work, p. 243). They are of primitive, although quite clever, construction, and hold up to 300 kg. of fresh fruit per charge. In some regions the natives break the fruit by hand, prior to distillation. The fruit (in most stills about 180 kg.) is placed into the retort, together with sufficient water to cover the material. The heating—by direct fire beneath the retort—has to be done slowly, to prevent the boiling over of the charge

[6] *Ibid.*
[7] *Bull. agence gén. colonies Paris* No. 222 (1927), 447.

and the escape of the vapors through the loose-fitting joints of the system. Steam and oil vapors pass, from the retort, through three small holes at the bottom of the vase-shaped still head, which also serves as a sort of condenser. The still head is covered with a flat bowl, through which cold water flows. Steam and oil vapors are condensed on the bottom of the flat bowl, the condensate dripping into the lower part of the still head, and from there, through a pipe, into the oil separator. Here, the oil and water separate; the distillation water flows back into the retort and is redistilled (cohobated). Because of the small condenser surface, distillation in these stills must be carried out slowly, a charge of fresh fruit requiring 48 hr., a charge of dried fruit 60 hr.

Yield of Oil.—It is difficult to state exact figures of yield, because this varies according to the maturity of the fruit, location and condition of the trees, region, climatic conditions, etc. Drouet [8] mentioned yields ranging from 3.1 to 3.5 per cent for fresh fruit from Langson and Diem-Her, and 2.2 to 2.5 per cent for material from Thât-Khê and Lôc-Binh.

Eberhardt [9] reported that 180 kg. of broken fruit, on distillation for 45 hr., yield approximately 5.5 kg. of oil (about 3 per cent).

In general, 100 kg. of fresh fruit, distilled for 48 hr., yield 2.5 to 3.0 kg. of oil; 100 kg. of dried fruit, distilled for 60 hr., give 8 to 9 kg. of oil.

Co-distilling dried star anise with glycol, *in vacuo*, Sabetay [10] obtained 8.25 per cent of volatile oil.

Native vs. Modern Distillation.—Primitive though it is, production of star anise oil in small native distillation posts offers nevertheless one great advantage, that of low cost. The trees, planted many years ago, grow semiwild near the villages and require no care. Harvesting, carried out by the village children, costs nothing. Any investment in the simple stills was amortized long ago. Distillation is a mere side occupation of the villagers; it goes along, at a leisurely pace, with their principal task, that of growing rice. The male members of the family tend to their rice plots, casting an occasional glance at the flow of the distillate in their stills, and separating the oil; the women cook food near the still and kindle the fire beneath it. In other words, the oil is produced at practically no cost, and simply means additional income for the family.

The erection of a modern distillery, on the other hand, would be relatively costly. Building and stills would have to be amortized, foreman and laborers hired at regular wages, fuel purchased, and taxes would have to be paid. True, in modern steam stills the time of distillation could be

[8] Conversation with Mr. L. Drouet in Langson, Tonkin, French Indo-China.
[9] *Bull. agence gén. colonies Paris* No. 222 (1927), 447.
[10] *Ann. chim. anal. chim. appl.* **22** (1940), 217.

shortened to perhaps 3 or 4 hr., but the saving in cost of fuel would not compensate for the other expenses involved. The only advantage would lie in the production of a very high grade of oil, free of all adulterants. Such an oil would consist almost entirely of anethole and possess an exceptionally fine odor and flavor. The cheapest way to produce an oil of very high quality is simply to submit native-distilled oils to rectification, a practice followed by a large European dealer of star anise oil in Langson, Tonkin, some years ago.

Physicochemical Properties.—Oil of star anise is a colorless to yellowish liquid, strongly refractive to light, and possessing a characteristic anise-like odor and sweet flavor. Gildemeister and Hoffmann [11] reported the following properties for oil of star anise:

Specific Gravity at 20°..........	0.98 to 0.99
Optical Rotation...............	Slightly laevorotatory, up to −2°. In recent years the oils have usually exhibited a slight dextrorotation (up to +0° 36′)
Refractive Index at 20°.........	1.553 to 1.557
Congealing Point...............	+15° to +18°, usually about 16°; in exceptional cases as low as +14°
Solubility.....................	Soluble in 1.5 to 3 vol. of 90% alcohol

Shipments of star anise oil, examined by Fritzsche Brothers, Inc., New York, over a number of years, have had properties varying within these limits:

Specific Gravity at 25°/25°......	0.978 to 0.987, occasionally as low as 0.975
Optical Rotation...............	−1° 46′ to +0° 34′
Refractive Index at 20°.........	1.5530 to 1.5582, occasionally as low as 1.5521
Congealing Point...............	+15.0° to +18.4°, in exceptional cases as low as +14.0°
Solubility.....................	Soluble in 1 to 2.5 vol. and more of 90% alcohol

Oils of acceptable quality should have a congealing point above +15°, a specific gravity above 0.978 at 25° C., and an index above 1.5530.

A genuine oil of star anise distilled exclusively from star anise fruit without any admixture of leaves, and in a native field still, under the author's supervision near Langson, at the end of 1938, had these properties:

Specific Gravity at 25°.....	0.984
Optical Rotation at 25°....	+0° 12′
Refractive Index at 20°....	1.5572
Congealing Point..........	+18.2°
Solubility................	Soluble in 1.5 vol. and more of 90% alcohol

[11] "Die Ätherischen Öle," 3d Ed., Vol. II, 568.

The oil had a very good odor and flavor; it was of exceptional quality, better than that of most commercial shipments. Note the high congealing point of this oil!

Because of its high content of anethole, oil of star anise congeals in the cold. However, under certain conditions, particularly on keeping in a closed container or on *slow* freezing, the oil can be supercooled considerably below its congealing point, and kept in the liquid form for a long time. In the course of analysis, congelation can be initiated by a slight tapping of the test tube, by scraping of the inner walls with a glass rod, or by "seeding" of the liquid with a crystal of anethole. The more the liquid has been supercooled, the more quickly congelation will take place.

Oils which have been kept in half-filled bottles for a long period of time, or which have otherwise been in contact with air—by frequent melting, e.g.—gradually lose their ability to congeal. This is probably the result of partial conversion of the anethole into anisaldehyde and anisic acid, or by polymerization (cf. Vol. II of this work, p. 509). Old, poorly stored oils no longer congeal on cooling.

The quality of star anise oil, like that of anise seed oil, can be evaluated by its congealing point. In fact, it is possible to estimate the anethole content of an oil directly from its congealing point (cf. the monograph on "Anise Seed Oil," Vol. IV of this work, p. 563). In commercial practice star anise oils may be judged as follows:

Congealing Point (+°C.)	Quality
18	Best
17	Very good
16	Good
15	Lowest limit
Below 15	Not acceptable

The congealing point of star anise oils (hence their anethole content) depends chiefly upon two factors:

1. The season. While studying production in the field, the author was told that the congealing point of the oils produced during the cool season averages 16° and higher, whereas that of the summer oils is only 15°, or slightly lower.

2. The composition of the distillation material. Native distillers, particularly in the interior of the Province of Kwangsi (China), frequently mix leaves and small terminal branches with the fruit, and then obtain oils of low congealing point. In some cases 85 to 90 per cent of the charge may consist of leaves and terminal branchlets, and only 10 to 15 per cent of

fruit. Oils distilled from such material will obviously exhibit abnormal properties but are used by the native dealers and intermediaries for bulking with lots of good quality. Oils of somewhat low congealing point, but still acceptable to the exporters, will result. As regards the properties of star anise *leaf* oil: its congealing point averages 13°. Details will be found in the following section (B. Oil of Star Anise Leaves).

In general, the quality (congealing point and anethole content) of the star anise oils from Tonkin and from the adjacent borders of Kwangsi is superior to that of the oils produced in the interior of Kwangsi. The congealing point of the former ranges from 15.5° to 18° (average 16°), while that of the latter averages 15.5°. There are Kwangsi oils with a congealing point as low as 13°, but such lots are rejected by the exporters and used by the native intermediaries for bulking. The export trade does not accept oils with a congealing point below 15°. In certain years—1937, e.g. —the quality of the Kwangsi oils declined to such an extent that the trade was almost forced to turn to the Tonkin oils.

In the Province of Kwangsi, the principal star anise producing region, there are three important trading centers, supplying somewhat different qualities of oil:

1. Long-Tcheou, nearest to the French Indo-Chinese frontier, renowned for a good quality of oil, the congealing point of which ranges from 16° to 17°.

2. Chun-On and Na-Por, with a fair quality, the congealing point averaging 15.5°.

3. Po-Seh, with a poor quality (congealing point about 15°).

All oils reaching the overseas markets are bulkings, consisting of numerous small lots.

As regards the optical rotation, most shipments were formerly slightly laevorotatory; for a time they were dextrorotatory, lately they have again been laevorotatory. According to Gildemeister and Hoffmann,[12] dextrorotation may be the result of admixture, to the fruit, of leaves and terminal branchlets, which yield slightly dextrorotatory oils. This explanation, however, does not appear quite satisfactory; the author observed slight dextrorotation in an oil that had been distilled in Tonkin, under his own supervision, and exclusively from fruit (see above). The reason for the dextrorotation, therefore, must be sought in other factors, the use of insufficiently matured fruit, perhaps.

[12] *Ibid.*

Adulteration.—The addition of star anise leaves to the distillation material, and of leaf oil to the fruit oil, has already been mentioned. These practices result in a lowering of the congealing point, and hence anethole content, of the oil. The congealing point is determined by the method described in Vol. I of this work, p. 253.

Another form of adulteration, occasionally practiced by the native producers and dealers, consists in the addition of small quantities of mineral oils (kerosene, etc.) or fatty oils. Added mineral oil lowers the congealing point, specific gravity, and solubility of the original, pure oil. A pure oil is clearly soluble in 3 volumes of 90 per cent alcohol, the solution remaining clear with added alcohol. In the presence of mineral oils, however, the solution will be turbid, drops of mineral oil separating on prolonged standing of the solution.

To what extent the addition of petroleum affects the physicochemical properties of star anise oil has been demonstrated by Schimmel & Co.: [13]

Quality of Oil	Specific Gravity at 15°	Congealing Point (°C.)	Solubility in 90% Alcohol
Pure Oil......................	0.986	+18°	1:2.2 and more
5 Per Cent Petroleum Added..	0.978	+16.25°	Not clearly soluble in
10 Per Cent Petroleum Added..	0.970	+14.75°	10 vol. of 90% alcohol

From these figures it appears that the congealing point is not the sole criterion of purity for a star anise oil; specific gravity and solubility, too, have to be determined carefully.

The most reliable test for the presence of mineral oils is the so-called "Oleum Test." For details, see Vol. I of this work, p. 332.

Local Trading and Exports.—The small lots of star anise oil produced by natives are sold, either directly or through field brokers, to buying agents in the nearest trading center. Frequently the oils pass through many hands before reaching the trading centers (Langson, Long-Tcheou, and Nanning). Much oriental shrewdness is displayed in trading between actual producers, field brokers, and local merchants. Although the cost of the oil is practically nil, producers are always intent on fetching the highest possible price, and are surprisingly well informed about the latest market quotations (which fluctuate continuously). Langson has a star anise market every fifth day; other centers have one every second day. Distillers or field brokers bring their oil to the markets in small vessels hung on both ends of a pole, which they carry on their shoulders. Whenever they cannot obtain sufficiently interesting prices, they simply store their oil away until the market improves. Thus their small stocks of oil represent more wealth than any

[13] *Ber. Schimmel & Co.*, April (1897), 42.

paper money, in which the natives have no confidence. European exporters were represented in the large markets of Langson, Long-Tcheou, and Nanning by their own purchasing agents, usually Chinese *compradores* who for centuries have been indispensable in the trading with native producers.

Passing through so many hands the oil is obviously exposed to all sorts of adulteration, particularly with kerosene, fatty oils, leaf oil, etc. The Chinese intermediaries and buying agents have no means of analyzing the small lots of oil. Lack of ice in these remote regions excludes any possibility of determining the congealing point. Nevertheless, the Chinese are most adept at evaluating the quality of an oil, which they do simply by shaking a sample and observing the formation and disappearance of bubbles, or by pouring a small quantity of oil from a certain height into a larger quantity of oil. It is quite amazing to watch how quickly and efficiently a Chinese expert can determine the quality of a lot by these simple means.

The purchasing agents in the larger trading centers then bulk the small lots, and ship the bulkings to their principals in the ports of Haiphong, Canton, and Hong Kong. Here each exporter makes up large standard lots (usually of 15.5° congealing point). The lowest congealing point accepted by the trade is 15°; but for the sake of certainty, the exporters standardize the lots at 15.5°. Oils of higher congealing point could be supplied by the exporters on demand from abroad, but such qualities would have to be made up specially. In general the exporters do not specialize in the handling of star anise oil, and to them the commodity, like cassia oil, represents simply another of the many domestic products which they export. Some exporters have in their warehouses one large tank filled with standardized star anise oil, and the quantities drawn from this tank for shipment are replaced with small lots arriving from inland. The only control exercised by the exporters is the determination of the congealing point; but in case of doubt a sample is submitted to the city analyst.

Because of much local trading and smuggling across the frontier, it is almost impossible to establish even approximate figures for the quantities of star anise oil produced in the two principal regions, viz., Kwangsi (South China) and Tonkin (French Indo-China). In normal times Kwangsi supplies about 75 per cent of the total amount, and Tonkin 25 per cent.

In the years before the Sino-Japanese War the oil produced in Tonkin was collected in Langson, and exported via the harbor of Haiphong, mostly to France. The oil produced in Kwangsi, on the other hand, was shipped from the important trading centers, Long-Tcheou and Nanning, down the West River to Canton and Hong Kong, and from there exported. River transport on Chinese junks was inexpensive.

The Sino-Japanese War, the occupation of the entire producing region by Japanese forces, the withdrawal of the Japanese, the insurrection of the natives in French Indo-China, and the civil war in China have brought about such fundamental changes that no clear picture of the state of affairs can be drawn at present. Even before the outbreak of World War II, and before the occupation of the producing regions by Japan, there was much local trading of oil across the frontier of China and French Indo-China. When Japan blocked the West River, many lots of oil were smuggled from the interior of Kwangsi overland to Pakhoi (South China), and from there shipped, on junks, either to Haiphong in French Indo-China, or to Macão (a Portuguese colony) and Hong Kong. Because of certain import duties, much Kwangsi oil has been smuggled into Tonkin; in short, there is so much illegal trafficking of oil between China and French Indo-China that official export figures are practically meaningless. To quote a few data: Cardot [14] reported the following quantities of star anise oil exported from French Indo-China, chiefly to France:

	Metric Tons
1924........	183
1925........	167
1926........	188
1927........	67
1928........	108

For 1930 to 1932 Schimmel & Co.[15] reported exports from French Indo-China, mostly to France, and to a lesser extent to Hong Kong:

	Metric Tons
1930........	240
1931........	182
1932........	82

As regards exports from Hong Kong the following statistics, issued by the U. S. Department of Commerce,[16] may be of interest:

	Pounds
1935...........	413,630
1936...........	465,485
1937...........	445,800 [17]

According to Schimmel & Co.,[18] Hong Kong imported these quantities of oil in 1938:

[14] *Rev. marques parfum. savon.* **8** (1930), 542.
[15] *Ber. Schimmel & Co.* (1934), 69; (1935), 75.
[16] *World Trade Notes Chem. Allied Products* **10** (1936), No. 17, 13, Washington, D. C.
[17] *Chem. Ind.* **61** (1938), 606.
[18] *Ber. Schimmel & Co.* (1939), 81.

377,288 lb. from South China
56,259 lb. from French Indo-China
_____
433,547 lb.

At the time of this writing it is impossible to predict what the future will bring. Most likely the natives in the interior of Kwangsi will continue producing star anise oil, and manage somehow to export it, either legally or by smuggling—the latter probably on junks down the West River to Hong Kong, or overland to Pakhoi, and from there by sea to Haiphong, Macão, or Hong Kong. So far as the Tonkin oil is concerned, its future availability will depend also upon the way the present chaotic situation resolves itself.

Compared with the export figures cited above, statistics of total production are even more unreliable; in fact they are nonexistent. Experts who have lived in the producing regions for many years estimate the total annual production of star anise oil, in normal years, at about 400 metric tons.

Chemical Composition.—The principal constituent of star anise oil is anethole. The quantity present in an oil can be determined from the congealing point of the oil. (For details see the monograph "Oil of Anise," Vol. IV of this work, p. 563.) According to Gildemeister and Hoffmann,[19] a good star anise oil, on repeated freezing, yields from 85 to 90 per cent of pure anethole; the actual content, therefore, is probably slightly higher. The balance (10 to 15 per cent) consists of almost twenty compounds. These are listed below. The following substances have been identified in the oil distilled from the fruit of the star anise tree:

Anethole. That the chief constituents of the oils of star anise (*Illicium verum*), anise seed (*Pimpinella anisum*), and fennel seed are one and the same substance was recognized more than a century ago by Cahours.[20] Shortly afterward Persoz,[21] on oxidation of star anise oil with chromic acid, obtained anisic acid, to which he assigned the name badianic acid. (Cf. Vol. II of this work, p. 508).

d-α-Pinene. The terpene b. 155°–158°, α_D +24° 31' was identified as *d*-α-pinene by Schimmel & Co.,[22] who prepared the nitrolbenzylamine m. 122°–125°.

Terpenes(?). The same authors [23] also investigated the fraction b. 163°–168°, d_{15} 0.8551, α_D +14° 7', which had a characteristic odor of terpenes. Repeated tests for β-pinene and sabinene, by oxidation with potassium permanganate in alkaline solution, gave negative results.

Δ[3]-Carene. It was only in 1931 that Duncan, Sherwood and Short [24] identified Δ[3]-carene (nitrosate m. 147°–148°) in the oil. In the course of their work, these re-

[19] "Die Ätherischen Öle," 3d Ed., Vol. II, 569.
[20] *Compt. rend.* **12** (1841), 1213. *Liebigs Ann.* **35** (1840), 313.
[21] *Compt. rend.* **13** (1841), 433. *Liebigs Ann.* **44** (1842), 311.
[22] *Ber. Schimmel & Co.*, April (1893), 57; April (1910), 99; October (1911), 86.
[23] *Ibid.*, Oct. (1911), 86.
[24] *J. Soc. Chem. Ind.* **50** (1931), 410T.

searchers also confirmed the presence of α-pinene, limonene, dipentene, and d-β-phellandrene, all of which had previously been reported as constituents of the oil (see below).

α- and β-Phellandrene. α-Phellandrene, as well as β-phellandrene, occur in the oil in both the dextro- and the laevorotatory forms. Schimmel & Co.[25] first observed l-α-phellandrene (nitrite m. 102°) in a fraction b. 170°–175°, $α_D$ −5° 40'. In a fraction with similar properties the same workers [26] later identified l-β-phellandrene by oxidation to tetrahydrocuminaldehyde (semicarbazone m. 202°–205°), and to cuminic acid m. 113°–115°. Shortly afterward they noted the presence of d-β-phellandrene,[27] and arrived at the conclusion that these three forms of phellandrene occur in the oil simultaneously, with one form predominating. More recently Goodway and West [28] succeeded in identifying the fourth isomer, viz., d-α-phellandrene, by preparing the maleic anhydride adduct m. 125°–126°, $[α]_D^{22}$ +9° 24' (CHCl₃, c = 8.655).

p-Cymene. Identified by Schimmel & Co.,[29] by oxidation to hydroxyisopropylbenzoic acid m. 156°–157°, and p-propenylbenzoic acid m. 159°–160°.

Cineole. Iodol compound m. 110°–111° (Schimmel & Co.[30]).

Dipentene. Tetrabromide m. 124°, nitrolpiperidine m. 153° (Schimmel & Co.[31]).

l-Limonene. Tetrabromide m. 102°–103° (Schimmel & Co.[32]).

α-Terpineol. According to Tardy,[33] the fractions b. 216°–218° and b. 218°–224° contain terpineol. However, definite proof of the presence of terpineol in star anise oil was furnished by Schimmel & Co.,[34] who conducted gaseous hydrogen chloride into two fractions b. 215°–218°, and b. 218°–220°, thus obtaining dipentene dihydrochloride m. 47°–48°, in good yield. Treatment with nitrosylchloride gave a nitrosochloride m. 113°, in poor yield. Schimmel & Co. also prepared a phenylurethane m. 110°–111° (no melting point depression with terpinyl phenylurethane m. 111°–112°). Oxidation of the terpineol with a dilute solution of potassium permanganate gave the corresponding glycerol; this, on treatment with chromic acid, yielded a ketolactone m. 62°–63°, the semicarbazone of which melted at 200°.

Methyl Chavicol (p-Methoxyallylbenzene). More than fifty years ago Schimmel & Co.[35] noted the presence of this phenolic ether in star anise oil, when, by repeated freezing, they eliminated so much anethole from the oil that no further crystals separated in a freezing mixture. The residual, mobile oil underwent profound changes on boiling with alcoholic potassium. The boiling point and the refractive

[25] *Ber. Schimmel & Co.,* April (1893), 56.
[26] *Ibid.,* April (1910), 99, 100.
[27] *Ibid.,* October (1911), 86.
[28] *J. Chem. Soc. Ind.* 56 (1937), 473T.
[29] *Ber. Schimmel & Co.,* April (1910), 99, 100.
[30] *Ibid.*
[31] *Ibid.,* October (1911), 86.
[32] *Ibid.*
[33] "Étude analytique sur quelques essences du genre anisique," Thèse Paris (1902). 22. Cf. *Ber. Schimmel & Co.,* October (1902), 83.
[34] *Ber. Schimmel & Co.,* April (1910), 99, 100.
[35] *Ibid.,* October (1895), 6.

index rose, and, on cooling, additional large quantities of anethole separated. This justified the conclusion that the additional anethole was formed from the originally-present methyl chavicol, by the action of alkali (Eykman's reaction).

Hydroquinone Monoethyl Ether. The presence of traces of this compound in star anise oil was noted by Schimmel & Co.,[36] who isolated colorless, shiny crystals m. 64° on shaking the oil with large quantities of alkali solutions.

Safrole. Years ago, Oswald [37] oxidized this compound with potassium permanganate, and obtained a substance m. 35°–36°, with a characteristic odor of piperonal (heliotropin). Years later, Schimmel & Co.[38] proved that the substance actually was piperonal by preparation of the semicarbazone m. 224°–225°.

p-Methoxyphenylacetone (Anisketone). According to Tardy,[39] oil of star anise contains a ketone b. 263°, the oxime of which melts at 72°, the semicarbazone at 182°. Tardy named this ketone *acétone anisique*. It reacts with bisulfite, and, on oxidation with an alkaline solution of potassium permanganate, yields acetic acid and anisic acid.

l-Bisabolene and *d*-Cadinene. The occurrence of sesquiterpenes in star anise oil was first reported by Tardy.[40] Years later Duncan, Sherwood and Short [41] noted the presence of dextrorotatory, bicyclic hydrocarbons, which, on dehydrogenation with sulfur, yielded cadalene. The chief constituent of the sesquiterpene fraction yielded a dihydrochloride m. 77° (optically inactive in chloroform solution).

More recently, Jackson and Short [42] investigated the sesquiterpene fraction, which amounts to about 4 per cent of the oil, and found that it consists chiefly of *l*-bisabolene (hydrochloride m. 79°–80°) and of small quantities of *d*-cadinene (hydrochloride m. 117°).

Farnesol(?). In 1929 Takens [43] reported the presence of farnesol in star anise oil, without, however, giving any proof.

Feniculin (*p*-Anol Prenyl Ether). In the last runs of star anise and of fennel seed oils the same author observed a fraction b_5 147°, d_{15} 0.967, α_D ±0°, congealing point 21.5°, which, on heating to 260°, decomposed with effervescence, yielding *p*-prenyl phenol m. 93°–94°. More recently, Späth and Bruck [44] showed that the parent substance of this phenol is *p*-anol prenyl ether $C_{14}H_{18}O$, which they named feniculin. (Cf. Vol. II of this work, p. 512.)

Anisoxide $C_{14}H_{18}O$. Heating the fraction b_{12} 135°–145° with potassium-sodium alloy, and submitting this fraction to further fractionation, Jackson and Short [45] obtained a highly unsaturated hydrofurano compound $C_{11}H_{13}O \cdot CH{=}CH \cdot CH_3$, which they named anisoxide. After recrystallization, the compound melted at 41° (b_{11} 140°, d_4^{50} 0.9604, α_D ±0°, n_D^{50} 1.5361). It had a faint, but characteristic odor and, on exposure to air, rapidly turned into a viscous yellow liquid of sharp

[36] *Ibid.*
[37] *Arch. Pharm.* **229** (1891), 98.
[38] *Ber. Schimmel & Co.,* April (1910), 101.
[39] "Étude analytique sur quelques essences du genre anisique," Thèse Paris (1902). 22. Cf. *Ber. Schimmel & Co.,* October (1902), 83.
[40] *Ibid.*
[41] *J. Soc. Chem. Ind.* **50** (1931), 410T.
[42] *Ibid.* **55** (1936), 8T.
[43] *Riechstoff Ind.* **4** (1929), 8.
[44] *Ber.* **71** (1938), 2708.
[45] *J. Chem. Soc.* (1937), 513.

odor. According to Jackson and Short the star anise oils investigated by them contained about 0.2 per cent of anisoxide.

Paraffins $C_{19}H_{40}$. In the course of their work these authors [46] also noted the presence, in the oil, of small quantities of paraffins $C_{19}H_{40}$, m. 45°–46°, and of anisic acid (cf. below).

On oxidation by air anethole is gradually converted to anisaldehyde and anisic acid. Old star anise oils, therefore, may contain these compounds, the quantity increasing with the age of the oil.

Use.—The most important use of star anise oil is for the technical isolation of anethole, which has a much finer odor and flavor than the oil itself (cf. Vol. II of this work, p. 511).

The oil is also employed for the flavoring of pharmaceutical—particularly oral—preparations, and as a mild expectorant in cough lozenges.

The flavor of anise is very popular in French, Italian, Spanish, Greek, and Turkish confectioneries, as well as in liqueurs and apéritifs, such as anisette and absinthe. In all of these cases, however, the employment of anethole, or still better, of anise oil (from *Pimpinella anisum*) is preferable.

Animals seem to relish food flavored with star anise oil; hence its wide use in all kinds of feed products.

The oil has its place also in the scenting of soaps, to which it imparts warm tonalities.

B. Oil of Star Anise Leaves

The leaves of the star anise tree (*Illicium verum* Hook. f.) contain a relatively high quantity (about 0.5 per cent) of an essential oil which can be isolated by steam distillation.

More than fifty years ago, Simon [47] reported that in the Pe-Se district of South China the natives had for quite some time been distilling oils from the leaves and terminal branches of the star anise tree. Shortly afterward Umney [48] investigated such leaf oils. Having determined the boiling range of the various oil fractions, Umney concluded that the leaf oils contained less anethole than those derived from the fruit, but more of the higher-boiling constituents (particularly anisaldehyde):

Boiling Range of Fraction	Fruit Oil (%)	Leaf Oil (%)
Below 225°.......	20	10
225°–230°........	65	60
Above 230°......	15	30

[46] *Ibid.*
[47] *Chemist Druggist* **53** (1898), 875.
[48] *Ibid.* **54** (1899), 323.

According to Gildemeister and Hoffmann,[49] however, it is doubtful whether anisaldehyde is an actual constituent of the leaf oil; the oil investigated by Umney may have been an old specimen, strongly oxidized by access of air.

More reliable data on star anise leaf oils were furnished a few years later by Eberhardt,[50] who distilled leaf material in Tonkin and obtained an oil with a congealing point of 13° C. (Oils distilled exclusively from fruit congeal at temperatures ranging from 16° to 18°.) From the congealing point observed by Eberhardt it can be concluded that the leaf oils contain less anethole than the fruit oils.

During his stay in the star anise oil producing regions, at the end of 1938, the author of the present work was able to gather additional information about the leaf oil from Mr. L. Drouet of Langson, an expert on the production of the oil. According to Drouet, the Agricultural Experiment Station of Phu had at various times distilled freshly harvested, green leaves and obtained an average yield of 0.5 per cent of oil. Experiments carried out by Drouet himself showed that the oils distilled from fresh (green) leaves exhibit congealing points as high as 17°, whereas the oils derived from fallen and wilted leaves (yield of oil, 0.55 per cent) have congealing points of about 13°. The oils from the fresh leaves, therefore, contain substantially more anethole than the oils from the wilted leaves; in fact, the former resemble the fruit oils as regards content of anethole. No wonder then that the native producers are tempted to mix leaf material with the fruit for purposes of distillation. (Cutting of the leaves deprives the tree of vital organs and does a great deal of harm. The natives in Tonkin, where the star anise oil industry is on a much higher level than in South China, therefore, abstain from cutting leaves for distillation; in Kwangsi, on the other hand, freshly cut, and even fallen leaves are often used for distillation. Hence the lower quality of the South Chinese star anise oils.)

It should also be mentioned here that the star anise leaf oils exhibit slight dextrorotation; for some time it was assumed, therefore, that dextrorotation in a star anise fruit oil indicates the presence of leaf oil. This position, however, cannot be maintained in the light of observations made by the author in French Indo-China (cf. the preceding section on "Oil of Star Anise Fruit").

SUGGESTED ADDITIONAL LITERATURE

Van den Driessen Mareeuw, "Unterscheidung von Anisöl und Sternanisöl," *Tech. Ind. Schweizer Chem. Ztg.* **27** (1934), 202. Cf. *Seifensieder-Ztg.* **62** (1935), 556.

[49] "Die Ätherischen Öle," 3d Ed., Vol. II, 575.
[50] *Compt. rend.* **142** (1906), 407.

M. Wagenaar, "Authentic and Poisonous Star Anise," *Pharm. Weekblad* **73** (1936), 1490. *Chem. Abstracts* **31** (1937), 211.

J. Small, "Star Anise," *Food* **12** (1943), 97, 107.

OIL OF STAR ANISE JAPANESE [1]

The Japanese star anise tree, *Illicium religiosum* Sieb. et Zucc. (fam. *Magnoliaceae*), known as "Shikimi" in Japan, is an evergreen tree with a trunk about 3 m. high, and pale, yellowish-white blossoms, which bloom from spring to summer. The tree grows wild in warm localities of southern and central Japan, on the Loochoo Islands, and in Formosa. It is grown extensively in the Prefecture of Nagasaki (chiefly on Goto Island), and to a smaller extent in the Prefectures of Kochi and Tokushima on the island of Shikoku.

For a long time the Japanese have been planting the star anise tree in temple compounds and in cemeteries in order to protect them from desecration by wild animals. This habit appears to have developed from the fact that the fruit of the tree is poisonous and that the leaves emanate a peculiar odor which is supposed to keep animals away. The custom developed to such an extent that even today altars during funeral services are decorated with leaves of *Illicium religiosum* Sieb.

The fruit of the Japanese star anise tree bears a striking resemblance to that of the Chinese tree (*Illicium verum* Hook. f.), except that the former is much smaller. In Japan the fruit is gathered for use in incense sticks, and for consumption as spice. It has stomachic and tonic properties, but owing to its toxicity can be consumed in very small quantities only. Some years ago a shipment of Chinese star anise, which contained admixed Japanese star anise, was exported to Germany and many people suffered from poisoning in Hamburg.

Of the total harvest of star anise in Japan (200,000 to 500,000 kin per year; 1 kin = 0.6 kg.), only about 10 per cent is used domestically, 90 per cent being exported. Prior to World War II, about 60 per cent of the exports went to India for use as incense in religious rites, while 40 per cent was shipped to China for admixture with the Chinese product and consumption as spice in cooking. From the end of hostilities to the end of 1949, about 60,000 kg. of Japanese star anise fruit were imported into the United States for unknown purposes.

The dried fruit of *Illicium religiosum* contains about 1 per cent of a volatile oil with an unpleasant odor, quite different from that of the Chinese product (*Illicium verum* Hook. f.). The oil is not produced commercially.

---

[1] The author is greatly obliged to Dr. Teikichi Hiraizumi, Tokyo, for much of the information contained in this monograph.

It was distilled experimentally years ago by Schimmel & Co.,[2] who reported these properties for a small number of oils derived from the fruit of the Japanese *Illicium religiosum:*

Specific Gravity at 15°......	0.984 to 0.985
Optical Rotation...........	−0° 50′ to −4° 5′
Acid Number.............	1.8
Ester Number.............	12.9
Solubility................	Soluble in 5 to 6 vol. of 80% alcohol, with separation of paraffins

One of the oils examined by Schimmel & Co. congealed at −18° with separation of *safrole.* The oil also contained *cineole,* and probably *linaloöl.* Anethole appeared to be absent.

Tardy [3] extracted dried and powdered fruits of *Illicium religiosum* with petroleum ether, and obtained 0.4 per cent of a concentrated extract. In it he noted the presence of *eugenol, cineole,* perhaps small quantities of *anethole* or *methyl chavicol,* and *safrole.* In the saponified distillation residue Tardy identified *palmitic acid* m. 62°.

Investigating the toxicity of the Japanese star anise fruit, Misaki [4] found that the poisonous principle is *hananomin,* a compound of the empirical molecular formula $C_{14}H_{22}O_{10}$. In addition, Misaki isolated from the fruit a phenolic and nontoxic substance $C_{14}H_{22}O_4$.

Takahashi [5] observed that the extract derived from the *bark* of *Illicium religiosum* Sieb. accelerates the coagulation of blood. He named the substance responsible for this effect "Illicin."

CONCRETE AND ABSOLUTE OF CHAMPACA

A. OIL OF *Michelia Champaca* L.
(True Champaca Oil)

The true champaca tree, *Michelia champaca* L. (fam. *Magnoliaceae*), occurs in many parts of tropical Asia, particularly in the Philippines and on the islands of the East Indian Archipelago. On Réunion Island and

[2] *Ber. Schimmel & Co.,* September (1885), 29; October (1893), 39 (Table), 46; April (1909), 52.

[3] "Étude Analytique sur Quelques Essences du Genre Anisique," Thèse, Paris (1902), 42.

[4] *J. Aichi Igakukai* **39** (1932), 1123.

[5] *J. Okayama Igakukai* **40** (1931), 1991.

in Nossi-Bé (Madagascar) it is cultivated, on a small scale, for the extraction of oil from its fragrant flowers. The flowers of the true champaca tree are orange-yellow. There are several other *Michelia* species—among them, *Michelia longifolia* Blume, which has white flowers (see below). These other species, however, are not commercially exploited.

The volatile oil contained in the flowers of the true champaca tree cannot be isolated by steam distillation for two reasons: the yield of oil is extremely small, and the odor of the distilled oils has no resemblance to that of the flowers. The first experimental extractions of the oil from the flowers were made in 1909 by Bacon,[1] who employed maceration in paraffin oil for this purpose. The oils thus obtained exhibited the fine odor characteristic of the flowers and had these properties:

Specific Gravity at 30°/30°...... 0.9543 to 1.020
Optical Rotation............... Too dark
Refractive Index at 30°......... 1.4550 to 1.4830
Saponification Number.......... 160 to 180

In these oils Bacon found 3 per cent of phenols (chiefly isoeugenol), 30 per cent of acids in ester form, 46 per cent of neutral substances with an odor reminiscent of bay oil, and a crystalline compound m. 165°–166°, to which he assigned the empirical molecular formula $C_{16}H_{20}O_5$. On saponification the oils lost their champaca odor.

Shortly afterward, Brooks [2] investigated champaca oils from Manila and reported the presence of *isoeugenol, benzyl alcohol, phenylethyl alcohol, benzaldehyde, cineole, p-cresol methyl ether,* and *a ketone* which was not identified. Its phenylhydrazone melted at 161°, the semicarbazone at 205°–206°. The ketone reacted quantitatively with bisulfite solution but could not be regenerated from the bisulfite compound. The same substance had been described somewhat earlier by Bacon (see above) who established the empirical molecular formula $C_{16}H_{20}O_5$, m. 165°–166°.

The oils described above were produced only experimentally; as a matter of fact, champaca oils from the Philippine Islands have never been made available in commercial quantities. There are at present only two sources of the genuine champaca oil, viz., Réunion Island and Nossi-Bé (Madagascar), where the oil is obtained, in the form of a concrete, by extraction with petroleum ether. (For the method, cf. Vol. I of this work, p. 200.) The yield of concrete varies from 0.16 to 0.20 per cent on Réunion Island, and from 0.13 to 0.15 per cent in Nossi-Bé. The concrete in turn yields about 50 per cent absolute.

[1] *Philippine J. Sci.* **4** (1909), A, 131; **5** (1910), 262.
[2] *Ibid.* **6** (1911), A, 333. *J. Am. Chem. Soc.* **33** (1911), 1763.

The absolute has a very smooth odor, floral and velvety—closely approximating that of the live flower. It recalls the fragrance of tea, orange blossoms, and ylang ylang.

Concrete of champaca flowers is produced in only small quantities (10 to 20 kg. per year in Nossi-Bé), but it constitutes one of the most exquisite raw materials for perfumery, being used in some of the finest French creations.

Igolen [3] submitted a concrete of champaca to co-distillation with glycol (method of Sabetay) and obtained an oil with these properties:

> Specific Gravity at 15°.................. 0.946
> Refractive Index at 20°................. 1.4895
> Acid Number......................... 6.2
> Ester Number........................ 70.1
> Carbonyl Number (Cold Method)........ 42.3
> Carbonyl Number (With Heating)........ 65

B. OIL OF *Michelia Longifolia* BLUME

This species of *Michelia* grows in tropical Asia, particularly in the Philippines and Java. The flowers are white, whereas those of true champaca (*Michelia champaca* L.) are yellow.

An oil distilled from fresh flowers more than fifty years ago in Java was investigated by Schimmel & Co.,[4] who noted these properties: d_{15} 0.883 and α_D $-12°$ 50′. The odor of the oil resembled that of basil.

In 1911, Brooks [5] examined an oil produced in the Philippine Islands and identified: *linaloöl, methyleugenol,* and the *methyl ester of methyl ethyl acetic acid,* to which, in the opinion of Brooks, the oil owes most of its odor. The oil also contained *a phenol* with a thymol odor.

C. COMMERCIAL SO-CALLED CHAMPACA OILS

The so-called "Champaca Oils" offered commercially about fifty years ago, but no longer on the market, appear to have been distilled from mixed flower material (*true* champaca and ylang ylang, or *true* champaca and *Michelia longifolia*). An oil distilled in Java, chiefly from the yellow flowers of *true* champaca, but with some mixture of the white blossoms of *Michelia longifolia*, and investigated by Schimmel & Co.,[6] had these properties:

[3] Cf. L. Girard, *Ind. parfum.* **2** (1947), 259.
[4] *Ber. Schimmel & Co.,* April (1894), 59.
[5] *Philippine J. Sci.* **6** (1911), A, 342. *J. Am. Chem. Soc.* **33** (1911), 1763.
[6] *Ber. Schimmel & Co.,* October (1906), 15; October (1907), 18.

Specific Gravity at 15°.............. 0.8861
Optical Rotation.................. −11° 10′
Acid Number..................... 10
Ester Number.................... 21.6
Ester Number after Acetylation...... 150.1
Solubility........................ Soluble in 2 vol. of 70% alcohol, tur-
bid in 4 and more vol.; soluble in 1
vol. of 80% alcohol, opalescence
and separation of paraffins in 7 and
more vol.

The oil contained about 60 per cent of *l-linaloöl*, a small quantity of *geraniol*, and *methyleugenol*. Only traces of terpenes were present. The first runs contained *esters of methyl ethyl acetic acid,* apparently the methyl and ethyl esters. The acid was present in the oil also in free form.

OIL OF MAGNOLIA

The most important of the various *Magnolia* species is undoubtedly *Magnolia grandiflora* L. (fam. *Magnoliaceae*). This stately tree, with its fragrant large flowers, is probably a native of the southern part of North America (perhaps South Carolina). It was introduced into Southern France and Italy at the end of the eighteenth century. Two types of magnolia oil have been described in literature, but neither is produced on a commercial scale.

A. Magnolia Flower Oil

Extracting fresh flowers of *Magnolia grandiflora* L., in the south of France, with highly purified petroleum ether (cf. Vol. I of this work, p. 200) Igolen [1] obtained from 1.2 to 1.63 per cent of a greenish-yellow concrete melting at 58°–60°, with an acid number of 28, and an ester number of 84. Two semisolid oils obtained by steam distillation of the concrete (yield 9.57 and 10.1) had these properties:

	I	II
Specific Gravity at 15°......................	0.900	0.903
Optical Rotation at 20° (in 25% benzene solution)	+0° 50′	+4° 36′
Refractive Index at 20°......................	...	1.5143
Acid Number.............................	11.84	9.33
Ester Number.............................	11.2	13.07

[1] *Rev. Marques Parfums France* **16** (1938), 33.

Both oils exhibited the characteristic odor of the flowers.
Nothing is known about the chemical composition of the flower oil.

B. Magnolia Leaf Oil

Steam distilling magnolia leaves gathered near Rome from April to September, Tommasi [2] obtained from 0.1 to 0.15 per cent of a volatile oil with a very agreeable odor; when exposed to air the oil became viscous. The properties were as follows:

Congealing Point	$-16°$
Boiling Range	Mostly between 170° and 265°
Specific Gravity at 15°	0.915 to 0.920
Specific Optical Rotation at 20°	$+1° 32'$ to $+1° 46'$
Refractive Index at 20°	1.5004 to 1.5020
Acid Number	1.9 to 2.3
Ester Number	26.9 to 30.3
Ester Content, Calculated as $C_{10}H_{17}OCOCH_3$	9.42 to 10.6%
Ester Number after Acetylation	51.0 to 55.5
Total Alcohol Content, Calculated as $C_{10}H_{18}O$	13.45%
Solubility	Soluble in 95% alcohol; soluble with turbidity in 30 vol. of 90% alcohol, and in 90 vol. of 80% alcohol

The oil contained about 3 per cent of phenols, 4 per cent of carbonyl compounds, cineole, and a mixture of probably sesquiterpenes and oxygenated compounds of unknown structure, in the fraction b. 230°–265°.

In this connection it should be mentioned that years ago Rabak [3] distilled the leaves of *Magnolia glauca* L., a species growing in North America, and obtained 0.05 per cent of a pale-yellow volatile oil with these properties:

Specific Gravity at 25°	0.9240
Specific Optical Rotation	$+3° 58'$
Refractive Index at 25°	1.4992
Acid Number	1.8
Ester Number	13.0
Ester Number after Acetylation	28.0
Solubility	Soluble in 3.5 vol. of 90% alcohol; insoluble in 80% alcohol

Suggested Additional Literature

St. Elmo Brady, "Phytochemical Study. Seed of the *Magnolia grandiflora.*" *J. Am. Pharm. Assocn.* **27** (1938), 407. *Chem. Abstracts* **32** (1938), 7668.

[2] *Rivista ital. essenze profumi* **10** (1928), 156.
[3] *Midland Druggist Pharm. Rev.* **45** (1911), 486.

CHAPTER XVI

ESSENTIAL OILS OF THE PLANT FAMILY *CAPRIFOLIACEAE*

CONCRETE AND ABSOLUTE OF HONEYSUCKLE

Quite a number of species of *Lonicera* grow wild in many parts of the world, or are cultivated in gardens. In the Grasse region of Southern France *Lonicera caprifolium* L. and *L. gigantea* L. (fam. *Caprifoliaceae*) are occasionally used for the extraction of their natural flower oils. Harvest takes place in mid-June. Treating the flowers of *Lonicera gigantea* with petroleum ether, Igolen [1] obtained 0.33 per cent of a dark green concrete, which on treatment with alcohol in the usual way gave 23.8 per cent of a viscous, olive green alcohol-soluble absolute. Steam distillation of the absolute yielded 9 per cent of a yellowish, thin liquid oil with these properties:

Specific Gravity at 15°	0.9012
Optical Rotation at 20°	±0°
Refractive Index at 20°	1.4613
Acid Number	25.2
Ester Number	145.6

The oil contained neither aldehydes, ketones, nor nitrogenous substances.

Nothing further is known about the chemical composition of the natural oil of honeysuckle.

In the past only very small quantities of absolute of honeysuckle have been produced in the Grasse region. The absolute is used in high-grade perfumes of the French type, to which it imparts alluring tonalities.

[1] *Parfums France* **15** (1937), 299.

CHAPTER XVII

ESSENTIAL OILS OF THE PLANT FAMILY *VIOLACEAE*

CONCRETE AND ABSOLUTE OF VIOLET

Of the numerous species and varieties of violet known to the horticulturist, there are only two which have been used for the extraction of their natural flower oils, viz., the Parma violet, and the Victoria violet. Both are probably derived from the sweet violet, *Viola odorata* L. (fam. *Violaceae*). In the course of the last 90 years the Parma has been replaced almost entirely by the Victoria. Both the violet flowers and the violet leaves contain a delightful and distinct perfume—the one quite different from the other—which years ago was recovered by maceration in hot fat, but is now extracted almost exclusively with volatile solvents. Today, as a matter of fact, only the leaves are used in the Grasse factories for the recovery of their perfume; the extraction of the flowers has been almost entirely abandoned. Several factors have been responsible for this development: the great amount of labor required in the harvesting of the flowers, the constantly increasing cost of labor during the last thirty years, the change in style of perfumes (away from violet), the introduction of new synthetics, by which the perfume of the violet flowers can be reproduced quite faithfully, etc. As a result, most violets grown today in Southern France are sold in the form of bouquets to florists in the principal cities of Europe, where they fetch much higher prices than the manufacturers of flower oils in Grasse could ever pay.

The Parma variety is characterized by compact, often double, pale blue, flowers. It is a tender plant, quite sensitive to cold, moisture, sun, and attacks by fungi and insects. A planting requires four years to get into full production. The flowers are collected twice a week, in the early morning after the dew has evaporated, or in the late afternoon. In 10 hr. one girl can pick from 3 to 4 kg. of flowers without peduncles, 1 kg. containing about 4,000 flowers. One hectare bears approximately 140,000 tufts, and when in full production yields about 800 kg. of flowers per season. The flowering season starts in early December, but the flowers gathered in December and January go to the florist trade, and it is only toward the end of the season, in February and March, that some of the flower material can be diverted to the extraction plants in Grasse.

Perfumers have always preferred the Parma variety over the Victoria for its very delicate and delightful scent. In fact, up to the turn of the century, the Parma was the only variety grown in Southern France. In 1900, for example, about 200 metric tons of Parma violet flowers were processed in Grasse, and this figure remained constant for several years. In addition, approximately 100 metric tons of leaves were treated every year in Grasse. Plantations were located from Vence to Spéracédès, near La Colle, Tourette, Le Bar, Magagnosc, and other villages of that picturesque and beautiful country. In the course of years, however, the plantings were attacked by fungi; many perished, and were never renewed. Today the only patch plantings remaining are those near Tourette, Le Bar, Magagnosc, and near Hyères. The small quantities of Parma flowers collected here are sold mostly to the florist trade; almost nothing is left for the extraction plants in Grasse. Prices have become exorbitant.

The Victoria violet, which is characterized by single flowers, larger than those of the Parma, and of deeper blue color, was introduced into Southern France about 1900. Being hardier and far more resistant to diseases than the Parma, and producing already in the second year after planting, the Victoria is much preferred by growers, although perfumers give preference to the Parma. Plantations of Victoria violets last six to seven years— much longer than those of the Parma. It is not surprising then that in the course of years the Parma variety was gradually, and almost entirely, replaced by the Victoria. The most important variety among the Victorias is the "Luxanne," distinguished by very fragrant, large flowers, of deep blue color, and with long peduncles. Plantations are located chiefly near Hyères, from where the flowers and leaves have to be transported to Grasse. The bulk of the Victoria blossoms goes to the florist trade, but from February on they are shipped also to Grasse. In March the *leaves* are cut and sold to Grasse. One hectare produces from 1,200 to 1,600 kg. of flowers per year. Their price is usually only one-fifth or one-sixth that of the Parmas.

In general, violets can be cultivated only in places sheltered from cold, as well as from exposure to the sun. For this reason plantings are usually located in olive groves, where they find the necessary shade. The soil should be rich in nitrogen and should perhaps contain some sulfurous compounds also.

Propagation is effected by means of stolons selected from older plantings, either at the beginning or at the end of winter. The young plants are set out 25 cm. apart, in rows 1 m. apart. Frequent irrigation is necessary. At the flowering stage the clumps have usually grown to a diameter of 25 to 35 cm.

As has been mentioned, the Parma violets are no longer used for the extraction of their natural perfume. Regarding the Victorias, substantial quantities of the *leaves* are still processed in Grasse for the preparation of the concrete and absolute. In 1939, for example, about 150 metric tons of leaves were worked up. During World War II the quantity diminished greatly; recently, however, it has been increasing again. On the other hand, production of concretes and absolutes of violet *flowers* (from the Parma as well as the Victoria) has declined almost to the vanishing point. Nevertheless, the various types of natural violet flower oils will now be discussed in more detail, because they offer considerable scientific interest.

A. Concrete and Absolute of Violet Flowers

Formerly, the natural perfume occurring in the flowers of violets was obtained exclusively by maceration with hot fat (cf. Vol. I of this work, p. 198). This process has been largely abandoned in favor of extraction with petroleum ether (Vol. I, pp. 200 ff.), whereby the so-called concrete and absolute of violet flowers are obtained. Naves and Mazuyer [1] reported yields of concrete (from Victoria violets) ranging from 0.09 to 0.13, in exceptional cases, up to 0.17 per cent. Parma violets yield from 0.07 to 0.12, rarely up to 0.17 per cent of concrete. In the experience of the author, from 1,100 to 1,150 kg. of Parma violets, or 1,200 to 1,400 Victoria violets are required for the production of 1 kg. of concrete.

Physicochemical Properties.—Concrete of violet flowers is a waxy, solid, yellow-greenish mass, only partly soluble in 95 per cent alcohol. The color of the concrete from Parma violets is somewhat lighter than that from the Victorias. The odor is most pleasant, and characteristic of the flowers. Victoria violets produce a concrete of harsher, more penetrating odor than the Parma; hence perfumers have always preferred the latter type for extraction, at least as long as they were available.

On treatment with alcohol in the usual way, concrete of Parma or Victoria violets yields from 35 to 40 per cent of alcohol-soluble absolute.

Walbaum and Rosenthal [2] (I), and Naves and Mazuyer [3] (II) reported for two concretes prepared from *Victoria* violets:

	I	II
Congealing Point	45.5°	36°–38°
Acid Number	11.2	64.4
Ester Number	95.2	49.8

Concrete (II) was of Italian origin.

Sabetay and Trabaud [4] examined a concrete from *Victoria* violets:

Drop Point (Ubbelohde's Apparatus)	49°–50°
Acid Number	62.6
Ester Number	53.7

[1] "Les Parfums Naturels," Paris (1939), 287.
[2] *Ber. Schimmel & Co., Jubiläums-Ausgabe* (1929), 193.
[3] "Les Parfums Naturels," Paris (1939), 288.
[4] *Ann. chim. anal. chim. appl.* [3], **23** (1941), 70. *Ber. Schimmel & Co.* (1942–43), 36.

Co-distillation of the concrete with ethylene glycol *in vacuo* yielded 4.4 per cent of a volatile oil. On treatment with strong alcohol, the concrete gave 35 per cent of absolute, which congealed at 16° to 20° and had these properties:

Refractive Index at 20°................ 1.4911
Acid Number........................ 120
Ester Number...................... 45.1

On distillation *in vacuo* the absolute yielded 8.35 per cent of a volatile oil (n_D^{20} 1.4637), which had a very fine odor of violet flowers and appeared to contain no eugenol (see below).

Sabetay and Trabaud [5] also examined a concrete extracted from *Parma* flowers (yield 0.12 per cent):

Drop Point (Ubbelohde's Apparatus).... 50°
Acid Number........................ 47.7
Ester Number...................... 58.6

This concrete yielded 43 per cent of a viscous, brownish absolute, which congealed below +15°.

Specific Gravity at 15°/15°............. 0.9539
Optical Rotation.................... +5° 2'
Refractive Index at 20°.............. 1.4932
Acid Number........................ 62
Ester Number...................... 42.9

On steam distillation *in vacuo* the concrete and the absolute gave 11.6 and 32.7 per cent, respectively, of volatile oil. Co-distillation of the concrete with ethylene glycol *in vacuo* yielded 11.7 per cent of volatile oil. These volatile oils had the following properties:

Specific Gravity at 15°/15°............. 0.9849
Optical Rotation.................... +10° 40'
Refractive Index at 20°.............. 1.5107
Methoxy Content.................... 8.06%

The odor of the volatile oil was peppery, "green," herb- and orris-like. (It should be mentioned here that these volatile oils are not produced commercially; they are prepared only for the purpose of analyzing concretes and absolutes—cf. Vol. I of this work, pp. 213, 215.)

Chemical Composition.—The chemical composition of the oil isolated from violet flowers was first investigated by Walbaum and Rosenthal,[6] who identified heliotropin by means of its semicarbazone m. 224°–225°. However, the occurrence of heliotropin in a natural flower oil appeared unusual

---

[5] *Compt. rend.* **209** (1939), 843.
[6] *Ber. Schimmel & Co., Jubiläums-Ausgabe* (1929), 193.

to Walbaum and Rosenthal, and they expressed doubt as to the purity of the concrete, although they had obtained it from a reputable house in Grasse.

The most thorough research on the chemical composition of violet flower oil is that of Ruzicka and his collaborators,[7] which extended over a period of more than ten years. Ruzicka and Schinz [8] steam-distilled 400 g. and 700 g. of two extracts from Victoria violet flowers, obtaining 23 g. and 50 g., respectively, of two volatile oils (I and II):

	I	II
Specific Gravity	d_4^{13} 0.956	d_4^{20} 0.896
Optical Rotation	$+8°\ 42'$	$+7°\ 36'$
Refractive Index	n_D^{16} 1.477	n_D^{20} 1.469

The following compounds were observed in the oil:

Benzyl Alcohol. Compared with the leaf oil, the oil from the flowers contains a little more benzyl alcohol.

2,6-Nonadien-1-ol. The so-called "violet leaf alcohol" (cf. Vol. II of this work, p. 164). Present in equal quantities in the flower oil and in the leaf oil (see below).

n-Hexanol, Heptenol(?), and Tertiary Octadienol(?). These three alcohols were observed in a fraction which had been hydrogenated. Therefore, it is not certain whether they actually occur as such in the original oil prior to hydrogenation, or whether they are the result of hydrogenation. Whatever the case, violet *flower* oil, in the opinion of Ruzicka, undoubtedly contains a number of alcohols, which are identical with those identified in violet *leaf* oil (see below).

2,6-Nonadien-1-al. The so-called "violet leaf aldehyde" (cf. Vol. II of this work, p. 322). Violet flower oil contains only $\frac{1}{10}$ the amount of "violet leaf aldehyde" that is present in the leaf oil.

Parmone. (Cf. Vol. II of this work, p. 495.) This ketone, isomeric with ionone, was isolated by means of reagent T of Girard and Sandulesco. Only 0.25 g. of parmone was obtained from 700 g. of absolute. Parmone has a very fine odor, characteristic of violet flowers. It does not occur in the leaf oil.

Eugenol. Investigating the volatile oil isolated by distillation of a concrete from *Parma* violets, Sabetay and Trabaud [9] found that it contained 21 per cent of eugenol. On the other hand, a small quantity of concrete from *Victoria* flowers was practically free of eugenol. The same authors [10] examined also violet *leaf* oils (from both Parma and Victoria violets) by odor tests and noted only trifling quantities of free eugenol.

Use.—When it was still produced on a sufficiently large scale, absolute of violet flowers was an esteemed and highly priced adjunct in expensive

[7] *Compt. rend. XVII Congr. Chim. Ind.* (1937), 915. *Drug Cosmetic Ind.* **41** (1937), 767. *Perfumery Essential Oil Record* **29** (1938), 174.
[8] *Helv. Chim. Acta* **25** (1942), 760.
[9] *Compt. rend.* **209** (1939), 843. *Ann. chim. anal. chim. appl.* **23** (1941), 70.
[10] *Perfumery Essential Oil Record* **31** (1940), 53.

perfumes, being used particularly in violet scents or in floral bouquets of the "heavier" and fancy type.

B. Concrete and Absolute of Violet Leaves

The perfume contained in the leaves of the violet plant is entirely different from that present in the flowers. Formerly it was isolated by submitting the leaves to maceration with hot fat at 60° to 80° (cf. Vol. I of this work, p. 198), but today extraction is accomplished with petroleum ether (Vol. I, p. 200), whereby the so-called concrete and absolute of violet leaves are obtained. Only Victoria violets are now used in Grasse; they have to be transported from the vicinity of Hyères, the principal violet growing region in Southern France.

In the author's own experience, about 1,000 to 1,100 kg. of violet leaves are required to yield 1 kg. of concrete. According to Naves and Mazuyer,[11] the yield of concrete varies greatly, depending upon the growth of the leaves. It ranges from 0.055 to 0.13 per cent, the average being 0.09 to 0.12 per cent.

Concrete of violet leaves contains from 4 to 12 per cent of volatile oil, which can be isolated by steam distillation. (These steam-volatile oils, however, are not produced commercially; they serve only for analytical examination of the concrete or absolute in question—cf. Vol. I of this work, pp. 213, 215.)

Physicochemical Properties.—Concrete of violet leaves is a solid, waxy, black-green, mass, only partly soluble in 95 per cent alcohol, and possessing an odor truly reminiscent of violet leaves.

Two concretes described by Naves and Mazuyer[12] (I), and by Girard[13] (II) had these properties:

	I	II
Melting Point	54°–55°	54.5°–55.5°
Congealing Point	∼50°	...
Specific Gravity at 60°	...	0.942
Acid Number	78.2	84
Ester Number	42.6	49.7

On treatment with strong alcohol in the usual way, concrete of violet leaves yields from 35 to 40 per cent, in exceptional cases up to 55 per cent, of absolute. Absolute of violet leaves is a viscous, dark green liquid, soluble in 95 per cent alcohol, and exhibiting the characteristic odor of violet leaves. The absolutes are marketed in concentrated form, or partly decolorized, in the latter case often diluted with solvents such as phthalates, glycols, and benzyl benzoate.

[11] "Les Parfums Naturels," Paris (1939), 288.　[13] *Ind. parfum.* **2** (1947), 219.
[12] *Ibid.*

Naves, Sabetay and Palfray [14] examined two volatile oils, which they had obtained by steam distillation *in vacuo* of two concretes of violet leaves:

	I	II
Specific Gravity at 15°.	0.925	0.913
Optical Rotation.	+16° 40'	+18° 12'
Refractive Index.	1.4808	1.4944
Acid Number.	12.4	...
Aldehyde Content, Calculated as Nonadienal	52.0%	...

Chemical Composition.—The chemical composition of violet leaf oil was first studied by Walbaum and Rosenthal,[15] later by Späth and Kesztler,[16] and particularly by Ruzicka and Schinz.[17] The following compounds have been reported in the steam-volatile oil derived from extracts of violet leaves:

Eugenol. Sabetay and Trabaud,[18] by olfactory tests, noted only traces of eugenol in violet leaf oil. (The *flower* oil derived from Parma violet contains 21 per cent of eugenol.)

2,6-Nonadien-1-al. The so-called "violet leaf aldehyde" (cf. Vol. II of this work, p. 322). This is the most important constituent of the oil, possessing a powerful odor typical of violet leaves. According to Ruzicka and Schinz, violet leaf oil contains from 30 to 50 per cent of this aldehyde, i.e., ten times more than the *flower* oil (see above).

n-Hexenol(?). In the neutral portions of the saponified oil Ruzicka and Schinz observed a primary hexenol, probably 3-hexen-1-ol (cf. Vol. II of this work, p. 158). These authors also reported in the saponified oil:

n-Heptenol(?). In optically active form.

n-Octenol(?). Also in optically active form.

2,6-Nonadien-1-ol. The so-called "violet leaf alcohol" (cf. Vol. II of this work, p. 164) occurs in violet leaf oil and violet flower oil in about the same proportion.

n-Hexanol(?) or *n*-2-Octen-1-ol(?). One of these two alcohols, probably the former, is present in the oil.

Benzyl Alcohol(?). Presence doubtful.

Violet leaf oil differs from other essential oils by the presence of the above listed unsaturated aliphatic aldehyde and alcohols. These alcohols occur in the oil free as well as esterified with:

Propionic Acid.

Enanthic Acid.

[14] *Perfumery Essential Oil Record* **28** (1937), 337.
[15] *Ber. Schimmel & Co., Jubiläums-Ausgabe* (1929), 211.
[16] *Ber.* **67** (1934), 1496.
[17] *Helv. Chim. Acta* **17** (1934), 1593, 1602; **18** (1935), 381.
[18] *Compt. rend.* **209** (1939), 843. *Perfumery Essential Oil Record* **31** (1940), 53.

An Octanoic Acid(?).   With a branched chain.

An Octenoic Acid(?).   Also with a branched chain.

Salicylic Acid.   Like the above named acids, present in ester form.

Palmitic Acid.   Also reported in the oil.

Use.—Absolute of violet leaves is a much esteemed and useful adjunct in the creation of high-grade perfumes of the French type.   It imparts unique tonalities not easily identified.   If skillfully blended, the absolute lends distinction and elegance to a perfume.   It can be used in many odor types, particularly "heavy" floral bouquets.

CHAPTER XVIII

ESSENTIAL OILS OF THE PLANT FAMILY *RESEDACEAE*

CONCRETE AND ABSOLUTE OF RESEDA

According to Louveau,[1] three varieties of *Reseda odorata* L. (fam. *Resedaceae*) were once used for the extraction of their perfume, viz., var. *gigantea, grandiflora,* and *pyramidalis.* Reseda (in France called "Mignonette") is a small plant with whitish flowers and brick colored anthers, possessing a peculiar, sweet and fresh odor.

Prior to World War I up to 40 metric tons of reseda flowers were processed annually in the Grasse region of Southern France. Production has now all but ceased. The center of reseda cultivation was in the fertile valley of the Siagne, near Pégomas, Mandelieu, and l'Abadie. For propagation, seed was planted in hotbeds in March; the harvest took place from May to July. One hectare yielded about 4,000 kg. of flowers.

In 1891, Schimmel & Co.[2] reported that, on steam distillation, fresh reseda flowers yielded 0.002 per cent of a volatile oil, which exhibited the characteristic reseda odor only in strong dilution. In the course of distillation, development of hydrogen sulfide took place. Because of the very low yield obtained by steam distillation, the flowers were then co-distilled with geraniol (500 kg. of flowers and 1 kg. of geraniol), and the product thus derived was marketed under the trade name "Reseda-Geraniol." Years later the reseda flowers were treated by the processes of *enfleurage,* or maceration with hot fat (see Vol. I of the present work, pp. 189, 198). More recently these processes have been replaced by that of extraction with petroleum ether (Vol. I, p. 200), whereby concretes and absolutes of reseda are obtained. In the author's experience, 1,150 to 1,200 kg. of reseda flowers are required to yield 1 kg. of concrete which, in turn, gives about 0.35 kg. of alcohol-soluble absolute. Naves and Mazuyer[3] reported yields ranging from 0.07 to 0.15, and in some cases even 0.26 per cent of concrete, which gives 30 to 35 per cent of absolute. The concrete contains 3.8 to 5.5 per cent of steam-volatile oil. According to Girard,[4] reseda flowers yield from 0.15 to 0.18 per cent of concrete.

Physicochemical Properties.—Four concretes described by Naves and Mazuyer (I and II), and by Girard (III and IV) had these properties:

[1] *Rev. marques* (1930), 139.
[2] Gildemeister and Hoffmann, "Die Ätherischen Öle," 3d Ed., Vol. II, 775.
[3] "Les Parfums Naturels," Paris (1939), 264.
[4] *Ind. parfum.* **2** (1947), 222.

	I	II	III	IV
Melting Point	46°	49°	53°	50°–51°
Specific Gravity at 60°	...	...	0.9178	0.9265
Acid Number	66	84.2	75.6	91
Ester Number	48.6	66.6	61.6	57.4
Content of Sulfur	...	...	1.66%	1.70%

The absolute is a viscous or semisolid mass of red-brown color, with a strong yet delicate odor, characteristic of the fresh flowers, but "heavier" and more fatty.

Naves and Mazuyer (V) and Girard (VI) reported the following values for two absolutes:

	V	VI
Melting Point	...	30.5°
Specific Gravity	...	0.9563
Acid Number	92.4	86.8
Ester Number	76.3	82.6
Sulfur Content	2.24%	1.62%

Steam-distilling an absolute of reseda, von Soden [5] obtained a volatile oil (yield: 0.003 per cent, calculated upon the flowers) with these properties:

Specific Gravity at 15°	0.961
Optical Rotation	+31° 20'
Acid Number	16.1
Ester Number	85

Naves, Sabetay and Palfray [6] described two steam-volatile oils, which they obtained from two concretes of reseda, with a yield of 3.8 and 5.5 per cent, respectively:

	I	II
Specific Gravity at 15°	0.972	0.966
Optical Rotation	+27° 42'	+37° 16'
Refractive Index at 20°	1.4944	1.4886
Acid Number	2.8	3.9
Ester Number	76.6	84.1

(Such volatile oils are not commercially available.)

Chemical Composition.—Little is known about the chemical composition of the reseda flower oil. As far back as 1913, Kerschbaum [7] reported the presence of *farnesol*, without giving any definite proof. More than twenty-five years later, Walbaum and Rosenthal [8] found that an extract from the

[5] *J. prakt. Chem.* **69** (1904), 264.

[6] *Perfumery Essential Oil Record* **28** (1937), 337.

[7] *Ber.* **46** (1913), 1732.

[8] *Ber. Schimmel & Co., Jubiläums Ausgabe* (1929), 221.

stalks, leaves, and flowers contained volatile sulfurous compounds, but failed to identify β-phenethyl isothiocyanate (see Vol. II of this work, p. 741), which had previously been found in reseda *root* oil by Bertram and Walbaum.[9] Sabetay, Igolen, and Monod [10] established the presence of *acetic, caprylic,* and solid *fatty acids, phenol* C_6H_5OH, *eugenol* and *paraffins* in the volatile oil derived by co-distillation with diethylene glycol from a concrete of reseda flowers.

Use.—In years past, when it was still produced in substantial quantities, reseda flower oil was used in high-grade perfumes of the French type, to which it imparted alluring tonalities.

[9] *J. prakt. Chem.* [2], **50** (1894), 555.
[10] *Ind. parfum.* **3** (1948), 85. Cf. Girard, *ibid.* **2** (1947), 222.

CHAPTER XIX

ESSENTIAL OILS OF THE PLANT FAMILY *SAXIFRAGACEAE*

CONCRETE AND ABSOLUTE OF *PHILADELPHUS CORONARIUS* L.

There are many species of *Philadelphus,* commonly called "Mock-Orange," sometimes also referred to as "Syringa," although the latter term should actually be restricted to the lilacs. *Philadelphus coronarius* L. (fam. *Saxifragaceae*), a native of Southern France, is called "Deutsches Jasmin," or "Pfeifenstrauch" in Germany, and "Seringat" in France. The bush occurs wild in the forests of Central and Southern Europe and is also cultivated for ornamental purposes in gardens. The flowers are white and emit a strong odor.

Extracting flowers of *Philadelphus coronarius* with petroleum ether, Treff, Ritter, and Wittrisch [1] obtained 0.237 per cent of a concrete which, on treatment with alcohol in the usual way, yielded 52.2 per cent of an alcohol-soluble absolute. Steam distillation of the concrete gave 2.5 per cent of a yellowish distillate with a powerful odor characteristic of the flowers. The oil had these properties:

Specific Gravity at 15°.............	0.947
Optical Rotation..................	±0°
Acid Number.....................	28
Ester Number....................	73
Ester Number after Acetylation......	224

The oil probably contained methyl anthranilate.

Farmiloe [2] obtained a yield of 0.25 per cent of concrete by extraction of the flowers with petroleum ether. The concrete gave 38.2 per cent of absolute.

In 1938 Igolen [3] carried out a number of experiments, extracting flowers of *Philadelphus coronarius* in the Grasse region of Southern France. Extraction with petroleum ether gave concretes with the most characteristic odor. Yields of concrete by extraction with petroleum ether ranged from 0.144 to 0.179 per cent. The concrete gave 25 to 27.2 per cent of absolute, a viscous, reddish-brown, liquid of strong fruity, somewhat harsh odor, reminiscent of the living flowers. Steam distillation of the absolute yielded

[1] *J. prakt. Chem.* [2], **113** (1926), 358.
[2] *Perfumery Essential Oil Record* **20** (1929), 321.
[3] *Parfums France* **16** (1938), 92.

9 per cent of a yellowish volatile oil with an odor characteristic of the flowers. The oil had these properties:

$$
\begin{array}{ll}
\text{Specific Gravity at } 15° & 0.912 \\
\text{Optical Rotation at } 24° & +3° \, 45' \\
\text{Refractive Index at } 20° & 1.4668 \\
\text{Acid Number} & 25.2 \\
\text{Ester Number} & 95.2
\end{array}
$$

The natural flower oil of *Philadelphus coronarius* is not produced commercially,[4] since the perfume of the flowers can be reproduced quite satisfactorily by blends of various synthetic aromatics, aromatic isolates, and essential oils.

[4] Recently Meunier [*Ind. parfum.* **5** (1950), 28] reported that a house in Grasse had succeeded in developing a satisfactory concrete, the solvent used for this purpose being butane.

CHAPTER XX

ESSENTIAL OILS OF THE PLANT FAMILY *CARYOPHYLLACEAE*

CONCRETE AND ABSOLUTE OF CARNATION
(Carnation Flower Oil)

Dianthus caryophyllus L. (fam. *Caryophyllaceae*), the common garden carnation, is cultivated on the French Riviera (Hyères, Vence, Antibes), and on the Italian Riviera (Ventimiglia, Bordighera, San Remo), chiefly for the cut flower trade. Surplus flowers which cannot be sold to florists, particularly at the end of the flowering season in June, are occasionally hauled to Grasse (A.M., France) and submitted to extraction with volatile solvents (petroleum ether), yielding the so-called concrete and absolute of carnation.[1] Steam distillation of the flowers gives such a poor yield of oil that the process cannot be applied.

The highest yield of flower oil is obtained from carnations that have been cut after a period of ample sunshine. The most common varieties, particularly those of lightest color, yield the most oil. According to Naves and Mazuyer,[2] carnation flowers, on extraction with petroleum ether, yield from 0.23 to 0.29, and in some cases even 0.33, per cent of concrete which, on treatment with alcohol in the usual way, gives 9 to 12 per cent of alcohol-soluble absolute. (On steam distillation, the concrete yields from 1 to 3.6 per cent of a volatile oil which, however, is not a commercial article.) Sabetay and Mane[3] reported yields of concrete ranging from 0.21 to 0.26 per cent; the concretes gave from 20 to 25 per cent of absolute. A sample of absolute, on steam distillation in the apparatus of Naves, yielded 7 per cent of a volatile oil. Treff, Ritter and Wittrisch[4] extracted carnation flowers in Riesa (Saxony) with petroleum ether, and obtained 0.282 per cent of concrete; the latter gave 32.9 per cent of absolute. The concrete contained 7.7 per cent of steam-volatile oil. Treff and Wittrisch[5] reported yields of 0.289 per cent of concrete, 0.088 per cent of absolute, and 0.00432 per cent of steam-volatile oil, all figures calculated upon the flower material.

Physicochemical Properties.—*Concrete* of carnation is a solid, waxy mass of light green color; the odor is "green," not characteristic of the flowers. Naves and Mazuyer[6] reported these properties for a concrete of carnation from Grasse:

[1] For details see Vol. I of this work, p. 200 ff.
[2] "Les Parfums Naturels," Paris (1939), 250.
[3] *Rev. chim. ind.* **48** (1939), 39.
[4] *J. prakt. Chem.* **113** (1926), 357.
[5] *Ibid.* **122** (1929), 332.
[6] "Les Parfums Naturels," Paris (1939), 250.

Melting Point.............. 56° to 62°
Congealing Point.......... 58° to 54°
Optical Rotation.......... ±0°
Acid Number.............. 15.2
Ester Number............. 23.8

Absolute of carnation is a viscous, dark brown oil, with a pleasant scent reminiscent of the live flowers. Girard [7] reported the following values for an absolute prepared by A. Chiris:

Specific Gravity at 15°...... 0.8908
Optical Rotation........... +5° 0'
Refractive Index........... 1.4712
Acid Number.............. 49
Ester Number............. 51.8

The steam-volatile oils obtained by Sabetay and Mane (I), and Treff and his collaborators (II and III) had these properties:

	I	II	III
Specific Gravity at 15°......	0.9726	1.010	1.0375
Optical Rotation...........	$[\alpha]_D$ −4° 0'	−0° 36'	−0° 39'
Refractive Index at 20°.....	1.4908	. . .	. . .
Acid Number.............. ...		28	16.8
Ester Number............. ...		132.0	131.6
Ester Number after Acetylation..................... ...		249.0	247.8
Phenol Content............	43%	. . .	. . .
Methoxy Content..........	7.2%	. . .	. . .

Adulteration.—In the past, when substantial quantities were produced, the concretes and absolutes of carnation were often adulterated with synthetic aromatics, such as eugenol, isoeugenol, and phenylethyl alcohol. These, when skillfully applied, imparted to the concretes and absolutes a more true-to-nature character than they actually possess. In point of fact, many so-called concretes and absolutes of carnation consisted largely of mixtures.

Chemical Composition.—The first investigation of the steam-volatile oil of carnation was carried out by Glichitch,[8] but to little effect. Heptacosane $C_{27}H_{56}$, m. 53°–54°, was identified in the oil.

A more thorough examination of the oil by Treff and Wittrisch [9] (see above) revealed the presence of the following compounds:

Eugenol (30%). Identified by means of its phenylurethane m. 95.5°–96°.

Phenylethyl Alcohol (7%). Phenylurethane m. 80°.

[7] *Ind. parfum.* **2** (1947), 188.   [9] *J. prakt. Chem.* **122** (1929), 332.
[8] *Bull. soc. chim.* [4], **35** (1924), 205.

Benzyl Benzoate (40%). Phenylurethane of the benzyl alcohol, m. 78°; benzoic acid m. 121°.

Benzyl Salicylate (5%). Salicylic acid m. 155°.

Methyl Salicylate (1%).

Total Production and Use.—About fifty years ago, up to 200 metric tons of carnation flowers were processed yearly in the Grasse region for the extraction of their flower oil. Since then this quantity has been diminishing steadily, and today only a few tons of carnations are treated.

Absolute of carnation is used in high-class perfumes of the floral as well as oriental types; it imparts a natural character to synthetic blends and acts as an excellent fixative.

CHAPTER XXI

ESSENTIAL OILS OF THE PLANT FAMILY *PRIMULACEAE*

CONCRETE AND ABSOLUTE OF CYCLAMEN

Cyclamen europaeum L. (fam. *Primulaceae*), a small, pretty plant, grows wild in several parts of Europe, particularly in the Alps and in the Jura. The flowers emit a delightful and sweet odor. However, the natural flower oil of cyclamen is not produced commercially, because the perfume of this plant can be reproduced quite satisfactorily by blends of synthetic aromatics, aromatic isolates, and essential oils.

Experimentally macerating cyclamen flowers of an unidentified species with hot fat, extracting the perfumed fat (*pommade*) with low boiling volatile solvents, and the concentrated extract with alcohol, Elze [1] obtained an absolute of cyclamen (d_{15} 0.94467, α_D $\pm 0°$) in which he identified *nerol* and *farnesol*. In addition, the absolute contained ketones, aldehydes, phenols and esters which, however, were not further characterized.

Gerhardt [2] extracted *Cyclamen europaeum* with petroleum ether, and obtained 0.18 per cent of a greenish-yellow concrete possessing a strong odor.

[1] *Riechstoff Ind.* **3** (1928), 91.
[2] *Perfumery Essential Oil Record* **20** (1929), 317.

CHAPTER XXII

ESSENTIAL OILS OF THE PLANT FAMILY *TILIACEAE*

CONCRETE AND ABSOLUTE OF LINDEN BLOSSOM

The blossoms of the linden tree, including the three species *Tilia cordata* Mill., *T. tomentosa* Moench., and *T. platyphyllos* Scop. (fam. *Tiliaceae*), emit a strong and delightful perfume which can be reproduced quite satisfactorily by mixtures of certain synthetic aromatics (hydroxycitronellal, etc.) and essential oils. The natural flower oil of the linden blossoms is, therefore, not produced commercially.

In 1938 Igolen[1] experimentally extracted linden blossoms in the south of France. The flowers from the region of Vaison-la-Romaine had a particularly suave odor. Extraction of fresh flowers with petroleum ether yielded 0.33 per cent of concrete, while extraction of dried flowers gave 0.915 per cent of concrete. Both concretes were hard, waxy masses of dark green color, with an herb-like odor reminiscent of dried hay, but not characteristic of linden blossoms. On treatment with alcohol in the usual way the two concretes gave 32 and 19 per cent, respectively, of viscous, greenish absolutes. Steam distillation of the absolute from the fresh flowers yielded 5.7 per cent of a yellowish, semisolid oil with these properties:

Specific Gravity at 15°............. 0.913
Optical Rotation at 28°............. −3° 20′
Refractive Index at 20°............. 1.4736
Acid Number.................... 44.8
Ester Number.................... 112.2
Ester Number after Acetylation..... 163.9
Carbonyl Number................ 25

The odor of the oil was only vaguely reminiscent of the linden blossoms and noticeable only in strong dilutions.

Years ago Kerschbaum[2] reported the presence of *farnesol* in natural linden blossom oil. Otherwise, nothing is known about its chemical composition.

[1] *Rev. marques parfums France* **16** (1938), 111.
[2] *Ber.* **46** (1913), 1732.

CHAPTER XXIII

ESSENTIAL OILS OF THE PLANT FAMILY *COMPOSITAE*

OIL OF *ACHILLEA MOSCHATA* L.
("Iva Oil")

Achillea moschata L. (family *Compositae*), the "Musk Yarrow," vernacularly called "Iva" in Europe, is an aromatic plant growing wild in the Alps at high altitudes. Because of its characteristic odor and flavor, caused by the presence of a volatile oil, *Achillea moschata* is used in the preparation of certain alcoholic liqueurs, particularly the well-known "Iva Liqueur." On steam distillation the dried flowering plant yields from 0.3 to 0.6 per cent of a yellowish-green—occasionally greenish-blue or dark blue—oil with a strong aromatic, somewhat narcotic odor, reminiscent also of valerian, cineole, thujone, and musk.

According to Gildemeister and Hoffmann,[1] the oil exhibits these properties:

Specific Gravity at 15°.............	0.925 to 0.959
Optical Rotation...................	−12° 30′ to −17° 0′
Refractive Index at 20°............	1.4730 to 1.4763
Acid Number......................	5 to 21
Ester Number.....................	18 to 44
Ester Number after Acetylation......	78 to 115.4
Solubility.........................	Soluble in about 1 vol. and more of 80% alcohol, usually with separation of paraffins

The presence of the following compounds has been reported in the oil:

Cineole.   Characterized by the iodol reaction (Schimmel & Co.[2]).

An Aldehyde.   Perhaps valeraldehyde (Schimmel & Co.[3]).

l-Camphor.   Identified by means of semicarbazone m. 237.5° (Schimmel & Co.[4]).

Palmitic Acid.   Reported by Haensel.[5]

Alcohols(?).   From the ester number after acetylation it appears that the oil also contains certain alcohols which, however, have not been identified.

"Ivaol"(?).   Von Planta-Reichenau[6] submitted the oil to fractional distillation and obtained a fraction b. 170°–210° to which he assigned the name "Ivaol," $C_{24}H_{40}O_2$. However, this substance was probably not a uniform compound, as can be concluded from the wide boiling range.

[1] "Die Ätherischen Öle," 3d Ed., Vol. III, 983. Cf. Gavalovski, *Pharm. Post* (1891), 153. *Ber. Schimmel & Co.,* October (1915), 16.
[2] *Ber. Schimmel & Co.,* October (1894), 27.   [5] *Chem. Zentr.* (1907), II, 1620.
[3] *Ibid.,* April (1912), 74.   [6] *Liebigs Ann.* **155** (1870), 148.
[4] *Ibid.*

Recently, Berk [7] distilled dried flowering *Achillea moschata* in Switzerland, and obtained 0.38 per cent of an oil (d_4^{18} 0.9436, α_D $-7°$ 42′ [2.5 cm.], r_D^{20} 1.4770, acid number 18, ester number 19.9, acetylation number 111.2), in which the presence of the following compounds was observed:

Esters, calculated as bornyl acetate. 6.9%

Free Alcohols, calculated as borneol. 30.9%

Volatile Acids of low molecular weight.

Palmitic Acid.

A Phenolic Compound.

Aldehydes.

l-Camphor.

Cineole. 22.9%

Chamazulene.

A Paraffin.

According to the author's knowledge, the oil is not produced today on a commercial scale.

OIL OF ARNICA

Arnica montana L. (fam. *Compositae*), is a herbaceous perennial, native to northern and central Europe, where it thrives on mountain meadows and upland moors. The flowers, leaves, and roots are employed in medicinal preparations.

According to Stockberger,[1] the plant requires a marshy soil, abundant rainfall, and a cool climate for its best development. It is propagated by divisions of the roots or from seeds sown in either the fall or the spring. Seed may also be sown in August in a seedbed and the plants transplanted the following spring to stand about 18 in. apart in the row. The flowers can be harvested the second year and the roots after three or four years.

Arnica is not produced commercially in the United States, and the small quantity imported annually is apparently sufficient to meet the market

[7] Thesis, l'Ecole Polytechnique Fédérale, Zurich, 1949. Copy kindly supplied by Dr. Asuman Berk, Istanbul.
[1] "Drug Plants under Cultivation," *U. S. Dept. Agr., Farmers' Bull.* No. 663 (1939), 13.

demands. Its cultivation presents many difficulties, and efforts to grow it in the milder sections of this country have generally proved unsuccessful.

A. OIL OF ARNICA FLOWERS

On distillation, the flowers of *Arnica montana* yield from 0.04 to 0.14 per cent of an essential oil with a reddish-yellow to brown color, and a strong aromatic odor and flavor. At room temperature the oil is of butter-like consistency, melting to a brownish liquid at 20° to 33°.

Gildemeister and Hoffmann [2] reported these properties for the volatile oil distilled from arnica flowers:

Specific Gravity at 30°.................	0.891 to 0.922
Acid Number.......................	62.5 to 127.3
Ester Number......................	22.5 to 32.2
Ester Number after Acetylation......	58.8 and 81.2 (two determinations)
Solubility.........................	Very difficultly soluble, even in absolute alcohol

Nothing is known about the chemical composition of the volatile oil derived from arnica flowers, except that it contains an acid m. 61° (Gildemeister and Hoffmann). On the other hand, much work has lately been done on the nonvolatile constituent of the arnica flowers (triterpenes and triterpenediols, etc.).[3] A discussion of these substances, however, falls beyond the scope of this book.

B. OIL OF ARNICA ROOT

On distillation, the freshly dried roots of *Arnica montana* yield from 0.5 to 1.5 per cent of an essential oil with a light yellow color that darkens on aging. The odor of the oil is reminiscent of that of radish, the flavor strong and aromatic.

Gildemeister and Hoffmann [4] reported these properties for the volatile oil distilled from arnica roots:

Specific Gravity at 15°..........	0.982 to 1.00
Optical Rotation...............	+0° 25' to −2° 38'
Refractive Index at 20°.........	1.507 to 1.509
Acid Number..................	2 to 10
Ester Number.................	56 to 100

[2] "Die Ätherischen Öle," 3d Ed., Vol. III, 1038.
[3] Cf. Dieterle and Engelhard, *Arch. Pharm.* **278** (1940), 225. Dieterle and Schreiber, *ibid.* **279** (1941), 312. Zimmermann, *Helv. Chim. Acta* **26** (1943), 642. Jeger and Lardelli, *ibid.* **30** (1947), 1020.
[4] "Die Ätherischen Öle," 3d Ed., Vol. III, 1039.

Ester Number after Acetylation... 82.1 (one determination)
Solubility...................... Soluble in 7 to 12 vol. of 80% alcohol,
                                 soluble in 0.5 to 6 vol. of 90% alcohol;
                                 in both cases occasionally with tur-
                                 bidity

The chemical composition of arnica root oil was investigated almost eighty years ago by Sigel,[5] who found that one-fifth of the oil consists of the *isobutyric ester of phlorol* (phlorol = o-ethylphenol), while four-fifths consists of *thymohydroquinone dimethyl ether* (cf. Vol. II of this work, p. 553) and small quantities of *phlorol methyl ether*.

Years later Kondakov [6] confirmed the presence of phloryl isobutyrate and thymohydroquinone dimethyl ether in the oil; in addition, he isolated a saturated hydrocarbon b. 176°–180°, a solid substance m. 69°, and a sulfur-containing compound.

Production and Use.—To the author's knowledge, very little (if any) of this oil is produced today. It could be used in fine liqueurs as a replacement for the tinctures of root and flowers of the herb.

SUGGESTED ADDITIONAL LITERATURE

Ludwig Kroeber, "*Arnica montana* L.," *Pharmazie* 4 (1949), 26.

OIL OF *ARTEMISIA TRIDENTATA* NUTT.
(Oil of American Sagebrush)

One of the most common of North American desert plants, growing in several parts of western United States, in the more arid sections, is the native "black" sagebrush, *Artemisia tridentata* Nutt. (fam. *Compositae*). Some thirty species of *Artemisia* have been distinguished in these regions, but *tridentata* is by far the most common. It furnishes browse for countless numbers of cattle, sheep, deer, and many smaller animals as well. This is particularly true in winter, when most desert plants have dropped their leaves or are covered with snow. Since the sagebrush retains its leaves through freezing weather, and is sufficiently tall to extend above the snow, desert range animals survive on the plant. In summer, even though an

[5] *Liebigs Ann.* **170** (1873), 345.   [6] *J. prakt. Chem.* [2], **79** (1909), 505.

abundance of feed is available, range animals continue to eat some sage. Consequently, the plant is of great economic importance.

Preliminary investigations by Adams and Billinghurst,[1] and Adams and Oakberg [2] have shown that the leaves and green twigs of *Artemisia tridentata* contain about 1 per cent of a volatile oil which can be isolated by steam distillation. The most thorough work on the chemical composition of this oil, however, was carried out by Kinney and his collaborators.[3]

Adams and Billinghurst found that the maximum yield of oil was obtained in the late summer, or in the fall; moreover, that the oil was most easily removed from material that had been air-dried prior to steam distillation. Kinney et al. reported yields of oil ranging from 0.45 per cent (from fresh plant material) to 1.26 per cent (from dried material).

Physicochemical Properties.—Several lots of oil of *Artemisia tridentata* distilled by Kinney from plant material growing wild in Utah (near Salt Lake City), and analyzed by Fritzsche Brothers, Inc., New York, had properties varying within these limits:

Specific Gravity at 15°/15°..........	0.928 to 0.940
Optical Rotation...................	+6° 56′ to +12° 30′
Refractive Index at 20°.............	1.4662 to 1.4729
Saponification Number..............	12.0 to 23.3%
Ester Number after Acetylation......	41.2 to 76.8%
Solubility........................	Soluble in 2 to 2.5 vol. of 70% alcohol and more

Oil of *Artemisia tridentata* is a liquid of powerful, camphoraceous, stinging, and lachrymatory odor, in great dilution slightly reminiscent of witch hazel.

Chemical Composition.—In 1934, Adams and Oakberg [4] reported that the oil consists of:

	Per Cent
"Artemisal"...........	5
α-Pinene.............	20
Cineole..............	7
l-Camphor...........	40
Sesquiterpenes........	12
Resins..............	16
	——
	100

With a view to elucidating the nature of the aldehyde "Artemisal," Kinney, Jackson, DeMytt and Harris [5] more recently investigated the oil, and identified the following compounds:

[1] *J. Am. Chem. Soc.* **49** (1927), 2895.
[2] *Ibid.* **56** (1934), 457.
[3] *J. Org. Chem.* **6**, No. 4 (1941), 612. Cf. *ibid.* **8**, No. 3 (1943), 290.
[4] *J. Am. Chem. Soc.* **56** (1934), 457.   [5] *J. Org. Chem.* **6**, No. 4 (1941), 612.

Methacrolein. The aldehyde named "Artemisal" by Adams and Oakberg was shown to be methacrolein (cf. Vol. II of this work, p. 318). It is responsible for the lachrymatory action of the oil.

α-Pinene. Identified by means of its nitrosochloride m. 100°, in the small fraction b. 151°–152°.

Terpenes(?). In addition to *α*-pinene, a lower and a higher boiling terpene were isolated, but not identified. The higher boiling terpene (in the small fraction b. 152°–153°) was perhaps *β*-pinene.

Cineole. Characterized in the fraction b. 167°–168° by its reaction with phosphoric acid and with resorcinol.

α-Terpinene. Identified in the decineolated residue of fraction b. 164°–168°, by preparation of the nitrosite m. 154°.

d-Camphor. Observed in the fractions boiling above 193°, from which large amounts of camphor crystallized.

Artemisol and Acetate. The fractions $b_{16.5}$ 115°–120° contained a liquid alcohol, isomeric with terpineol; it was named "artemisol" (cf. Vol. II of this work, p. 754). The acetate of artemisol appeared to be present in this fraction also.

Kinney and collaborators obtained negative results when testing the oil for the presence of limonene, dipentene, terpinolene, sylvestrene, and phellandrene.

Production and Use.—Although enormous quantities of wild growing herb material are available for distillation, oil of sagebrush cannot be produced economically in the United States, because of the high cost of labor. The oil could be used in certain types of deodorants, insect sprays, and perhaps for oil flotation but, in respect to this last, would have to compete with eucalyptus oil and low-priced camphor fractions, natural and synthetic.

OIL OF *ARTEMISIA VULGARIS* L.

Oil of *Artemisia vulgaris* L., the common mugwort, is a much-branched perennial shrub which occurs wild in Europe and Asia, and has also become naturalized in the eastern areas of North America. All parts of the plant contain a volatile oil, which can be isolated by steam distillation. According to Gildemeister and Hoffmann [1] the yield of oil from the roots averages about 0.1 per cent, that from the herb ranges from 0.026 to 0.2 per cent.

---

[1] "Die Ätherischen Öle," 3d Ed., Vol. III, 1018.

(*Top Left*) Close-up of a tuberose plant. (*Top Right*) Harvest of tuberose flowers near Khémisset, Morocco. *Photos Mr. Pierre Chauvet, Seillans (Var), France.* (*Bottom*) Branch of a mimosa tree in full bloom. *Photo Fritzsche Brothers, Inc., New York.*

(*Top*) Production of star anise oil in French Indo-China. A native field distillery near Langson. (*Bottom*) Star anise oil, French Indo-China. Two native stills located on the edge of a rice field. The retorts are imbedded in the hearth. The still tops serve as condensers. Between the two stills are two oil separators arranged in cascade form. *Photos Fritzsche Brothers, Inc., New York.*

The root oil is a yellowish-green, crystalline mass of buttery consistency, and has a disagreeable, bitter, at first burning, and then cooling, flavor. The odor of the herb oil is not strongly characteristic.

Little is known about the physicochemical properties of the oils derived from the roots and the overground parts of the plant, probably because these oils have no commercial importance.

As regards the chemical composition, Schimmel & Co.[2] years ago noted the presence of *cineole* in the *herb* oil.

More recently Stavholt and Sörensen[3] steam distilled 5.5 kg. of *roots* of *Artemisia vulgaris*, obtaining 1.1 g. of essential oil, in which they noted the presence of *dehydromatricaria ester* (methyl *n*-decenetriynoate), $C_{11}H_8O_2$, m. 113°. This ester is assumed to possess either of the following structural formulas:

$$CH_3 \cdot CH{=}CH \cdot C{\equiv}C \cdot C{\equiv}C \cdot C{\equiv}C \cdot COOCH_3$$

or

$$CH_3 \cdot C{\equiv}C \cdot C{\equiv}C \cdot C{\equiv}C \cdot CH{=}CH \cdot COOCH_3$$

So far as the author knows, the oil has never been produced commercially.

[2] *Ibid.*
[3] *Acta Chem. Scand.* **4** (1950), 1567. *Chem. Abstracts* **45** (1951), 7005.

OIL OF *BLUMEA BALSAMIFERA* DC.

Blumea balsamifera DC., a shrub-like plant of the family *Compositae*, is a native of India; it occurs wild from the Himalaya Mountains to Singapore, in Burma, Tonkin, South China, Formosa, Hainan, in the Malayan Archipelago, and the Philippine Islands.

The stalks and leaves of the plant contain from 0.1 to 0.4 per cent of an essential oil which can be isolated by hydrodistillation.[1,2] Experiments carried out by the Indian Forest Department yielded as much as 1.88 per cent of oil.[3]

According to Gildemeister and Hoffmann,[4] the so-called "Ngai-camphor"

[1] Bacon, *Philippine J. Sci.* **4A** (1909), 127.
[2] Cayla, *J. Agr. Tropicale* **8** (1908), 30; **9** (1909), 251. *Ber. Schimmel & Co.,* April (1908), 154. Cf. Gildemeister and Hoffmann, "Die Ätherischen Öle," 3d Ed., Vol. III, 956.
[3] *Perfumery Essential Oil Record* **3** (1912), 341.
[4] "Die Ätherischen Öle," 3d Ed., Vol. III, 956.

obtained by distillation of the plant in the Province of Kwangtung (China) and on the Island of Hainan (China) is used locally for ritual and medicinal purposes. As much as 15,000 lb. per year are said to have been exported at one time from Hainan, whereas in the south of China (Kwangtung) up to 18,000 lb. were produced. The crude product was refined in Canton.[5,6]

In 1895, Schimmel & Co.[7] examined a sample of Ngai-camphor and found that the yellow-white, crumbly, crystalline mass consisted almost entirely of pure *l*-borneol. That Ngai-camphor is actually identical with laevorotatory borneol had already been reported years ago by Plowman,[8] and by Flückiger.[9]

In 1909, Jonas [10] examined a dark brown Ngai-camphor oil of characteristic borneol-like odor; most of the borneol, however, had probably been already removed from the oil by steam distillation. In this oil (d_{15} 0.950, α_D $-12°$ 30', n_D^{20} 1.48151, acid number 23.35, ester number 1, ester number after acetylation 198, total alcohol content, calculated as $C_{10}H_{18}O$, 63.95 per cent) Jonas [11] established the presence of the following compounds:

Cineole.

Limonene.

l-Borneol.

l-Camphor.

Phloracetophenone Dimethyl Ether.

Sesquiterpenes (b. ~280°).

Sesquiterpene Alcohols (b. ~280°).

Palmitic Acid (presence probable).

Myristic Acid (presence probable).

For details regarding the phloracetophenone dimethyl ether, see Vol. II of the present work, p. 543.

In 1910, Schimmel & Co.[12] investigated a Ngai-camphor which had been distilled in Dehra Dun (India) from air-dried leaves and found that it contained 75 per cent of *l*-camphor, 25 per cent of *l*-borneol, and minute quantities of a yellowish oil. Schimmel & Co. could not explain why some lots of Ngai-camphor consisted exclusively of *l*-borneol, and other lots of

[5] Holmes, *Pharm. J.* [3], **21** (1891), 1150.
[6] Cayla, *J. Agr. Tropicale* **13** (1913), 317. Cf. Gildemeister and Hoffmann, "Die Ätherischen Öle," 3d Ed., Vol. III, 956.
[7] *Ber. Schimmel & Co.*, April (1895), 74.
[8] *Pharm. J.* [3], **4** (1874), 710.
[9] *Ibid.*, 829.
[10] *Ber. Schimmel & Co.*, April (1909), 149.
[11] *Ibid.*
[12] *Ibid.*, April (1910), 149.

a mixture of *l*-borneol and *l*-camphor. The reason may be that the product was sometimes distilled exclusively from *Blumea balsamifera* DC., and sometimes from a mixture with other *Blumea* species—*lacera* DC., for example.

In 1925, Chiris [13] analyzed a distillate of *Blumea balsamifera* DC., obtained near Langson (Tonkin), where the plant is called "Dai-bi." The distillate contained chiefly *d*-camphor and only 1.54 per cent of borneol. In the opinion of Gildemeister and Hoffmann [14] the presence of *dextrorotatory* camphor in Ngai-camphor is most unusual and may be explained probably by ample addition of ordinary *d*-camphor to the product in question.

OIL OF CHAMOMILE
(Roman or English Chamomile Oil)

Essence de Camomille Romaine   *Aceite Esencial Manzanilla Romana*
Römisch-Kamillenöl   *Oleum Chamomillae Romanae*
Oleum Anthemidis

Anthemis nobilis L. (fam. *Compositae*), the so-called Roman or English chamomile, is an herbaceous composite with a perennial root. The stems are from 6 to 12 in. long, trailing, and divided into branches which turn upward at their extremities. The flowers are solitary, possessing a yellow, convex disc and white rays. The plant, native to Europe, is cultivated in Belgium, France, and England. Belgium is the principal producer of chamomile, with areas of cultivation located chiefly near Lessines and Flobecq in the province of Hainaut, and near Grammont in the province of East Flanders. In the United States chamomile occurs wild in certain regions, or serves as an ornamental plant in gardens, especially along borders. It blossoms from mid-summer until killed by frost.

Roman chamomile grows well in almost any good, fairly dry soil, with full exposure to the sun. It prefers temperate climates and must be protected from the rigors of adverse weather.

Dafert and Brandl [1] showed that *Anthemis nobilis* gives a maximum yield of flowers and oil when supplied with artificial fertilizers (125 g. Chile

[13] *Parfums France* (1925), 257.
[14] "Die Ätherischen Öle," 3d Ed., Vol. III, 959.
[1] *Angew. Botan.* **12** (1930), 212.

saltpeter, 395 g. superphosphate, and 100 g. potassium salt per 10 sq. m.), or with ample organic fertilizers. The Roman chamomile behaves similarly to the German chamomile (*Matricaria chamomilla* L.): great sensitivity to unbalanced phosphatic fertilizers and preference for potassium. Yield of oil appears to depend on the growth of the plant as a whole.

Anthemis nobilis is a difficult crop, and requires careful handling. In Belgium, much of the work of cultivation and harvesting is done by women and children as a sort of family industry. The plant is set out in the fields in the first warm days of spring. Propagation may be by seed or by root cuttings. Root divisions are planted in straight lines, with spacing of about 50 cm. between the plants, and 60 cm. between the rows. As the herb grows, it develops numerous clustered, curved stalks about 40 cm. high, which almost completely cover the ground, so that weeding is rarely necessary. The ends branch out and bear the flowers, which are gathered during dry, clear weather as they mature. The young flower buds continue to develop into full-grown flowers, to be gathered later in the season.

The flowers are harvested about every two weeks, by women and young girls, who kneel between the rows and with both hands skillfully pick them. The larger ones are first collected in wicker baskets and from there emptied into sacks standing along the fields. A season's harvest consists of about five or six pickings in each field. The flowers of the second and third pickings are usually the most beautiful and contain the most essential oil. In view of the great number of small fields, there are fresh flowers available almost every day of good weather. In Belgium the harvest lasts about two and a half months, from the end of July to the middle of October.

Harvesters are paid according to weight of fresh flowers gathered. Growers sell the fresh flowers to herb exporters, whose trucks daily collect the sacks of chamomile flowers harvested during the day. The material is transported as quickly as possible to the kilns, where large quantities of flowers can be dried with hot air within a few hours.

Five kilograms of fresh chamomile flowers yield about 1 kg. of dried flowers, but the ratio varies considerably with the season. In dry, sunny weather, the flowers are very beautiful, large and pure white, whereas during rainy, damp weather they appear grayish or rusty.

The cultivation of Roman chamomile is a relatively important industry in Belgium. It extends over a territory comprising more than twenty communities, and is carried out by several hundred families, all of whose members participate, since a good deal of work is involved.

Before World War II, Belgium produced from 200 to 250 metric tons of dried flowers annually. These were exported for distillation in other countries (chiefly France). Within the last few years, the Belgians have made attempts to carry out distillation locally; the new essential oil industry does not appear, as yet, to have reached major proportions. However, with time and experience Belgium ought to be able to produce several oils, of excellent quality, particularly from domestically grown aromatic plants such as chamomile, angelica, valerian and hops.

It appears doubtful whether *Anthemis nobilis* could be successfully cultivated on a large scale in the United States, as the crop requires much hand labor. Under normal conditions, such labor is expensive in the United States, and competition with the low-priced European labor does not seem feasible.

According to Gildemeister and Hoffmann,[2] steam distillation of the whole plant yields from 0.2 to 0.35 per cent of oil. Flowering heads alone give a higher yield (up to 1 per cent). Dried flowers distilled in France, under the author's supervision, yielded from 0.32 to 1 per cent of oil.

Physicochemical Properties.—Freshly distilled Roman chamomile oil has a light blue color which, on prolonged standing and exposure to air and light, gradually changes first to green and later to yellow-brown. The odor of the oil is strong, aromatic, and characteristic of the flowers; the flavor is slightly burning.

Gildemeister and Hoffmann [3] reported these properties for Roman chamomile oil:

Specific Gravity at 15°...............	0.905 to 0.918
Optical Rotation....................	−1° 0′ to +3° 0′; in most cases too dark for determination [4]
Refractive Index at 20°..............	1.442 to 1.457
Acid Number.......................	1.5 to 14
Ester Number......................	210 to 317
Solubility.........................	Soluble in 5 to 10 vol. of 70% alcohol, occasionally with turbidity. Soluble in 1 to 2 vol. of 80% alcohol, occasionally with turbidity and separation of paraffins

Oils distilled under the author's supervision in Seillans (Var), France, from Belgian and French flower material, exhibited properties varying within the following limits:

Specific Gravity at 15°...............	0.904 to 0.912
Optical Rotation....................	−0° 40′ to +0° 48′
Refractive Index at 20°..............	1.4410 to 1.4461
Acid Number.......................	4.2 to 11.2

[2] "Die Ätherischen Öle," 3d Ed., Vol. III, 976.
[3] *Ibid.*
[4] The use of sodium vapor lamps is recommended (see Vol. I of the present work, p. 242).

Ester Number...................... 272 to 293.5
Solubility........................... Soluble in 1 vol., and more, of
80% alcohol. Occasionally
with separation of paraffins

Stafford Allen & Sons Ltd.[5] observed the properties quoted below for Roman chamomile oils produced in England from domestically grown flower material:

Specific Gravity at 15°................ 0.901 to 0.911
Optical Rotation...................... ...
Refractive Index at 20°............... 1.443 to 1.449
Acid Number....................... 0.6 to 3.8
Saponification Number................ 267 to 298
Solubility in 70% Alcohol at 15.5°...... Soluble in 7 vol. with turbidity.
Deposit of a slight precipitate,
on standing, in 10 vol.

Chemical Composition.—The first investigation of Roman chamomile oil was undertaken more than a century ago by Gerhardt,[6] who identified the principal constituent as angelic acid. About thirty years later Demarçay[7] found that angelic acid occurs in the oil not in free form, but esterified with butyl and amyl alcohols. More extensive work on the chemistry of Roman chamomile oil was carried out by Fittig and Kopp,[8] Fittig, Kopp and Köbig,[9] van Romburgh,[10] and Blaise.[11] The following compounds have been identified as constituents of Roman chamomile oil:

Methacrylic Acid (Free and as Ester). In the lowest boiling fractions b. 150°–160° of the acids (after separation from the alcohols) Fittig et al. observed the separation of a white, amorphous powder, which could not be isolated in pure form, but appeared to be methacrylic acid (cf. Vol. II of the present work, p. 583). The acid occurs in the original oil not only in free, but probably also in esterified, form.

n-Butyl Isobutyrate. Fittig and Köbig observed an ester of isobutyric acid in the oil; this they believed to be the ester of isobutyl alcohol. Later, however, Blaise identified the alcohol as n-butyl alcohol by means of its phenylurethane m. 55°–66°.

n-Butyl Angelate and Isoamyl Angelate. As was mentioned above, Demarçay identified these two esters many years ago as the chief constituents of the oil. His findings were confirmed by Fittig et al., and by Blaise in 1903.
Fittig and his collaborators also reported that Roman chamomile oil contains esters of tiglic acid. However, Blaise later proved that tiglic acid is not an original

[5] Private communication from Mr. R. K. Allen, London.
[6] *Compt. rend.* **26** (1848), 225. *Ann. chim. phys.* [3], **24** (1848), 96. *Liebigs Ann.* **67** (1848), 235. *J. prakt. Chem.* **45** (1848), 321.
[7] *Compt. rend.* **77** (1873), 360; **80** (1875), 1400.
[8] *Ber.* **9** (1876), 1195; **10** (1877), 513.
[9] *Liebigs Ann.* **195** (1879), 79, 81, 92.
[10] *Rec. trav. chim.* **5** (1886), 219; **6** (1887), 150.
[11] *Bull. soc. chim.* [3], **29** (1903), 327.

constituent of the oil, but is formed from angelic acid on treatment of the latter with alkali, or by heating.

3-Methyl-1-pentanol (β,β-Methylethylpropyl Alcohol) as Ester. Fittig and Köbig noted, in the oil, a hexyl alcohol, esterified probably with angelic acid. On oxidation this hexyl alcohol yielded a caproic acid. Later van Romburgh showed that the latter was methylethylpropionic acid, and that the alcohol present in the oil was β,β-methylethylpropyl alcohol b. 154°, d_{15} 0.829, $[\alpha]_D$ +8° 12' (cf. Vol. II of the present work, p. 148). The oil contains about 4 per cent of this alcohol.

Anthemol(?) $C_{10}H_{16}O$. According to Fittig and Köbig, the fraction b. 213.5°–214.5° is a colorless, viscous liquid of camphoraceous odor, which at atmospheric pressure cannot be distilled without some degree of decomposition. The substance in question appears to be an alcohol of the empirical molecular formula $C_{10}H_{16}O$; to it Fittig and Köbig assigned the name anthemol. The acetate boiled at 234°–236°; on saponification of the ester the alcohol could be regenerated. Treatment of anthemol with chromic acid resulted in complete degradation; oxidation with dilute nitric acid yielded p-toluic acid, terephthalic acid, and a third, more readily soluble, acid.

Chamazulene. The highest boiling fractions of the oil are blue, and contain chamazulene. (See the monograph "Oil of German Chamomile," "Chemical Composition," and Vol. II of the present work, p. 127.)

As regards the quantitative composition of Roman chamomile oil, Blaise [12] saponified 500 g. of oil at room temperature and obtained 190 g. of crude acids, of which 90 g. were angelic acid and 25 g. isobutyric acid. On distillation of the acids, a considerable quantity of polymethacrylic acid remained as a colorless powder in the distillation residue. The neutral products of saponification, on the other hand, consisted of 30 g. of n-butyl alcohol, 25 g. of isoamyl alcohol, 80 g. of optically active hexyl alcohol (3-methyl-1-pentanol), and 33 g. of anthemol.

From a petroleum ether extract of Roman chamomile flowers, Naudin [13] isolated a paraffin m. 63°–64°, which he named *anthemene*. Years later Klobb, Garnier and Ehrwein [14] found that this paraffin has the empirical molecular formula $C_{30}H_{62}$. In the opinion of Gildemeister and Hoffmann,[15] this triacontane is perhaps a constituent also of the *distilled* oil of Roman chamomile, from which it can probably be isolated by proper means.

Use.—Oil of Roman chamomile finds the same uses as German chamomile oil (cf. the following monograph).

[12] *Ibid.*
[13] *Ibid.* [2], **41** (1884), 483.
[14] *Ibid.* [4], **7** (1910), 940.
[15] "Die Ätherischen Öle," 3d Ed., Vol. III, 979.

OIL OF GERMAN CHAMOMILE

Essence de Camomille (Matricaire)   Kamillenöl
Aceite Esencial Manzanilla Alemana   *Oleum Chamomillae*

Matricaria chamomilla L. (fam. *Compositae*), the so-called German, Hungarian or small chamomile, is an annual, asteraceous plant, native to Europe, and cultivated in Germany, Hungary, and Russia. Introduced into North America some time ago, it has now become naturalized and is occasionally cultivated in gardens. In certain areas it has escaped cultivation and can be found growing along highways and paths.

Matricaria chamomilla L. possesses a branching stem, 1 or 2 ft. high, bearing very green, smooth and alternate leaves. The flower heads appear singly at the ends of stems and branches. The flowers are smaller than those of *Anthemis nobilis* L., the so-called Roman chamomile.

The plant thrives in moderately heavy soil, rich in humus and rather moist. It can withstand considerable cold. In Hungary, it also grows abundantly in great patches in the clayey lime soil of the plains, where the depleted and almost barren condition of the earth offers little support for other flora. In these areas the plants are of almost uniform size—about 9 in. tall. Chamomile grows also on farmland, and in grain fields around houses, along roads, and elsewhere; under these conditions it attains a greater height, and its flowers are larger.

Chamomile blooms from the end of April to the end of May, about eight weeks after the seed is sown. A crop may thus be raised from seed sown early in the spring, or from seed planted later, after early vegetable crops. Seed may be sown in drills and barely covered, or it may be broadcast, since the plants soon occupy the entire ground, and exclude weeds.

Chamomile thrives in the poorest areas of Hungary, and its harvest represents a considerable source of income to impoverished inhabitants of these regions. The chamomile grown on farmland is gathered by hand, which permits removal of the flowers without the stem. In other sections, however, it must be collected by means of flower scoops or strippers. When using these implements, the harvesters must gather the flowers as carefully and with as little stem material and extraneous matter as possible. A single worker with a "flower comb" can collect from 60 to 100 kg. of fresh flowers per day; by hand he can gather only 8 to 10 kg.

The harvested flowers are sifted in a suspended sieve (mesh diameter, 7–11 mm.), to separate the flower heads from the flowers with attached stems, and from clinging bits of weed or grass. Thus sifted, the flowers are spread out on the floor or on sheets, in thin layers, to dry. Artificial dryers may also be employed. The delicate material must not be turned while drying, since it is easily damaged.

Five kilograms of fresh flowers give 1 kg. of air-dried flowers. These are sifted once more, and packed into boxes or bales for shipment. If packed in bags, the flowers are likely to be damaged by handling.

According to Stockberger,[1] one acre yields about 400 to 600 lb. of dried flowers under favorable conditions.

It is impossible to estimate figures of production at the present time. Before World War II, Hungary exported about 600,000 kg. of dried flowers, and about 130,000 kg. of "chamomile dust," largely to Germany. German production of dried flowers at that time averaged about 35,000 kg.

It seems doubtful that chamomile can be grown successfully in the United States. Harvesting and handling entail a great deal of hand labor, which is costly. In Europe the work is done by women and children, or by an entire family—in other words, on a garden-crop basis. The United States cannot compete with such an industry.

As regards the various grades of chamomile flowers exported from Hungary, the drug was marketed, before World War II, in six grades—under strict government control. The "Extra" grade, assorted by hand, was of beautiful appearance and usually three times higher in price than other grades. Type V, the so-called "bath chamomile," consisted of flowers with attached stems as they remained in the sieve after sifting of the dried material. This quality was used in pharmacy, chiefly for external application. Type VI, the so-called "chamomile dust," consisted of sifted-out dust.

Distillation.—High-grade chamomile flowers are much too costly to use for essential oil production. For this purpose, therefore, only the lower grades, including flowers with stalks ("bath chamomile"), siftings, and dust are employed in Hungary and Germany. Dust alone, however, cannot easily be distilled with direct steam, because it has a tendency to "bake" in the retort and to form lumps, through which the steam cannot penetrate. To prevent this, dust is first thoroughly mixed with an inert material, such as straw, or, still better, with flowers and stalks.

Because oil of chamomile consists chiefly of high boiling constituents, including paraffins, steam of relatively high pressure (7 atmospheres per square centimeter, in the steam generator) is applied. Distillation of one charge requires from 7 to 13 hr. Chamomile oil may separate crystals even at $+15°$, and has a tendency to form a deposit on the cool walls inside of the condenser tubes. It is, therefore, necessary to stop the flow of the cooling water from time to time, until the temperature of the condenser rises

[1] "Drug Plants Under Cultivation," *U. S. Dept. Agr., Farmers' Bull.* No. 663 (1935), 16.

sufficiently for the deposit to reliquefy and flow off. Great care must be exercised in doing this.

Certain constituents of the oil are soluble in large quantities of warm water; moreover the oil has a high specific gravity and tends to flow off with the distillation water in the form of a milky emulsion. For these reasons the distillation waters must be redistilled (cohobated); by this means as much as 30 per cent of the total oil distilled over may be recovered.

Yield of Oil.—Foreign, particularly German, chemical literature contains numerous references to investigations of the various methods of assaying the volatile oil content of chamomile flowers, and of the yield of oil obtained from different qualities of flower material. To cite only the more recent works:

Kaiser, Eggensperger and Bärmann [2] found that whole German flowers contained from 0.6 to 0.67 per cent of a deep blue oil; whole Hungarian flowers from 0.3 to 0.35 per cent.

Bergmann [3] reported 0.72 to 0.78 per cent of oil from German flowers, and 0.38 to 0.63 per cent from Hungarian material.

Rom [4] obtained these yields of oil from Hungarian flowers:

(a) 0.35 to 0.38 per cent from poorly dried material,
(b) 0.41 to 0.84 per cent from medium to high-grade material.

Gstirner [5] suggested an improvement in the official German assay (Deutsches Arzneibuch VI). He found that pulverized flowers yield more oil than whole flowers.

Liebisch [6] observed these yields of oil from:

	Per Cent
Yugoslavian flowers	0.39–0.56
Hungarian flowers	0.43–0.54
German flowers	0.70–1.05

Improvements in the official German assay were suggested also by Kleinert and Zimmermann [7] (who reported yields of 0.75 to 1.05 per cent from 12 samples of flowers), and by Peyer.[8]

Will [9] reported 0.5 to 1.5 per cent of oil from German flowers, and widely varying yields from Hungarian material. Most of the Yugoslavian flowers gave a poor yield. Two Greek samples gave 0.56 and 0.44 per cent of oil

[2] *Süddeut. Apoth. Ztg.* **68** (1928), 284. *Chem. Zentr.* (1928), II, 1234.
[3] *Pharm. Zentr.* **71** (1930), 785.   [7] *Pharm. Ztg.* **81** (1936), 1091.
[4] *Pharm. Post* **66** (1933), 109.   [8] *Deut. Apoth. Ztg.* **52** (1937), 247.
[5] *Apoth. Ztg.* **48** (1933), No. 70.   [9] *Ibid.* **53** (1938), 1479.
[6] *Chemistry Industry* **57** (1934), 762.

with an abnormal odor. Spanish flowers yielded 0.31 per cent of a colorless oil with a fruity odor.

Determining the volatile oil content of 84 chamomile samples originating from 10 different regions, Heeger and Rosenthal [10] obtained yields ranging from 0.25 to 1.35 per cent, the average being 0.48 per cent.

As regards the yield of oil from large-scale industrial distillation, Gildemeister and Hoffmann [11] claim that the most suitable flower material comes from Hungary, and that the yield of oil ranges from 0.2 to 0.38 per cent. According to de Bittera,[12] Hungarian producers obtain from 0.2 to 0.5 per cent, on the average 0.3 per cent, of oil. Distilling dried Hungarian chamomile flowers in Seillans (Var), France, the author of the present work observed yields varying between 0.1 and 0.4 per cent. German chamomile flowers distilled by Schimmel & Co.[13] yielded 0.25 to 0.425 per cent of oil.

Physicochemical Properties.—The oil derived from *Matricaria chamomilla* L. is a deep blue liquid of strong and characteristic odor, and bitter aromatic flavor. Depending upon the temperature, it is more or less viscous. Under the influence of light and air the deep blue color of the oil gradually changes to green, and finally to brown.

Gildemeister and Hoffmann [14] reported these properties:

Specific Gravity at 15°.............. 0.917 to 0.957 (0.9586 in the case of one oil distilled from whole flowers)
Acid Number...................... 5 to 50
Ester Number..................... 3 to 39
Ester Number after Acetylation...... 117 to 155
Solubility........................ Even in 95% alcohol soluble only with more or less pronounced separation of paraffins

Some oils are quite viscous even at 15° and at that temperature commence separating crystals; the specific gravity of such oils has to be determined in their superfused state. According to Gildemeister and Hoffmann,[15] oils of chamomile, despite literature reports to the contrary, do not assume a buttery consistency on cooling, and do not congeal, at 0°, to a solid mass. Oils distilled by Schimmel & Co. were only viscous even at −20°; none solidified, none had even a butter-like consistency at this temperature. However, a sample of pure German chamomile oil, examined by Fritzsche Brothers, Inc., congealed to a solid mass when cooled to 0° C.

---

[10] *Pharmazie* **4** (1949), 385.
[11] "Die Ätherischen Öle," 3d Ed., Vol. III, 985.
[12] Private information of Dr. Jules de Bittera, Budapest.
[13] *Ber. Schimmel & Co.* (1939), 39.
[14] "Die Ätherischen Öle," 3d Ed., Vol. III, 987.
[15] *Ibid.*

Chamomile oils produced by Schimmel & Co.[16] from flower material of German origin exhibited the following properties:

Specific Gravity at 15°.............. 0.9326 to 0.9459
Acid Number...................... 18.7 to 31.7
Ester Number.................... 1.9 to 12.1
Ester Number after Acetylation...... 66.3 to 115.7
Solubility........................ Soluble in 90% alcohol, with separation of
                                    paraffins

Chamomile oils, distilled under the author's supervision in Seillans (Var), France, from imported flower material of Hungarian origin, had properties varying within these limits:

Specific Gravity at 15°.............. 0.919 to 0.955 (0.974 in an oil distilled from
                                      chamomile dust)
Acid Number...................... 14 to 43.4
Ester Number.................... 6.1 to 12.6 (49.0 in the oil distilled from cham-
                                  omile dust)
Solubility........................ In most cases soluble only in 95% alcohol,
                                    with separation of paraffin crystals. In ex-
                                    ceptional cases soluble in 2.5 to 3 vol. of
                                    80% alcohol, with separation of paraffin
                                    crystals

The oils normally are too dark in color to permit the determination of the optical rotation or refractive index.

An experimental lot of chamomile oil distilled from Hungarian flower material by Fritzsche Brothers, Inc., New York, showed the following properties:

Specific Gravity at 15°/15°...... 0.912
Acid Number................ 35.0
Ester Number............... 7.0
Solubility.................. Waxy separation in 10 vol.
                             of 80% alcohol

Chemical Composition.—Early investigations of the chemical composition of the volatile oil derived from *Matricaria chamomilla* L., carried out by Bornträger,[17] Bizio,[18] Gladstone,[19] and Piesse [20] yielded no positive results; therefore they do not require further discussion.

[16] *Ber. Schimmel & Co.* (1939), 39.
[17] *Liebigs Ann.* **49** (1844), 243.
[18] *Wiener akadem. Ber.* **43** (1861), 2 Abtlg., 292. *Jahresber. Chem.* (1861), 681.
[19] *J. Chem. Soc.* **17** (1864), 1.
[20] *Compt. rend.* **57** (1863), 1016.

In 1871 Kachler [21] fractionated a pure oil of his own distillation and obtained these fractions:

b. 105°–180°...... 4.5 per cent of a slightly bluish oil, with a strong chamomile odor;

b. 180°–255°...... 8.3 per cent of oil;

b. 255°–295°...... 42 per cent of oil, with development of a deep blue vapor;

b. above 295°..... 25 per cent of a very viscous oil, with development of a violet color;

Residue.......... 20 per cent of a brown, peat-like mass.

Distillation was accompanied by decomposition, because all fractions were acidic. Treatment with potassium hydroxide solution yielded an acid which, on analysis of the silver salt, was shown to be *capric acid* $C_{10}H_{20}O_2$. Almost sixty years later, Ruhemann and Lewy [22] confirmed the presence of this acid. Most likely capric acid occurs in the oil as an ester.

(In this connection it should be mentioned that more recently Sörensen and Stene [23] distilled the flowers of *Matricaria inodora* L., which in Scandinavia are used as a substitute for *Matricaria chamomilla* L. From the volatile oil thus obtained—it had a "heavy," unpleasant odor—these authors isolated a highly unsaturated ester $C_{11}H_{10}O_2$, m. 37°, which they named *matricaria-ester*. It has this constitution: $H_3C \cdot CH = CH \cdot C \equiv C \cdot C \equiv C \cdot CH = CH \cdot COOCH_3$. Sörensen and Stene expressed the opinion that the occurrence of capric acid in certain essential oils—*Matricaria chamomilla*, e.g.—may perhaps be explained by perhydrogenation of highly unsaturated esters, such as matricaria-ester.)

The most important constituent of the essential oil derived from *Matricaria chamomilla* is an azulene which has been named *chamazulene*, in order to differentiate it from the azulenes contained in other essential oils. (For details regarding the azulenes, chamazulene in particular, the reader is referred to Vol. II of the present work, pp. 127 and 132.) Chamomile oils contain from 1 to more than 15 per cent of chamazulene, the average being about 6 per cent. The content of chamazulene in the various chamomile oils depends upon the origin and the age of the flower material; it decreases during storage of the flowers. The plant itself does not seem to contain the azulene in free form, but as a precursor, the latter being colorless or yellowish, easily soluble in chloroform, ether, methyl alcohol, and ethyl alcohol, also in warm water, but only difficultly in petroleum ether.

[21] *Ber.* **4** (1871), 36.
[22] *Ber.* **60** (1927), 2463.
[23] *Liebigs Ann.* **549** (1941), 80. Cf. Sörensen et al. (*Acta Chem. Scand.* **4** [1950], 416, 1080. *Chem. Abstracts* **45** [1951], 2398) who confirmed the structural formula of matricaria ester given above, and also isolated hexahydromatricaria ester $C_{11}H_{16}O_2$ from the oil of *Matricaria inodora*.

The precursor is resistant to ammonia, but decomposes on heating, or in an acid medium, yielding the azulene. Thus it appears that the chamazulene present in the essential oil is formed during steam distillation of the plant material.

As regards the other constituents of the oil, only a few have been identified. The exact nature of the sesquiterpenes and sesquiterpene alcohols occurring in the oil is not yet known.

Ruhemann and Lewy [24] isolated two sesquiterpenes (which they named "B" and "C") from the fractions b_{10} 120°–130°, and b_{10} 130°–140°, respectively, after the chamazulene had been removed by treatment with ferrocyanic acid.

Sesquiterpene "B" b_{10} 124°–125°, appeared to be monocyclic, because on hydrogenation it yielded a saturated hydrocarbon $C_{15}H_{30}$, and on treatment with hydrogen chloride it formed a hydrochloride $C_{15}H_{24} \cdot 3HCl$, m. 45°.

Sesquiterpene "C" b_{10} 129°–131°, was present in such small quantities that it could not be characterized. To judge by its molecular refraction, it was a mixture of a monocyclic sesquiterpene (probably "B") and a bicyclic sesquiterpene.

Ruzicka and Rudolph,[25] in their own investigation, arrived at the conclusion that oil of chamomile contains 10 per cent of (chiefly) monocyclic sesquiterpenes which, on dehydrogenation with sulfur, yield cadalene.

The *sesquiterpene alcohols* contained in the oil were first investigated by Ruhemann and Lewy [26] who, after removal of the azulene and the esters, obtained an alcohol fraction b_{10} 170°–180° Efforts to isolate the alcohol in free or in ester form were unsuccessful. Treatment of the fraction with potassium bisulfate yielded a hydrocarbon $C_{15}H_{24}$, b_{13} 137°–139°, which Ruhemann and Lewy named sesquiterpene "A." Its molecular refraction indicated a mixture of monocyclic and bicyclic sesquiterpenes. Shortly after Ruhemann and Lewy had published their findings, Ruzicka and Rudolph [27] reported that the oil contains a mixture of tertiary, chiefly bicyclic, sesquiterpene alcohols. Like the sesquiterpenes, they yield cadalene when treated with sulfur.

So far as the *paraffins* are concerned, Schimmel & Co.[28] noted almost sixty years ago that, in the purified form, they are snow white, melting at 53°–54°. On distillation of the oil, the paraffins remain in the residue as a dark mass, easily soluble in ether, but sparingly soluble in alcohol. The paraffins tend to retain the blue color of the oil most tenaciously. Extracting dried chamomile flowers with petroleum ether, Klobb, Garnier and Ehrwein [29] isolated a paraffin m. 52°–54°, to which they ascribed the em-

[24] *Ber.* **60** (1927), 2463.
[25] *Helv. Chim. Acta* **11** (1928), 253.
[26] *Ber.* **60** (1927), 2463.
[27] *Helv. Chim. Acta* **11** (1928), 253.
[28] *Ber. Schimmel & Co.*, April (1894), 13.
[29] *Bull. soc. chim.* [4], **7** (1910), 940.

pirical molecular formula $C_{29}H_{60}$. Power and Browning Jr.[30] identified, in chamomile flowers, a paraffin m. 63°–65°, viz., triacontane $C_{30}H_{62}$.

Steam-distilling an alcoholic extract of chamomile flowers, the same authors noted that the volatile oil thus obtained contained still other constituents. (In the opinion of Gildemeister and Hoffmann,[31] however, it is questionable whether these substances occur also in the essential oil produced by steam distillation of the flowers.) Furfural was characterized by color reaction. The oil investigated also contained a fatty acid m. 61°. On standing, the oil deposited small quantities of a substance m. 110° which, on heating, developed a coumarin odor, and which in concentrated sulfuric acid displayed blue fluorescence. To all appearance the substance was umbelliferone methyl ether (cf. Vol. II of the present work, p. 667).

Use.—Mild infusions and teas made from chamomile flowers have been for centuries employed as a popular remedy against fever, stomach troubles, indigestion, intestinal pain, and as a mild tonic and antispasmodic. Externally, the flowers are applied in the form of fomentations against inflammation or irritation. The active principle of the flowers is the chamazulene. In many medicinal preparations the essential oil, therefore, replaces extracts of the flowers.

The oil is also used, but sparingly, as a flavoring agent in fine liqueurs, particularly those of the French type. The oil finds further application in perfume compositions, to which it imparts pleasing and warm tonalities that are difficult to trace.

SUGGESTED ADDITIONAL LITERATURE

W. Wolf, "The Essential Oil of Camilla and Azulene, the Substance Responsible for the Blue Color." *Fette u. Seifen* **47** (1940), 122. *Chem. Abstracts* **34** (1940), 8180.

Charlotte Pommer, "Action of Some Azulenes on Inflammation." *Arch. exptl. Path. Pharmakol.* **199** (1942), 74. *Chem. Zentr.* (1942), I, 3227. *Chem. Abstracts* **37** (1943), 3828; cf. *ibid.* **27** (1933), 5816.

E. F. Heeger, K. H. Bauer, and W. Poethke, "*Matricaria chamomilla*, True Chamomile." *Pharmazie* **1** (1946), 210. *Chem. Zentr.* (1947), I, 64. *Chem. Abstracts* **41** (1947), 6021.

[30] *J. Chem. Soc.* **105** (1914), 2280.
[31] "Die Ätherischen Öle," 3d Ed., Vol. III, 991.

OIL OF COSTUS ROOT

Saussurea lappa Clarke (*Aplotaxis lappa* Dec.; *A. auriculata* DC.; *Aucklandia costus* Falc.), fam. *Compositae,* is a tall, sturdy, herbaceous perennial, up to 8 ft. high, with large radical leaves and a robust stem bearing a cluster of several bluish-black flower heads. Indigenous to the northwestern regions of the Himalaya Mountains (Kashmir and Hazara), it has lately been introduced also to Chamba and Tehri. *Saussurea lappa,* locally called "Kuth," grows abundantly in the Kishengang Valley of Kashmir, and in the higher elevations of the Chenab Valley. It thrives in shady, moist places beneath birch and dwarf willow, at altitudes ranging from 9,000 to 11,000 ft.

The most useful part of the plant is its root, usually called costus root.[1] Since ancient times this root has been used in the Orient as an aromatic stimulant against all kinds of diseases, as an aphrodisiac, insect repellent, incense, and to perfume and preserve woolen fabrics. Large quantities of costus root are exported from Kashmir to China and Japan (in normal times) for use in temples, and to Europe for distillation purposes. The industry is a state monopoly in Kashmir, the principal trading center being Baramulla. Prior to World War II, about 2 million pounds of costus root were said to be produced yearly, but only a small part of this quantity went to Europe for extraction of the essential oil. Shipping ports for export are Bombay and Calcutta.

Stewart and Raynor [2] experimented with the cultivation of *Saussurea lappa* in the Garhwal district (India), and reported that best results were obtained by sowing seed in nursery beds (instead of directly in the forest), and by transferring nursery plants to intensively prepared and manured beds on sheep and goat steadings. The best time for extracting the roots from the earth was found to be the end of October or early November, when they are well matured. At this period of the year the moisture content of the roots is lowest. After having been dug out, the roots are immediately cut into short lengths of about 2 to 4 in. and cleaned of earth before being placed on shelves for drying. Thus the extracted roots are partially dried on the spot before the close of the season in early November, after which they have to be transported down to a drying "godown" at a lower altitude for further drying and subsequent export to the plains.

In Kashmir the roots are dug up soon after fructification in September

---

[1] Not to be confused with the root of *Costus speciosus* Smith = *Amomum hirsutum* Lam. (fam. *Zingiberaceae*), which is odorless, and serves as a native food.

[2] *Perfumery Essential Oil Record* **33** (1942), 341.

and October, cut into pieces 3 to 4 in. long, gently roasted to prevent sprouting, and hauled to the "godowns" in Baramulla for complete drying and export.

On steam distillation the dried and triturated root material yields a viscous oil of light yellow to brown color, and most peculiar, slightly animal-like, very lasting odor, reminiscent of violet, orris, and vetiver.

Gildemeister and Hoffmann [3] reported yields of oil ranging from 0.3 to 1.0 per cent (in one case 2.78 per cent [4]). Ghosh, Chatterjee and Dutta [5] obtained 1.5 per cent of oil on steam distillation. In the author's own experience, the yield of oil in large-scale production ranges from 0.98 to 1.5 per cent.

According to Naves,[6] the yield of oil is affected by the nature of the metal of the still, the size of the batch of root material, and the form of the gooseneck and cooling coil. The distillate is generally extracted from the aqueous phase with benzene or toluene.

The oil can also be produced, according to the same author, by first preparing a resinoid and then treating the resinoid with a current of superheated steam at reduced pressure, or simply by distilling it in a high vacuum. These are delicate operations. Yield of oil thus obtained ranges from 3.3 to 4.8 per cent, which is much higher than that obtained by direct steam distillation of the root itself. The essential oil derived from the resinoid contains more lactonic products, particularly deshydrocostus lactone (see below), than that resulting from steam distillation of the root.

Physicochemical Properties.—In 1931 Gildemeister and Hoffmann [7] reported these properties for oil of costus root obtained by direct steam distillation:

Specific Gravity at 15°..............	0.940 to 1.009
Optical Rotation..................	+13° 0′ to +27° 0′; in the case of one oil (roots from Madras) +47° 15′
Acid Number.....................	8 to 36
Ester Number....................	55 to 114 (in the titration of the excess alkali, the liquid must be kept cool to prevent regeneration of the lactones; otherwise the ester number will be lowered to 18 and 45, respectively)
Ester Number after Acetylation......	105 to 162
Solubility........................	Clearly soluble in 90% alcohol, opalescent to turbid after addition of 2 to 5 vol. of 90% alcohol, or of larger amounts of 95% alcohol

[3] "Die Ätherischen Öle," 3d Ed., Vol. III, 1044.
[4] D. Hooper, "Board of Scientific Advice for India" (1911 to 1912), 31.
[5] *J. Indian Chem. Soc.* **6** (1929), 517.
[6] *Mfg. Chemist* **20** (1949), 318.
[7] "Die Ätherischen Öle," 3d Ed., Vol. III, 1044.

Because of the presence of acids and lactones, 33 to 52 per cent of the oil was soluble in a 5 per cent aqueous solution of sodium hydroxide.

Two oils distilled from roots originating from the Punjab were semi-solid and solid at room temperature and were permeated with crystals of naphthalene (see Chemical Composition).

In 1936, Chiris [8] reported the following properties of two steam-distilled costus root oils:

	I	II
Specific Gravity at 15°....	1.0067	1.012
Optical Rotation.........	+7° 15′	+9° 10′
Refractive Index.........	1.5177	1.5188
Acid Number...........	34.23	30.8
Ester Number..........	88.9	89.78
Solubility..............	Soluble in 10 vol. of 90% alcohol; soluble in 4 vol. of 95% alcohol	Soluble in 10 vol. of 90% alcohol; soluble in 1 vol. of 95% alcohol

Of oil (I) 46 per cent was soluble in a 3 per cent aqueous solution of sodium hydroxide; of oil (II) 44 per cent was soluble.

Oils of costus root distilled in the same manner under the author's supervision in Seillans (Var), France from imported root material exhibited properties which varied within these limits:

Specific Gravity at 15°..............	1.000 to 1.045
Optical Rotation...................	+12° 0′ to +14° 14′
Refractive Index at 20°............	1.5159 to 1.5280
Acid Number.....................	29.4 to 40.6
Ester Number....................	90.5 to 121.3
Ester Number after Acetylation......	173.3 to 176.1
Solubility at 20°...................	Soluble in 0.5 vol. of 90% alcohol, cloudy and occasionally with separation of paraffin crystals on addition of more alcohol

In 1949, Naves [9] reported the following properties for distilled oils analyzed by himself:

Specific Gravity at 15°.......	0.9985 to 1.043
Optical Rotation...........	+12° 12′ to +18° 32′
Refractive Index at 20°......	1.5120 to 1.5230
Acid Number..............	18.0 to 33.0
Ester Number.............	86.2 to 126.6

Comparing the data supplied by Chiris, Guenther, and Naves with those of Gildemeister and Hoffmann, it appears that the physicochemical proper-

[8] *Parfums France* **14** (1936), 271.   [9] *Mfg. Chemist* **20** (1949), 318.

ties of costus root oil are now somewhat different from what they were 20 years ago. The higher specific gravity and acid number, and the lower optical rotation permit the conclusion that the root material imported now may be of greater age than it was years ago.

Adulteration.—Because of its high price, oil of costus root is often adulterated, particularly with oil of vetiver, the physicochemical properties of which do not differ too much from those of a genuine costus root oil. In the evaluation of a costus root oil by odor tests, special attention should be paid to the last notes which appear on a blotting paper after several days of standing.

Chemical Composition.—The chemical composition of costus root oil was investigated first by Semmler and Feldstein,[10] and more recently by Ukita,[11] Crabalona,[12] and Naves.[13]

In the course of their work Semmler and Feldstein established the presence of the following compounds:

Phellandrene (0.4%). Identified by means of the bisnitrosite m. 106°–108°.

Camphene (0.4%). Characterized by conversion to isoborneol.

A Terpene Alcohol(?) (0.2%). Semmler and Feldstein also found in the oil what they believed to be a terpene alcohol of the empirical molecular formula $C_{10}H_{16}O$.

Aplotaxene (20%). In the fraction b_{11} 160°–175°, α_D +14°, the same authors observed a hydrocarbon $C_{17}H_{28}$ (?) to which they assigned the name aplotaxene. This compound b_{11} 154°–156°, d_{21} 0.8604, could not be obtained free of oxygenated substances. Reduction of aplotaxene with sodium and alcohol yielded dihydro-aplotaxene $C_{17}H_{30}$, b. 154°–157°, d_{21} 0.8177, which, on treatment with hydrogen and platinum black was converted into octohydroaplotaxene b_{11} 159°–163°, d_{16}^{21} 0.7805. The latter was identical with *n*-heptadecane. Semmler and Feldstein thus arrived at the conclusion that aplotaxene is an aliphatic hydrocarbon with a normal carbon chain possessing four double bonds, two of which are in conjugated position.

α-Costene (6%) and β-Costene (6%). In the fractions b_{11} 100°–130° and b_{11} 130°–150° Semmler and Feldstein noted the presence of two sesquiterpenes, which they named α-costene and β-costene, respectively. Details regarding these two compounds will be found in Vol. II of the present work, p. 747.

Costus Acid (14%). This bicyclic acid $C_{15}H_{22}O_2$, belonging to the sesquiterpene series, was observed in the fraction b_{11} 190°–200° (cf. Vol. II of this work, p. 612).

Costol (7%). A bicyclic, primary sesquiterpene alcohol $C_{15}H_{24}O$, with two double bonds, present in the fraction b_{11} 175°–190° (cf. Vol. II of this work, p. 756).

[10] *Ber.* **47** (1914), 2433, 2687.
[11] *J. Pharm. Soc. Japan* **59** (1939), 231—Abstracts in German (1939), 80.
[12] *Bull. soc. chim.* (1948), 357.
[13] *Helv. Chim. Acta* **31** (1948), 1172.

Costus Lactone (11%). A bicyclic sesquiterpene lactone $C_{15}H_{20}O_2$, containing two double bonds, present in the fraction b_{11} 200°–210° (cf. Vol. II of this work, p. 767).

Dihydrocostus Lactone (15%). The dihydro derivative, $C_{15}H_{22}O_2$, of costus lactone, present in the fraction b_{11} 190°–200° (cf. Vol. II of this work, p. 767).

The above are the compounds identified by Semmler and Feldstein.

Deshydrocostus Lactone. Twenty five years later, Ukita [14] extracted 4 kg. of costus root with petroleum ether and obtained 200 g. of oil. Distilling this oil at 6 mm. (175°–190°), Ukita obtained palmitic acid m. 63°, an acid $C_{15}H_{22}O_3$, m. 118.5°, and a lactone $C_{15}H_{18}O_2$, m. 60.5°.

In 1948, Crabalona [15] investigated the crude crystals m. 58°, which had precipitated from a costus root oil after several years of standing. After recrystallization from petroleum ether, they melted at 61.5°, $[\alpha]_D$ −12° 32′ (alcohol; c = 15). From various reactions Crabalona concluded that the compound was a bicyclic lactone $C_{15}H_{18}O_2$, containing three double bonds. The corresponding acid melted at 122°.

Shortly afterward, Naves [16] published the results of his own work on this lactone, m. 63.5°–64°, $[\alpha]$ −13° 42′ (alcohol; c = 5), to which he assigned the name deshydrocostus lactone. Reduction with sodium and alcohol yielded a dihydrolactone $C_{15}H_{20}O_2$, while catalytic hydrogenation gave a saturated lactone, viz., tetrahydrocostus lactone $C_{15}H_{24}O_2$. On treatment with selenium at 300°–340° the latter lactone yielded guaiazulenes and bicyclic dihydrosesquiterpenes.

According to Naves,[17] his research on the structure of deshydrocostus lactone indicates a hydroguaiazulenic skeleton, which would make this lactone a close relative of the hydronaphthalene lactones of the santonin group, differing from the latter only in the position of the cyclic bridge.

d-α-Ionone. With a view to investigating the ketones present in costus root oil, Naves [18] treated the fraction b_2 80°–110° (about 15 per cent of the oil) with Girard and Sandulesco's reagent P and found that the oil contains about 0.8 to 1 per cent of a mixture of ketones. Among them he identified *d-α*-ionone by means of its phenylsemicarbazone m. 182°–183°, and dinitro-2,4-phenylhydrazone m. 132°–133°.

cis-Dihydroionone. Along with *d-α*-ionone, Naves identified *cis*-dihydroionone by preparation of its phenylsemicarbazone m. 174°–175°, and dinitro-2,4-phenylhydrazone m. 177°–178°.

β-Ionone. The above two ketones were accompanied by small quantities of *β*-ionone, which Naves characterized by means of its thiosemicarbazone m. 157°–158° (mixed melting point determination).

Ketones(?). Aside from the three ketones mentioned above, Naves also observed the presence, in small quantities, of other isomeric ketones which, on ozonolysis and reduction of the ozonides, gave formaldehyde. In the opinion of Naves, it is possible that oil of costus root also contains some *γ*-ionone or dihydro-*γ*-ionone or a related compound.

[14] *J. Pharm. Soc. Japan* **59** (1939), 231—Abstracts in German (1939), 80.
[15] *Bull. soc. chim.* (1948), 357.   [17] *Mfg. Chemist* **20** (1949), 318.
[16] *Helv. Chim. Acta* **31** (1948), 1172.   [18] *Helv. Chim. Acta* **32** (1949), 1064.

In two semisolid and solid costus root oils originating from the Punjab, and submitted by D. Hooper, Schimmel & Co.[19] identified *naphthalene* m. 79° (complex with picric acid, m. 149°).

Use.—Oil of costus is used in high-grade perfumes of heavy, oriental type. It blends well with sandal, vetiver, patchouly, rose, and violet, imparting to them unique and alluring tonalities that are difficult to trace. Because of its very strong and tenacious odor the oil has to be dosed most carefully; otherwise the effect will be unpleasant. The oil is expensive and not always readily available.

OIL OF DAVANA

Davana (or Davanam) is an aromatic herb, much prized in Southern India for its delicacy of odor. Though probably native to this area, it apparently does not now grow wild to any large extent. It is, however, cultivated here and there, in gardens, for its leaves and flowers, which are used as decorative garlands and chaplets, and in religious offerings. As regards the essential oil of davana, it is produced by only one distiller (near Mysore), in very limited quantities.

Scientific literature contains few references to the plant or its volatile oil—a rather surprising fact, in view of the popularity of the herb. The botanical identity of the plant, long a matter of dispute, has only recently been established as *Artemisia pallens* Wall.[1]

Davana is an annual herb (fam. *Compositae*), requiring about four months to reach maturity, at which time it attains a height of about $1\frac{1}{2}$ ft. According to Sundera Rao,[2] two distinct varieties of the plant occur. This last fact is of considerable importance in the production of davana oil.

Planting, Cultivating, and Harvesting.—Propagation is by seed, the seed being sown in beds in December. About one and a half months later the young plants are transplanted into fertile, sandy, loamy soil, which has been previously plowed several times and amply fertilized with manure. Planting is done in long rows, about 6 in. apart. Davana, a delicate plant, cannot withstand much rain and should be artificially irrigated—i.e., by

[19] Gildemeister and Hoffmann, "Die Ätherischen Öle," 3d Ed., Vol. III, 1047.
[1] Cf. S. G. Sastry, *Indian Soap J.* **11** (1946), 242.
[2] *Ibid.*

sprinkling with ordinary garden hose. Two months after transplanting, the plant develops flowers, light yellow in color. Best yield and quality of oil are obtained from plants cut just before the flowers open completely.

Davana can be grown throughout the year, but optimum conditions as regards oil hold for plants set out in December, and harvested before the end of April, i.e., toward the end of the summer (which in southern India lasts from February to April). For harvesting, the entire plant is cut with a sickle. After harvesting, the plantation must be replanted with seedlings.

Yield and Distillation.—As was noted above, Sundera Rao indicates that there are two varieties of *Artemisia pallens* and that it is important, in planting, to distinguish the two, since the oil of one is superior to that of the other. Unfortunately, Sundera Rao has not detailed his own experiments to support this observation. However, he does state, after prolonged work upon the proper type of fertilizer, correct stage of maturity for harvest and best type of soil for planting, that the oil obtained from plants grown in severe summer conditions, without rainfall, and harvested at a particular stage of development is of the best quality. Rain and cold have an adverse effect upon the yield and quality of the oil.

Prior to distillation, the plants are dried for about one week in the shade. Distillation is carried out with direct steam, one charge requiring about 12 hr.

According to Sastry,[3] yield of oil varies from 0.13 to 0.58 per cent. The author of the present work, while surveying production of this oil in Mysore, was told that the yield averages 0.2 per cent.

Physicochemical Properties.—Oil of davana is a brownish, viscous liquid of peculiar, very aromatic, somewhat balsamic and persistent odor. Very little material on the properties of the oil has been published.

B. S. Rao et al.[4] recorded these data:

Specific Gravity at 30°/30°. 0.9833
Specific Optical Rotation at 30°. . . . . . −25° 48'
Refractive Index at 30°. 1.4898
Acid Number. 2.6
Ester Number. 19.1
Ester Number after Acetylation. 78.0

The laevorotation reported by Rao seems strange. According to Sastry,[5] an oil distilled in Mysore had the following properties:

Specific Gravity at 15.5°. 0.9605
Optical Rotation. . . . . +35° 0'
Refractive Index at 20°. 1.4880

[3] *Ibid.*
[4] *Perfumery Essential Oil Record* **28** (1937), 411.
[5] *Indian Soap J.* **11** (1946), 242.

Acid Number..................... 2.4
Ester Number.................... 52.9
Solubility...................... Not always clearly soluble in 10 vol.
of 70% alcohol

Shipments of davana oil from Mysore, examined by Fritzsche Brothers, Inc., New York, had properties varying within these limits:

Specific Gravity at 15°/15°......... 0.961 to 0.990
Optical Rotation.................. +42° 0' to +54° 28', often the oil is too dark to determine the optical rotation
Refractive Index at 20°............ 1.4874 to 1.5003
Acid Number..................... 1.0 to 5.6
Saponification Number............. 31.7 to 58.2
Solubility...................... Not always clearly soluble up to 10 vol. of 80% alcohol

Oils received prior to 1937 occasionally exhibited optical rotations as low as +5°.

Nothing is known of the chemical composition of this oil.

Use.—The only use of davana oil is in fine perfumes; however, because of its high price, it has not been widely applied. Unless its cost is reduced by large-scale production, the oil can be employed only in expensive perfume compositions.

OIL OF ELECAMPANE

Essence de Racine d'Aunée   *Aceite Esencial Enula Campana*
Alantöl   *Oleum Helenii*

Elecampane (*Inula helenium* L.), fam. *Compositae*, is a large, coarse perennial, native to Central Asia, but now cultivated on a limited scale in Central Europe. Introduced into the Western Hemisphere, it grows wild along roadsides and in fields throughout the northeastern sections of the United States. The root is used in medicinal preparations; owing to the presence of alantolactone (see below) it possesses anthelminthic properties.

According to Stockberger,[1] elecampane will grow in almost any soil, but thrives best in deep clay loam well supplied with moisture. The ground on which this plant is to be cultivated should be deeply plowed and thoroughly

[1] "Drug Plants under Cultivation," *U. S. Dept. Agr., Farmers' Bull.* No. 663 (1939), 22.

prepared before planting. It is preferable to use divisions of old roots for propagation, and these should be set in the fall about 18 in. apart in rows 3 ft. apart. Plants may also be grown from seeds, which may be sown in the spring in seedbeds, and the seedlings transplanted later to the field and set in the same manner as the root divisions. Plants grown from seed do not flower the first year. Cultivation should be sufficient to keep the soil in good condition and free from weeds.

The roots are dug in the fall of the second year, thoroughly cleaned, sliced, and dried in the shade. The available data on yield indicate that a ton or more of dry root per acre may be expected.

On steam distillation the triturated root yields from 1 to 3 per cent of a solid, crystalline mass, permeated with a very small quantity of (liquid) oil of brown color.

Physicochemical Properties.—The normal (total) oil, from which none of the alantolactone (its chief constituent) has been removed, melts at 30°–45° to a brown liquid. The oil possesses a peculiar odor, reminiscent of labdanum.

According to Gildemeister and Hoffmann,[2] the physicochemical properties of the oil have been determined in relatively few cases only, and not always under the same conditions, in the superfused state, e.g.:

Specific Gravity at 15°	1.0320 (one determination)
30°	1.015 and 1.038 (two determinations)
40°	1.0438 (one determination)
45°	1.0355 to 1.0405 (three determinations)
Optical Rotation	+124° 0' to +155° 0' (four determinations)
Refractive Index at 20°	1.5221 and 1.5250 (two determinations)
40°	1.5181 (one determination)
45°	1.5140 and 1.5167 (two determinations)
Acid Number	6 to 10 (five determinations)
Ester Number	160 to 202 (five determinations)
Ester Number after Acetylation	199 to 211 (three determinations)
Solubility	Soluble in 2 to 3 vol. of 90% alcohol; opalescent with more alcohol

In 1931 Schimmel & Co.[3] reported these properties for an oil of their own distillation (yield 0.896 per cent):

Congealing Point	+41.6° (supercooled to +41°)
Specific Gravity at 50°	1.0347
Optical Rotation	Too dark
Refractive Index at 50°	1.51430
Acid Number	13.1

[2] "Die Ätherischen Öle," 3d Ed., Vol. III, 964.
[3] *Ber. Schimmel & Co.* (1931), 3.

Ester Number....................... 194.1
Ester Number after Acetylation...... 205.3
Solubility......................... Soluble in 90% alcohol, up to 1.5
vol., then slight turbidity; floc-
culent precipitation on standing

The high congealing point of this oil indicated a high content of alanto-lactone.

Chemical Composition.—By far the greatest part of oil of elecampane consists of a mixture of alantolactone and isoalantolactone, the former pre-dominating (cf. Vol. II of the present work, p. 691). In this connection it should be mentioned that the position of the double bond in the lactone ring of the isoalantolactone molecule has not been definitely established. An alternative structural formula for the lactone ring of isoalantolactone, first suggested by Hansen,[4] is reproduced below:

Separating alantolactone from elecampane oil, Ruzicka and van Melsen [5] observed the presence, in the oil, of a monocyclic sesquiterpene b_{12} 135°–138°, d_4^{15} 0.8864, n_D^{15} 1.500, which yielded eudalene upon dehydrogenation with selenium.

A liquid with the same molecular formula as that of alantolactone and isoalantolactone, $C_{15}H_{20}O_2$, was also isolated from the oil.

The highest boiling portions of the oil contain a compound of blue color.

Production and Use.—According to Harms and Schneider,[6] substantial quantities of elecampane were grown in Thuringia (Germany) during World War II, but nothing is known about the condition of these plantings since the end of the war. Years ago small quantities of the root came from Austria where it was grown by farmers in the mountainous sections of Styria.

The oil, when produced on a commercial scale, is used for the isolation of alantolactone which, in its physiological action, resembles santonin, but

[4] *Ber.* **64** (1931), 67, 943, 1904. *J. prakt. Chem.* **136** (1933), 185. Cf. Ruzicka, Pieth, Reichstein and Ehmann, *Helv. Chim. Acta* **16** (1933), 268.
[5] *Helv. Chim. Acta* **14** (1931), 397.
[6] *Die Deutsche Heilpflanze* **7** (1941), 142.

has only a slightly bitter flavor, and does not act as an emetic. Alanto-lactone is an antiseptic, expectorant, diuretic, and above all an anthel-mintic.[7]

SUGGESTED ADDITIONAL LITERATURE

R. Jaretzky, "Microscopic Identification of Alantolactone in the Roots of *Inula helenium* L.," *Arch. Pharm.* **280** (1942), 236.

OIL OF ERIGERON

Erigeron canadensis L. (fam. *Compositae*), known also as fleabane, horseweed, or butterweed, is a common weed, growing abundantly in fields, waste places, and along roadsides. It is widely distributed in the Old World, in South America, Canada, and the northern and central sections of the United States. The height of the plant depends upon the kind of soil on which it grows; it may vary from a few inches to several feet. From June to November the weed produces numerous heads of small, inconspicuous white flowers, followed by an abundance of seed. Because of its pronounced aroma, the plant is most troublesome on peppermint plantations, becoming intermixed with the mint material, and imparting an objectionable by-odor to the distilled peppermint oil.

The erigeron oil industry constitutes, actually, little more than a side line in the production of peppermint and spearmint oils. In the United States the oil is distilled in northern Indiana and southern Michigan, with South Bend as the principal marketing center for the oil. Yearly production ranges from 1,000 to 2,000 lb.

The plants are not cultivated: invading abandoned corn fields, they grow prolifically. Harvesting takes place during the flowering period, in the latter part of July. The plants are simply mowed down with a wheat binder, allowed to dry for 24 hr., loaded on wagons or trucks, and hauled to the distilleries.

The same type of still employed to process peppermint, spearmint, wormwood, or tansy, is used to produce oil of erigeron. Live steam, generated

---

[7] Ozeki, Kotake and Hayashi, *Proc. Imp. Acad. Tokyo* **12** (1936), 233. Chevalier, *Presse méd.* (1939), 445. *Merck's Jahresbericht* **53** (1939), 187. Chinsho Go, *Japanese J. Med. Sci.* IV, *Pharmacology* **11** (1938), 110. *Merck's Jahresbericht* **53** (1939), 188. Harms and Schneider, *Die Deutsche Heilpflanze* **7** (1941), 142. *Ber. Schimmel & Co.* (1942/43), 5, 103.

in a separate steam boiler, is blown through the herb material, and the oil is collected in an oil separator. The charge per still consists of about 2,500 lb. of dried plant material; such a charge yields from 8 to 10 lb. of oil. Several acres are required to make up one charge of herb material. Since the plants are naturally quite dry, distillation proceeds rapidly, only 40 min. being required to complete exhaustion of each batch.

Physicochemical Properties.—When freshly distilled, oil of erigeron is a colorless or slightly yellow, mobile liquid possessing a peculiar persistent odor, and a slightly stinging flavor. On exposure to air the oil resinifies rapidly, turning viscous and dark. According to Rabak,[1] crystals may occasionally separate. Old or improperly stored resinified oils exhibit an abnormally high specific gravity. It is possible, however, to recondition such oils by redistillation.

Gildemeister and Hoffmann [2] reported these properties for erigeron oil:

Specific Gravity at 15°.............. 0.8565 to 0.868
Optical Rotation................... +52° 0' to +83° 0'
Acid Number...................... 0
Ester Number.................... 39 to 108
Ester Number after Acetylation...... 67 to 108
Solubility........................ Occasionally soluble in an equal volume of 90% alcohol; often the solution remains turbid even on addition of several volumes of alcohol

Genuine oils of erigeron examined by Fritzsche Brothers, Inc., New York, had properties varying within the following limits:

Specific Gravity at 15°/15°...... 0.857 to 0.877
Optical Rotation.............. +48° 58' to +68° 40'
Refractive Index at 20°......... 1.4820 to 1.4931
Saponification Number.......... 21.5 to 49.5
Solubility at 20°.............. Opalescent to turbid in 10 vol. of 90% alcohol

The abnormally high specific gravity and poor solubility of some of these oils indicated partial resinification. The oils in question were, therefore, redistilled. Several such reconditioned lots had these properties:

Specific Gravity at 15°/15°...... 0.843 to 0.859
Optical Rotation.............. +56° 48' to +77° 8'
Refractive Index at 20°......... 1.4751 to 1.4799
Saponification Number.......... 6.5 to 18.7
Solubility at 20°.............. Soluble in 4.5 to 6.5 vol. and more of 90% alcohol

[1] *Pharm. Rev.* **23** (1905), 81; **24** (1906), 326.
[2] "Die Ätherischen Öle," 3d Ed., Vol. III, 954.

Schimmel & Co.[3] distilled two oils from *Erigeron canadensis* growing wild in the environments of Miltitz, Germany, and in both cases obtained a yield of 0.26 per cent. Oil (I) was derived from herb material including roots, but few flowers; oil (II) was distilled from herb without roots, but with many flowers. The two oils had the following properties:

	(I)	(II)
Specific Gravity at 15°........	0.8720	0.8836
Optical Rotation..............	+53° 56'	+50° 4'
Refractive Index at 20°........	1.49922	1.50624
Acid Number................	0.3	0.3
Ester Number................	About 63.5	About 70.9
Ester Number after Acetylation	70.3	81.9
Solubility in 90% Alcohol......	Soluble in 5.5 vol. and more	Soluble in 4 vol. and more, with slight turbidity

Oil (II) appeared more prone to resinification than oil (I). Both oils had a slightly aromatic odor, on dilution somewhat reminiscent of neroli.

Chemical Composition.—The following compounds have been identified in oil of erigeron:

d-Limonene. After earlier work by Beilstein and Wiegand,[4] and Wallach,[5] Power,[6] in 1887, reported that by far the largest part of the oil boils at 175°, and that the oil consists chiefly of *d*-limonene. In 1893 Meissner [7] definitely proved the presence of *d*-limonene in the oil by means of the nitrosochlorides; the α-nitrosochloride yielded a benzylamine compound m. 90°–92°.

dl-Terpineol. In the fraction b. 205°–210° Hunkel [8] identified *dl*-terpineol by preparation of the nitrolpiperidine m. 159°–160°. Examining a sample of erigeron oil, which showed distinct evidence of resinification, Foote and Matthews [9] found that it contained 16.82 per cent of terpineol.

Terpinyl Acetate. The same authors also reported the presence of 5.94 per cent of terpinyl acetate in their oil.

Citral(?). The oil gives a positive aldehyde reaction with Tollen's and Schiff's reagents. Investigating the aldehydes of the oil, Foote and Matthews obtained very small quantities of a semicarbazone m. 149°–152.5°. Upon saponification this semicarbazone yielded a residue with an odor suggestive of citral. Since no bisulfite compound could be prepared, the amount of aldehyde present in erigeron oil must be very small.

[3] *Ber. Schimmel & Co.* (1922), 20.
[4] *Ber.* **15** (1882), 2854.
[5] *Liebigs Ann.* **227** (1885), 292.
[6] *Pharm. Rundschau* (New York) **5** (1887), 201.
[7] *Am. J. Pharm.* **65** (1893), 420.
[8] Cf. *Pharm. Rundschau* (New York) **13** (1895), 137
[9] *J. Am. Pharm. Assocn.* **28** (1939), 1031.

More recently Sörensen and Stavholt,[10] using chromatographic and spectrographic methods, observed, in oil of *Erigeron canadensis,* the presence of several acetylene compounds, among them *matricaria ester* m. 31.5°. (Cf. the monographs on "Oil of German Chamomile," and "Oil of *Artemisia vulgaris.*")

The oil derived by steam distillation of the *root* of *Erigeron canadensis,* on the other hand, contained:

Dehydromatricaria Ester (Methyl *n*-Decenetriynoate), $C_{11}H_8O_2$, m. 112.5°–112.7°.

Matricaria Ester and "Centaur X."

Use.—Oil of erigeron is employed, but only rarely, in certain medicinal preparations. Its effect is said to be hemostatic.

OIL OF ESTRAGON
(Oil of Tarragon)

Essence d'Estragon   *Aceite Esencial Estragon*   *Esdragonöl*
Oleum Dracunculi

Oil of estragon (tarragon) belongs among those lesser-known oils which, in the hands of the expert perfumer and flavorer, yield unique results.

Artemisia dracunculus L. (fam. *Compositae*), the estragon plant, is a Eurasian perennial, which grows to a height of about 2 ft. It is cultivated in Southern France, the main area of production being in the Département Vaucluse, as well as near Pégomas and Grasse in the Département Alpes-Maritimes. The latter two districts produce perhaps half of the crop.

For planting, a cluster of estragon roots is divided into several segments; these segments are planted into the fields in February or March. A first harvest takes place in July; a second in September. The plants are cut in one day, flowers and leaves together, only very short stumps being left aboveground. These stumps grow sufficiently to form the second (September) harvest. A planting lasts about three years, after which it must be renewed. Deel and Deel [1] found that the optimum yield of herb material and essential oil is obtained with soil with a pH of 6.2. (With such soil

---

[10] *Acta Chem. Scand.* **4** (1950), 1575. *Chem. Abstracts* **45** (1951), 7005.
[1] *Bull. soc. chim* [4], **45** (1929), 175.

the oil may contain as much as 66.3 per cent of methyl chavicol, its chief constituent.)

Distillation of the herb is carried out in direct steam stills, from 1 to 1½ hr. being required to complete a charge. Care must be exercised during distillation not to "burn" the plant material in the stills, since this adversely affects the odor of the oil, destroying its characteristic and delicate estragon note.

According to Gildemeister and Hoffmann,[2] dried German herb material yields from 0.25 to 0.8 per cent, fresh material from 0.1 to 0.45 per cent of oil. Chiris[3] reported a yield of 1.4 per cent for fresh herb grown in Abadie (A.M., France). When dried for 24 hr. the herb yielded only 1.04 per cent of oil (calculated upon the fresh herb). In the author's own experience, plant material that has been cut during the flowering stage and reduced to clover dryness yields from 0.3 to 0.35 per cent of oil. According to Heeger,[4] estragon yields from 1.35 to 2.40 per cent of volatile oil. The so-called German estragon contains more oil than the Russian. The plant should be cut at the beginning of the flowering stage; otherwise it loses some of its aroma.

Physicochemical Properties.—Oil of estragon is a light yellow to greenish liquid of peculiar, strongly aromatic odor, characteristic of the plant, and reminiscent of sweet basil and anise.

Gildemeister and Hoffmann[5] reported these properties for estragon oil:

Specific Gravity at 15°.............	0.900 to 0.945
Optical Rotation.................	+2° 0′ to +9° 0′
Refractive Index at 20°...........	1.504 to 1.516
Acid Number....................	Up to 1
Ester Number...................	1 to 9
Ester Number after Acetylation....	15 to 22
Solubility.......................	Soluble in 6 to 11 vol. of 80% alcohol, eventually with slight turbidity. Soluble in 0.6 to 1.5 vol., and more, of 90% alcohol

An oil distilled by Chiris[6] (see above) from estragon grown in France exhibited the following properties:

Specific Gravity at 15°.............	0.9467
Optical Rotation at 20°...........	+2° 44′
Refractive Index at 20°...........	1.51572
Acid Number....................	0.28

[2] "Die Ätherischen Öle," 3d Ed., Vol. III, 1,000.
[3] *Parfums France* (1923) No. 8, 28.
[4] *Pharm. Ind.* **7** (1940), 372.
[5] "Die Ätherischen Öle," 3d Ed., Vol. III, 1,000.
[6] *Parfums France* (1923) No. 8, 28.

Ester Number.................... 2.8
Ester Number after Formylation in
 the Cold...................... 12.32
Free Alcohol Content............. 2.63%
Methyl Chavicol Content.......... 65%
Solubility...................... Soluble in 13.5 vol. of 75% alcohol;
  soluble in 6.5 vol. of 80% alcohol

Four estragon oils (I, II, III and IV) distilled under the author's supervision in Seillans (Var), France, from cultivated herb had these properties:

	Specific Gravity at 15°	Optical Rotation	Refractive Index at 20°	Saponification Number
No. I...........	0.926	+3° 46'	1.5112	3.7
No. II..........	0.934	+4° 12'	1.5130	2.8
No. III.........	0.941	+3° 34'	1.5141	4.7
No. IV..........	0.943	+3° 25'	1.5152	7.5
No. V..........	0.966	+3° 20'	1.5201	17.7

Oil No. V originated from a reliable source in Southern France but was more than one year old when analyzed. The somewhat abnormal properties of this oil are characteristic of older estragon oils (see below).

All five oils were soluble in 0.5 and more volumes of 90 per cent alcohol. Oils I, II, III and IV were not clearly soluble in 80 per cent alcohol; oil V was soluble in 80 per cent alcohol.

Pure estragon oils produced in Southern France and examined in the laboratories of Fritzsche Brothers, Inc., New York, exhibited properties varying within the following limits:

Specific Gravity at 15°/15°......... 0.929 to 0.966, seldom higher than
  0.956
Optical Rotation................... +2° 0' to +4° 12'
Refractive Index at 20°............. 1.5121 to 1.5201
Saponification Number.............. 2.8 to 18.4
Ester Number after Acetylation...... 11.2 to 36.4
Solubility....................... Soluble in 0.5 to 1 vol. of 90% al-
  cohol and more

An oil of estragon distilled in Seillans (Var), Southern France, from Italian plant material had similar characteristics:

Specific Gravity at 15°/15°......... 0.925
Optical Rotation................... +3° 30'
Refractive Index at 20°............. 1.5098
Saponification Number.............. 1.9
Ester Number after Acetylation...... 15.9
Solubility....................... Hazy in 10 vol. of 80% alcohol

In recent years oil of estragon has also been produced in the United States, chiefly near Fremont, Ohio. Such oils, analyzed by Fritzsche Brothers, Inc., New York, had these properties:

Specific Gravity at 15°/15°...... 0.930 to 0.944
Optical Rotation............... +1° 53' to +4° 37'
Refractive Index at 20°......... 1.5120 to 1.5159
Saponification Number.......... Up to 4.2
Solubility..................... Soluble in 0.5 to 1.5 vol. of
            90% alcohol and more

The odor and flavor of the American estragon oils were not quite as fine and characteristic of the estragon plant as those produced in Southern France.

Oil of estragon has to be stored in well-filled bottles, protected from light. When exposed to air the chief constituent of the oil, methyl chavicol, is oxidized to an aldehyde, with accompanying increase in the specific gravity and refractive index of the oil. The solubility in alcohol improves to such an extent that an aged oil is soluble in 80 per cent alcohol. Gildemeister and Hoffmann [7] reported that the physicochemical properties of an oil (I) after standing six months in an open bottle underwent considerable changes (II):

	I	II
Specific Gravity at 15°....	0.9168	1.0475
Optical Rotation.........	+3° 10'	+3° 10'
Refractive Index at 20°...	1.50847	1.52493
Solubility...............	Soluble in 1 and more vol. of 90% alcohol	Soluble in 1 and more vol. of 80% alcohol

The aged oil (II) gave a positive reaction on treatment with bisulfite solution, which indicates that an aldehyde had been formed by air oxidation of the oil.

Chemical Composition.—Like most essential oils, oil of estragon probably contains numerous constituents, but only very few have been identified:

Ocimene(?). In the terpene fraction (15 to 20 per cent of the oil) Daufresne [8] noted the presence of an aliphatic hydrocarbon b. 173°–175°, d_{15} 0.812, α_D +29° 26', n_D^{15} 1.48636, mol. refr. 48.04, which appeared to contain three double bonds, and was perhaps ocimene. Reduction of this hydrocarbon yielded dihydromyrcene b_{12} 66°, d_{15} 0.7972, n_D^{15} 1.45782, mol. refr. 47.19; hydration according to the method of Bertram and Walbaum gave a substance b_{15} 95°–100°, which had an odor similar to linalyl acetate.

[7] "Die Ätherischen Öle," 3d Ed., Vol. III, 1001.
[8] Thèse, Paris (1907). *Compt. rend.* **145** (1907), 875. *Bull. soc. chim.* [4], **3** (1908), 330.

(*Top*) German chamomile (*Matricaria chamomilla*) growing wild in Farmos, Com. Pest, Hungary. (*Middle*) A plantation of Roman chamomile (*Anthemis nobilis*) near Deáki (Pozsony County), Hungary. *Photos Dr. Jules de Bittera, Budapest, Hungary.* (*Bottom*) Harvest of Roman chamomile (*Anthemis nobilis*) in Belgium. *Photo Fritzsche Brothers, Inc., New York.*

(*Top*) Production of oil of wormwood (*Artemisia absinthium*) in southern Michigan. Harvesting of the plant material. (*Bottom*) Oil of wormwood. A field distillery in southern Michigan. *Photos Fritzsche Brothers, Inc., New York.*

Phellandrene(?). In the terpene fraction Daufresne [9] observed another hydrocarbon, identical perhaps with phellandrene.

Methyl Chavicol. More than a century ago Laurent [10] oxidized estragon oil and obtained an acid m. 175°, which he named dragonic acid. A few years later Gerhardt [11] arrived at the conclusion that this acid was nothing other than anisic acid, and reported that oil of estragon in many ways reacted like oil of anise seed. On the basis of Gerhardt's work, it was believed for almost fifty years that anethole formed the principal constituent of estragon oil. It was only in 1892 that Schimmel & Co.[12] proved that the chief component of the oil is actually methyl chavicol, and that no anethole is present. One year later Grimaux [13] investigated the chemical composition of estragon oil and confirmed the conclusion of the Schimmel chemists. Grimaux assigned the name estragole to the compound, although there was no justification for assigning a new name to the substance already well known as methyl chavicol.

As regards the quantity of methyl chavicol present in estragon oil, Chiris [14] found 65 per cent in an oil distilled in Southern France (Abadie, A.M.) from cultivated plants (cf. above—Physicochemical Properties).

p-Methoxycinnamaldehyde. Examining a ten-year-old estragon oil of German origin Daufresne [15] noted the presence of 4.5 per cent of p-methoxycinnamaldehyde b_{13} 170°, d_0 1.137, which he identified by means of its semicarbazone m. 222°, and oxime m. 154°. In another German oil (probably not so old) Daufresne found only 0.4 per cent, and in a French oil, 0.5 per cent of p-methoxycinnamaldehyde. Oxidation of the aldehyde with permanganate in acidic solution gave anisic acid m. 184°; oxidation with silver oxide yielded p-methoxycinnamic acid m. 170°.

An Aldol(?). Investigating the high boiling fractions of the oil, Daufresne [16] observed that they were laevorotatory and that splitting off of water and resinification occurred. He concluded that the fractions probably contained an aldol.

Use.—Oil of estragon is used widely, but in small dosage, as a flavoring agent in vinegars, table sauces, salad dressings, canned soups, and liqueurs. It is also employed in perfumes, particularly of the chypre type.

[9] *Ibid.*
[10] *Liebigs Ann.* **44** (1842), 313.
[11] *Compt. rend.* **19** (1844), 489. *Liebigs Ann.* **52** (1844), 401.
[12] *Ber. Schimmel & Co.,* April (1892), 17.
[13] *Compt. rend.* **117** (1893), 1089.
[14] *Parfums France* (1923), 28.
[15] Thèse, Paris (1907). *Compt. rend.* **145** (1907), 875. *Bull. soc. chim.* [4], **3** (1908), 330.
[16] *Ibid.* Cf. Daufresne and Flament, *Bull. soc. chim.* [4], **3** (1908), 656.

OIL OF SWEET GOLDENROD

In the United States there are about 75 species of *Solidago* (fam. *Compositae*), commonly known as "Goldenrod." In Texas alone, 39 species have been reported by Cory and Parks.[1] Some of the *Solidago* species grow so profusely that they are considered weeds. Many possess a characteristic odor and flavor. *Solidago odora* Ait., the so-called sweet goldenrod, e.g., has a sweet, long-lasting flavor resembling licorice.

According to Holland, Johnson and Sorrels,[2] production of oil of sweet goldenrod is economically feasible. Through cultivation, large quantities of the disease and insect resistant plant could be made available. Little care is required during the growing period; harvesting can be accomplished by means of row binders.

Botany and Occurrence.—The sweet goldenrod (*Solidago odora* Ait.) flourishes from New Hampshire to Florida, and as far west as Missouri and Texas. It grows in open woods, along the edges of hedgerows, in thickets or woods and old abandoned fields. The typical wild plant, when mature, is from 2 to 4 ft. tall; the cultivated plant becomes larger—often reaching a height of 5 ft. or more.

Solidago odora has a perennial root and crown, and an annual top. It spreads by sending up several new stems from the crown and root each spring, and also by seed. Propagation may also be by division of old clumps —probably the fastest method for establishing a new planting. Division of old clumps is best done in the spring, after the new growth has reached a height of about 6 in.

Cultivating and Harvesting.—The plant is put out in rows, the width of these being determined by the amount of space required for cultivating and harvesting equipment. A traction-operated rowbinder, for example, requires $3\frac{1}{2}$- to 4-ft. rows. Second and third year plantings may have a spread of from 6 to 10 in. in the row, and a width of 4 ft. between rows may be necessary for cultivating and harvesting operations. The principle object in cultivating goldenrod is to obtain pure stands, which can be harvested at the least cost. Further care beyond control of weeds and grass is not necessary.

According to Holland, Johnson and Sorrels,[3] goldenrod should be harvested when most of the plants are in bloom, since the oil yield in relation to

[1] "Catalogue of the Flora of Texas," *Texas Agr. Expt. Sta., Bull.* No. 550, July (1937).
[2] "Essential Oil Production in Texas. II. Sweet Goldenrod," *Texas Eng. Expt. Sta., Bull.* No. 107, September (1948).
[3] *Ibid.*

quality is highest at that stage of growth. Delaying the harvesting past full bloom may increase the yield slightly, but the quality of the oil suffers. Harvesting normally occurs from September 20th to the 25th, although there may be seasonal variations of from ten to fourteen days. Method of harvesting should be adapted to the size and manner of planting: small areas are easily harvested by hand—by breaking or cutting with pruning shears; larger areas, grown in rows, may be economically harvested with a rowbinder. Broadcast plantings can be harvested with a grain binder or mower. Rate of harvesting should be commensurate with rate of distillation, as a lapse of two days between cutting and distilling adversely affects yield of oil. Close stacking or tramping of piles, to prevent drying, is not advisable, as the material will become heated, with a further loss of oil.

Distillation and Yield.—The herb is steam distilled in the usual way, a charge of 150 to 250 lb. of fresh plant material requiring 2 to 3 hr. for completion. The freshly cut goldenrod should be distilled at once; certainly not later than two to three days after cutting.

Holland, Johnson and Sorrels reported yields of oil ranging from 1.89 per cent on September 23, to 0.58 per cent on September 30. Highest yields have consistently been obtained from fresh material harvested when the plant was in full bloom.

Physicochemical Properties.—When freshly distilled, oil of sweet goldenrod is a colorless, refractive liquid, with an odor and flavor reminiscent of estragon and anise.

Holland, Johnson and Sorrels [4] reported these properties for a fresh oil:

Specific Gravity at 30°	0.939
Optical Rotation	+14° 0′
Refractive Index	1.5095
Acid Number	0.18
Ester Number	4.98
Saponification Number	5.16
Solubility	Soluble in 6.6 vol. of 80% alcohol

Samples of oils analyzed by Fritzsche Brothers, Inc., New York, exhibited the following properties:

Specific Gravity at 15°/15°	0.950 to 0.956
Optical Rotation	+10° 38′ to +14° 21′
Refractive Index at 20°	1.5123 to 1.5153
Saponification Number	2.7 to 5.8
Ester Number after Acetylation	5.7 to 7.7
Solubility	Soluble in 0.5 to 1 vol. of 90% alcohol and more. Some oils soluble in 5 to 6 vol. of 80% alcohol and more

[4] *Ibid.*

Gildemeister and Hoffmann [5] reported values varying within these limits:

Specific Gravity at 15°.....................	0.94 to 0.96
Optical Rotation..........................	+9° 20' to +13° 12'
Refractive Index at 25°...................	1.506 to 1.514
Saponification Number....................	7 to 9
Ester Number after Acetylation............	19.4
Solubility................................	Soluble in 0.4 vol. of 90% alcohol

On long storage in clear glass bottles, exposed to light, the color of the oil changes to a deep amber, the optical rotation decreases; the refractive index, acid number and ester number, on the other hand, increase.

Chemical Composition.—The chemical composition of oil of *Solidago odora* was investigated years ago by Miller and Moseley,[6] and more recently by Holland.[7] The former two authors concerned themselves chiefly with the oxygenated compounds present in the oil, whereas Holland examined the terpenes. The following compounds have been found in the oil:

d-Limonene. Identified by preparation of the β-nitrolaniline m. 153°, and of the tetrabromide m. 103.5° (Holland).

Dipentene(?). The optical rotation and refractive index of the limonene fraction indicate that the oil probably contains dipentene (Holland).

l-α-Pinene. Characterized by means of the nitrolpiperidine m. 118.5° (Holland). The quantity of *l*-α-pinene found by distillation represented 0.2 per cent of the oil.

Methyl Chavicol. This phenolic ether, the chief constituent of the oil, was identified by Miller and Moseley, who oxidized it to anisic acid m. 184°, and homoanisic acid m. 85°–86°. The oil investigated by Miller and Moseley contained 76 per cent of methyl chavicol.

l-Borneol. The same workers also noted the occurrence of borneol in the oil (m. 203°–204°, phenylurethane m. 138°–139°). Years later, Holland reported that the borneol present is probably the laevorotatory form.

Acids(?). According to Miller and Moseley, the saponification lye of their oil probably contained three volatile acids, and a nonvolatile acid.

The investigation carried out by Miller and Moseley showed that oil of sweet goldenrod contains no phenols, aldehydes, or ketones. Phellandrene, camphor, and anethole are absent.

[5] "Die Ätherischen Öle," 3d Ed., Vol. III, 951. Cf. *Ber. Schimmel & Co.*, October (1891), 40. Miller and Moseley, *J. Am. Chem. Soc.* **37** (1915), 1286.

[6] *J. Am. Chem. Soc.* **37** (1915), 1285.

[7] *Ibid.* **70** (1948), 2597.

According to Holland, Johnson and Sorrels,[8] freshly distilled oil of *Solidago odora* has the following composition:

	Per Cent
Terpenes (chiefly *d*-Limonene)	15
Methyl Chavicol	75
l-Borneol	3
Esters (calculated as Bornyl Acetate)	3
Volatile Fatty Acids	Traces
	—
	96

Use.—At the time of this writing, oil of sweet goldenrod is not yet produced on a commercial scale. Because of its aromatic, licorice-like flavor the oil could be used perhaps in chewing gums and candies. Holland, Johnson and Sorrels suggest insecticides and deodorants as another possible outlet for this oil.

OIL OF *HELICHRYSUM ANGUSTIFOLIUM* DC.

Helichrysum angustifolium DC. (*H. italicum* G. Don), fam. *Compositae* —the common "Everlasting"—grows wild in the southern part of Europe, particularly in areas bordering the Mediterranean—e.g., the Dalmatian Islands, the Italian Riviera (near Portofino), and the French Riviera (Estérel Mountains). On distillation, the flowering plants yield an essential oil of peculiar, aromatic odor, reminiscent of rose and chamomile (cf. the following monograph on "Oil of *Helichrysum stoechas*").

The principal producing regions for the essential oil are the islands of Cherso and Lussino in Dalmatia. For the production of the oil, the flowering tops, including the leaves, of wild growing plants are cut with sickles, and distilled as soon as possible. If kept longer than 24 hr., the plant material will start to fade and to ferment, and the oil will be of poor quality. Production of helichrysum oil in Dalmatia was started on the island of Cherso, in 1908. Only small quantities of oil are produced annually.

Physicochemical Properties.—An oil distilled in San Marino di Cherso, under the author's supervision, had these properties:

[8] "Essential Oil Production in Texas. II. Sweet Goldenrod," *Texas Eng. Expt. Sta., Bull.* No. 107, September (1948).

Specific Gravity at 15°.............. 0.904
Optical Rotation................... −0° 30′
Refractive Index at 20°............. 1.4741
Saponification Number.............. 122.3
Ester Number after Acetylation...... 147.5
Solubility........................ Soluble in 0.5, and more,
    vol. of 90% alcohol

Shipments of pure helichrysum oils from Dalmatia analyzed by Fritzsche Brothers, Inc., New York, exhibited properties varying within the following limits:

Specific Gravity at 15°/15°......... 0.901 to 0.911
Optical Rotation................... −2° 20′ to +0° 10′
Refractive Index at 20°............. 1.4735 to 1.4759
Acid Number...................... 0.9 to 2.8
Ester Number..................... 115.7 to 122.3
Ester Number after Acetylation...... 141.9 to 150.3
Solubility........................ Soluble in 0.5 vol. of 90%
    alcohol and more

Dalmatian helichrysum oils examined by Schimmel & Co.[1] had these properties:

Specific Gravity at 15°.............. 0.8971 to 0.9101
Optical Rotation................... −14° 10′ to +3° 4′
Refractive Index at 20°............. 1.47750 to 1.49011
Acid Number...................... 0.6 to 1.9
Ester Number..................... 56.9 to 101.7
Ester Number after Acetylation...... 78.4 to 121.3
Solubility........................ Soluble from 3.5 to 9 vol.
    90% alcohol

Oils distilled by the same firm from dried herb (yield 0.075 per cent), imported from Dalmatia, exhibited the following values:

Specific Gravity at 15°.............. 0.892 to 0.920
Optical Rotation................... −9° 20′ to +4° 0′
Refractive Index at 20°............. 1.4745 to 1.4849
Acid Number...................... Up to 15
Ester Number..................... 39 to 134
Solubility........................ Soluble in 3.5 to 10 vol.
    of 90% alcohol, occa-
    sionally with separation
    of paraffins

[1] *Ber. Schimmel & Co.*, October (1903), 80; April (1909), 51; October (1911), 48; April (1914), 62; (1927), 54; (1932), 36. Cf. Gildemeister and Hoffmann, "Die Ätherischen Öle," 3rd Ed., Vol. III, 962.

Rovesti [2] reported these properties for helichrysum oils from various parts of Italy:

Specific Gravity at 15°.............. 0.890 to 0.922⟩
Optical Rotation................... −7° 47′ to +10° 0′
Refractive Index at 20°............ 1.4756 to 1.4825
Acid Number..................... 1 to 4
Ester Number.................... 9 to 75
Ester Number after Acetylation...... 35 to 159
Solubility........................ Soluble in 4 to 11 vol. of
90% alcohol

The dextrorotation of the oils from plants in southern areas is higher than that of plants originating in more northerly areas.

Distilling fresh plant material, Burger [3] obtained an oil (yield 0.2 per cent) exhibiting these values:

Specific Gravity at 20°......... 0.89
Optical Rotation.............. −5° 30′ to −5° 48′

Fractionating an oil, Burger reported:

Per Cent
b_{10} 55° to 65°....... 25
b_5 125° to 130°...... 50 (oil with an odor characteristic of tea roses)

Chemical Composition.—Except for the fact that oil of helichrysum contains *neryl acetate* as chief constituent, literature reports relatively little about the chemical composition of the oil. The following compounds have been identified:

d-α-Pinene. The presence of pinene in the oil was reported by Rovesti,[4] and later confirmed by Crabalona and Teisseire,[5] who oxidized this terpene to pinonic acid m. 70°. This acid was identified by means of its semicarbazone.

Nerol (Free and as Ester). The oil contains from 30 to 50 per cent of nerol, partly in the free form, partly esterified—chiefly with acetic, perhaps also with isovaleric and caprylic acids. (Regarding the isolation of nerol from oil of helichrysum, see Vol. II of this work, p. 175.)

Use.—Oil of *Helichrysum angustifolium* is produced in small quantities only. The principal use of the oil is for the isolation of nerol.

[2] *Rivista ital. essenze profumi* **17** (1935), 19.
[3] *Riechstoff Ind. Kosmetik* **13** (1938), 10.
[4] *Rivista ital. essenze profumi* **17** (1935), 19.
[5] *Bull. soc. chim.* (1948), 270.

CONCRETE AND ABSOLUTE OF *Helichrysum Angustifolium*

According to Girard,[6] *Helichrysum angustifolium* is also extracted with volatile solvents, whereby the so-called concretes and absolutes of "Immortelle" are obtained. (In 1939 about 5,000 kg. of plant material were processed with solvents in Grasse.) Extraction with petroleum ether yields about 1 per cent of a waxy, light brown concrete which, on treatment with alcohol in the usual way, gives 85 per cent of an alcohol-soluble, viscous, dark colored absolute. Sabetay [7] submitted the concrete, as well as the absolute, to steam distillation at reduced pressure, and obtained 4.9 per cent and 6.35 per cent, respectively, of a volatile oil. That from the concrete had these properties:

Specific Gravity at 15°/15°...... 0.9239
Refractive Index at 20°......... 1.5005 to 1.5046
Acid Number................. 14
Ester Number................ 28.1
Methoxy Content............. 1.25%

In the volatile oil thus obtained Sabetay found:

Caprylic Acid.

Eugenol.

Acetic Acid. In esterified form.

Valeric Acid(?). In esterified form.

An Azulogenetic Sesquiterpene $C_{15}H_{24}$.

It is interesting to note that Sabetay did not find any nerol in the volatile oils distilled from concrete and absolute of helichrysum. The plant material originated from the Estérel Mountains in Southern France.

Similar extraction experiments were carried out by Rovesti,[8] who treated plant material from the Abruzzi Mountains (Italy) in August and obtained 0.92 per cent of concrete by means of petroleum ether, and 1.13 per cent by means of benzene. The former concrete yielded 68.3 per cent of absolute, the latter 78.7 per cent.

Steam-distilling the two absolutes at ordinary pressure, Rovesti obtained 21.8 per cent and 15.8 per cent, respectively, of volatile oil with the following properties:

[6] *Ind. parfum.* **2** (1947), 218.
[7] *Ann. chim. anal.* [3], **22** (1940), 89.
[8] *Rivista ital. essenze profumi* **17** (1935), 23.

	Petroleum Ether	Benzene
Specific Gravity at 15°.	0.8874	0.9061
Optical Rotation.	+2° 53'	+3° 42'
Refractive Index at 20°.	1.4815	1.4791
Ester Number.	89.60	35.47
Ester Content.	31.36%	12.41%
Ester Number after Acetylation.	136.24	112

Naves, Sabetay and Palfray [9] distilled two concretes of helichrysum produced in Grasse (A.M., France) and obtained volatile oils with these properties:

	I	II
Specific Gravity at 15°.	0.939	0.935
Optical Rotation.	+5° 35'	+7° 40'
Refractive Index at 20°.	1.5021	1.4968
Acid Number.	20.2	13.1
Ester Number.	33.6	42.0

OIL OF *HELICHRYSUM STOECHAS* DC.

The various species of *Helichrysum* are very polymorphous, and cannot always be distinguished easily. *Helichrysum angustifolium* DC. (*H. italicum* G. Don), *H. arenarium* DC., and *H. stoechas* DC. (fam. *Compositae*), for example, grow wild and abundantly in Mediterranean countries, particularly in Dalmatia, Italy, Southern France, Spain, and North Africa. In addition to *Helichrysum angustifolium* (see the preceding monograph), *H. stoechas* is occasionally used for the production of its essential oil.

Steam-distilling flowering *Helichrysum stoechas* DC., collected in July on the Italian Riviera, Rovesti [1] obtained 0.107 per cent of an orange colored oil, which had an odor reminiscent of rose and chamomile. The oil exhibited these properties:

Specific Gravity at 15°.	0.9184
Optical Rotation at 20°.	−4° 31'
Refractive Index at 20°.	1.4774
Acid Number.	2.61
Ester Number.	158

[9] Cf. Naves and Mazuyer, "Les Parfums Naturels," Paris (1939), 221.
[1] *Rivista ital. essenze profumi* **12** (1930), 149.

<pre>
Ester Number after Acetylation...... 162.4
Solubility........................ Soluble in 3.3 vol.
                                    of 90% alcohol
</pre>

In the oil, Rovesti observed the presence of *neryl acetate* as chief constituent (nerol tetrabromide m. 118°), traces of *furfural*, probably some *l-α-pinene* (nitrosochloride m. 100°–101°), other hydrocarbons, and compounds of high molecular weight.

Rovesti also found that older plants, collected at the end of August and containing fully opened flowers, yielded only 0.051 per cent of essential oil.

Extracting *Helichrysum stoechas* with petroleum ether, the same author [2] obtained a concrete which, on treatment with alcohol in the usual way, gave 76.8 per cent of an alcohol-soluble absolute. Steam distillation of this absolute yielded 17.9 per cent of a volatile oil with these properties:

<pre>
Specific Gravity at 15°...... 0.9178
Optical Rotation........... −3° 17′
Refractive Index at 20°..... 1.4749
Ester Number............. 169.87
</pre>

From the above it appears that oil of *Helichrysum stoechas* resembles oil of *H. angustifolium* in regard to physiochemical properties and chemical composition. In fact, the two plants are often processed together, for production of oil of helichrysum (as it is called commercially).

OIL OF MILFOIL
(Oil of Yarrow)

Achillea millefolium L. (fam. *Compositae*), commonly known as milfoil or yarrow, is a widely distributed weed, official in the pharmacopoeias of several countries. Like chamomile, it finds extensive use in popular medicines and teas.

On steam distillation, the plant yields an essential oil of blue color, deeper than that of chamomile oil; this is due to the presence of azulene. The latter, however, does not appear to occur as such in the plant, but in the form of a precursor which, on treatment with steam, produces the azulene (cf. the monograph "Azulenes," Vol. II of this work, pp. 127 ff.). Extracting

2 *Ibid.*

powdered flowers of *Achillea millefolium* with petroleum ether, Graham [1] obtained an oil which did not contain any azulene. Steam distillation of the *extracted* flowers, however, yielded 0.01 per cent of azulene. The precursor of the azulene can be extracted from the flowers by means of chloroform. On digestion of the chloroform extract with petroleum ether, the essential oil dissolves, whereas the azulene-precursor remains in the residue. Steam distillation of this residue yields a blue distillate containing the azulene. Rosenthal [2] found that yarrow blossoms of varying origin contained up to 0.199 per cent of azulene; environment does not seem to influence the azulene content of the plant.

As regards the essential oil content of the plant, Gildemeister and Hoffmann [3] reported that, on distillation, fresh yarrow flowers yield from 0.07 to 0.25 per cent, dried and semidried flowers 0.24 to 0.5 per cent of oil. A part of the oil (about 21 per cent) dissolves in the distillation waters and has to be recovered by cohobation; the oil of cohobation must be added to the direct (main) oil.

Physicochemical Properties.—The following values have been reported in literature:

Specific Gravity at 15°..............	0.900 to 0.936 (oils produced in Germany—Gildemeister and Hoffmann [4])
Specific Gravity...................	0.8687 to 0.8935 (oils distilled by Sievers [5] in the United States)
Specific Gravity at 22°.............	0.9217 (Aubert [6])
Specific Gravity at 17° of Direct Oil..	0.913 to 0.915
Specific Gravity at 25° of Cohobation Oil...........................	0.939 to 0.959 (Kremers,[7] and Miller [8])
Optical Rotation..................	Oil usually too dark. In one case −1° 39', determined in a solution of absolute alcohol (1:200), using a 50 mm. tube (Haensel [9])
Refractive Index at 20°.............	1.48645 (one determination—Gildemeister and Hoffmann)
Acid Number.....................	1 to 16 (Gildemeister and Hoffmann)
Ester Number....................	10 to 27 (Gildemeister and Hoffmann)
Ester Number after Acetylation......	66 to 79 (Gildemeister and Hoffmann)
Saponification Number.............	29.3 to 37.7 (American oils—Sievers)
Solubility.......................	Soluble in 0.2 to 1 vol., often in all vol. of 90% alcohol, occasionally with separation of paraffins

[1] *J. Am. Pharm. Assocn.* **22** (1933), 819.
[2] *Arch. Pharm.* **279** (1941), 344.
[3] "Die Ätherischen Öle," 3d Ed., Vol. III, p. 980.
[4] *Ibid.*
[5] *Pharm. Rev.* **25** (1907), 215.
[6] *J. Am. Chem. Soc.* **24** (1902), 778.
[7] *J. Am. Pharm. Assocn.* **10** (1921), 252.
[8] *Wisconsin Univ. Bull., Science Ser.* **4**, No. 9 (1916).
[9] *Pharm. Ztg.* **47** (1902), 74.

Chemical Composition.—The chemical composition of the essential oil derived from *Achillea millefolium* was investigated by Schimmel & Co.,[10] Miller,[11] Kremers,[12] and Ruzicka and Rudolph,[13] who established the presence of the following compounds:

d-α-Pinene. Identified by means of its nitrosochloride m. 102°–103°, and nitrolpiperidine m. 119° (Miller, Kremers).

β-Pinene. Oxidation to nopinic acid m. 126°; nopinone semicarbazone m. 188° (Kremers).

l-Limonene. Not identified, but presence possible (Miller).

l-Borneol. M. 202°–203°; acetate m. 29°; phenylurethane m. 139° (Miller, Kremers).

Bornyl Acetate and Other Esters of Borneol. (Miller, Kremers).

l-Camphor. Oxime m. 118°; semicarbazone m. 235°–236° (Miller, Kremers).

Thujone. β-Thujaketonic acid m. 78°–79°; semicarbazone m. 191°–192° (Kremers).

Cineole. Resorcinol compound; iodol derivative m. 106°–107° (Miller, Kremers).

Azulene. From 100 g. of the fraction $b_{1.1}$ 140°, Kremers obtained 35.5 g. of a blue oil, viz., azulene $b_{1.1}$ 135°–136°, and 59 g. of oil that was not blue. The azulene readily formed a picrate on treatment with an alcoholic solution of picric acid.

Caryophyllene. After removal of the azulene by means of phosphoric acid, the oil was fractionated into eight fractions $b_{0.1-5.0}$ 80°–180°. When purified by distillation over metallic sodium, one of these fractions ($b_{0.5}$ 105°–110°, d_{20} 0.916, α −13° 45') yielded a nitrolbenzylamine m. 172°–173°, and a hydrate m. 94°–95°, indicating that the oil contains caryophyllene.

A Sesquiterpene(?) and a Sesquiterpene Alcohol(?). In the high boiling portions of the oil Ruzicka and Rudolph noted the presence of what was probably a sesquiterpene (cadinene type), and a mixture of (chiefly bicyclic) primary and secondary sesquiterpene alcohols.

Aside from the compounds listed above, Miller also found aldehydes (probably two), formic, acetic, butyric(?), and isovaleric acids, and a nonvolatile acid or lactone.

According to Kremers, the distillation waters of the oil contain formaldehyde, methyl alcohol, ethyl alcohol, acetone, furfural and borneol m. 203°–204°.

Production and Use.—Oil of milfoil is not produced in any appreciable quantity, if indeed at all. It has been reported as an adulterant of chamomile oil; any such practice would be absurd today in view of the high price of milfoil oil.

[10] *Ber. Schimmel & Co.*, October (1894), 55.
[11] *Wisconsin Univ. Bull., Science Ser.* **4**, No. 9 (1916).
[12] *J. Am. Pharm. Assocn.* **10** (1921), 252; **14** (1925), 399.
[13] *Helv. Chim. Acta* **11** (1928), 258.

SUGGESTED ADDITIONAL LITERATURE

W. Ripperger, *"Achillea millefolium* L. The Weed Yarrow as Drug Plant." *Pharm. Zentralhalle* **78** (1937), 641. *Chem. Abstracts* **32** (1938), 723.

A. Bänninger, "Investigations of the Influence of Mountain Climates on the Content of Active Components of Pharmaceutical Plants." *Ber. schweiz. botan. Ges.* **49** (1939), 239.

W. Peyer, "Yarrow Oil." *Deut. Apoth. Ztg.* **55** (1940), 1. *Chem. Abstracts* **34** (1940), 3016.

K. Koch, "Azulene Content of Yarrow and Its Colorimetric Estimation." *Deut. Apoth. Ztg.* **55** (1940), 758.

OIL OF *SANTOLINA CHAMAECYPARISSUS* L.

Santolina chamaecyparissus L. (family *Compositae*), the so-called "Cypress Lavender-Cotton," is a native of southern Europe, where it grows wild on clay and lime-containing soils, and often is also cultivated in gardens. The plant, a much branched evergreen subshrub, possesses a strong and penetrating aromatic odor. Because of its antispasmotic and anthelmintic properties, the plant was formerly official in the pharmacopoeias of various countries, and even today is used in old-fashioned medicines of Europe.

The essential oil, to which the plant owes its aroma, can be isolated by steam distillation. The yield of oil varies considerably with the stage of vegetation, being highest shortly before the flowering period. According to Francesconi and Scarafia,[1] it varies between 0.198 and 1.15 per cent. Schimmel & Co.[2] obtained 0.47 per cent from plants of Italian (Torino) origin. The yield is highest shortly before the flowering period. Like the yield, the chemical composition of the oil varies greatly with the stage of vegetation.

Physicochemical Properties.—The oil distilled by Schimmel & Co. (see above) exhibited a dark blue color and had an odor reminiscent of wormwood and tansy. It had the following properties:

Specific Gravity at 15°.............. 0.9065
Optical Rotation.................. Color of the oil too dark
Refractive Index at 20°............. 1.50040

---

[1] *Atti accad. Lincei* [5], **20** (1911), II, 255, 318, 383. *Chem. Zentr.* (1912), I, 344, 345. Cf. *Gazz. chim. ital.* **41** (1911), II, 180.

[2] *Ber. Schimmel & Co.,* October (1911), 107.

Acid Number...................... 6.6
Ester Number..................... 16.4
Ester Number after Acetylation...... 74.2
Solubility........................ Soluble in 0.5 vol. and more of 90%
alcohol with separation of paraf-
fins. Insoluble in 80% alcohol

In their oil, Francesconi and Scarafia (see above) noted: d_{15} 0.8732, $[\alpha]_D^{25}$ $-11°$ 44'.

Pellini and Morani [3] distilled plants growing in gardens of Palermo, Sicily, and obtained 0.41 per cent and 0.327 per cent, respectively, of orange-yellow and greenish-yellow oils with these properties:

	I	II
Specific Gravity at 15°......................	0.8868	0.9060
Optical Rotation.............................	$-24°$ 38'	$-20°$ 27'
Refractive Index.............................	1.4769	1.4807
Acid Number................................	0.9	2.7
Ester Number...............................	7.2	13.6
Ester Content, Calculated as $C_{10}H_{17}OCOCH_3$.....	2.52%	4.76%
Ester Number after Acetylation.................	33.2	56.1
Free Alcohol Content, Calculated as $C_{10}H_{18}O$.....	7.29%	12.08%
Total Alcohol Content.........................	9.27%	15.82%
Solubility....................................	Soluble in 0.37 vol. of 90% alcohol	

Massera [4] steam-distilled herb material of *Santolina chamaecyparissus* originating from Cyrenaica (Africa), obtaining 0.133 per cent of a greenish volatile oil with an agreeable odor; it exhibited the following values:

Specific Gravity at 15°........... 0.9275
Optical Rotation at 20°.......... $-5°$ 50'
Refractive Index at 20°.......... 1.4632
Acid Number.................. 5.64
Ester Number................. 114.4
Ester Number after Acetylation... 164.4
Total Alcohol Content, Calculated
as $C_{10}H_{18}O$.................. 51.56%
Solubility at 25°................ Soluble in 3 vol. of
70% alcohol

A test for aldehydes by means of Schiff's reagent was positive; phenols were absent from the oil.

More recently, Chiris [5] steam-distilled the overground parts of *Santolina chamaecyparissus* during the flowering period; these plants came from clay

[3] *Ann. chim. applicata* **13** (1923), 126.
[4] *Rivista ital. essenze profumi* **6** (1924), 136. Cf. *Chem. Zentr.* (1925), I, 915.
[5] *Parfums France* **15** (1937), 126.

and lime-containing soils in southern France. The oil obtained by Chiris had these properties:

Specific Gravity at 15°........... 0.9546
Optical Rotation................ −6° 26′
Refractive Index at 20°.......... 1.4908
Acid Number.................... 9.12
Ester Number.................... 22.41
Ester Number after Acetylation... 126.25
Solubility...................... Soluble in 0.1 vol.
of 95% alcohol

Years ago, Chiris had examined also a dextrorotatory santolina oil (α_D +10° 50′).

Chemical Composition.—The volatile oil derived from *Santolina chamae-cyparissus* L. contains a terpene(?), b. 165°–170°, in the first fractions, and perhaps a phenolic ether. The chief constituent is a mixture of two unsaturated ketones $C_{10}H_{16}O$, viz., α- and β-*santolinenone*, which have been investigated by Francesconi and his collaborators. (For details see Vol. II of the present work, p. 409.) The third isomer appears to be a saturated ketone $C_{10}H_{16}O$, which is probably of the camphor type.

Use.—According to the author's knowledge, the essential oil of *Santolina chamaecyparissus* L. is not produced on a commercial scale.

OIL OF TAGETES

The common term "marigold" embraces a diversity of plants with golden flowers, most of which belong to the family *Compositae*. Prominent among the marigolds are various species of *Tagetes*,[1] particularly *Tagetes glandulifera* Schrank, an annual herb, which, according to the "Handlist of Herbaceous Plants,"[2] is synonymous with *Tagetes minuta* L.[3]

[1] For a detailed discussion, see *Parfums France* **14** (1936), 6.

[2] Royal Botanic Gardens, Kew (1925).

[3] In 1924 the Imperial Institute, London [*Bull. Imp. Inst.* **22** (1924), 279] examined an essential oil distilled in Africa, supposedly from *Tagetes minuta* L., and found these properties: d_{15} 0.9369, α_D +1° 42′, n_D^{20} 1.496, Acid Number 1.5, Ester Number 44.5, Ester Number after Acetylation 116.5; soluble in 1.5 volumes of 90% alcohol, turbid on dilution. The oil polymerized rapidly on standing. The Imperial Institute also reported the presence, in this oil, of olefinic terpenes (myrcene or ocimene?), small amounts of phenols, linaloöl, and carvone, without, however, indicating how

Tagetes glandulifera Schrank, the so-called Mexican marigold or "orina," is a native of Central America but now grows abundantly and widely, as a weed, from Canada to Argentina. According to information gathered by the author while surveying the production of essential oils in East Africa, the plant was introduced into South Africa probably during the Boer War in 1900, when a great number of horses and a large quantity of fodder were imported from the Argentine. After the Boer War, Australian troops returning from South Africa brought the plant to Australia where it now grows most profusely as a weed. During the East African campaign (1914 to 1918) the marigold was introduced from South Africa to Kenya Colony, where it took root and is still spreading over wide areas—a real nuisance to planters. Locally known as "khaki bush" it thrives along roadsides and on any land that has been plowed up. In Africa it is claimed that the plant acts as a fly and vermin repellent. The natives of East Africa hang tagetes plants in their huts to keep out the swarms of flies which are a terrible nuisance in those parts of Africa. It has been demonstrated that common houseflies and blow flies avoid baits scented with tagetes oil, whereas they are attracted by, and readily lay their eggs on, control baits not scented with the oil. Attempts have also been made to develop an effective larvicide that would kill maggots in wounds. An emulsion of water, carbon tetrachloride, some wool fat, five per cent of tagetes oil, and a preserving agent was found to be very effective. On the other hand, worm-killing experiments with tagetes oil were negative; the same is true in regard to ticks and other external parasites.

The plant contains an essential oil which can be isolated by steam distillation. In order to obtain the maximum of oil, the plant should be harvested and distilled during the period of maturity (formation of seed), after the full flowering stage. In East Africa natives cut the plants with long bush knives (machetes). Because of transportation difficulties, only limited quantities of tagetes oil can be produced from the widely scattered, wild growing plants. To obtain larger quantities of oil the plant would have to be cultivated—a very easy task, however.

In Kenya Colony distillation is carried out with direct steam, one batch requiring 3 to 4 hr. The yield of oil ranges from 0.3 to 0.4 per cent.[4] Distilling flowering plants, collected near Brisbane (Australia), Jones and Smith[5] obtained 0.5 per cent of oil. Igolen[6] reported yields ranging from

these compounds were identified. It appears quite possible that the ketone (carvone) reported by the Imperial Institute was in reality tanacetone (see below), and that the oil investigated was actually distilled from *Tagetes glandulifera*, the principal source of the commercial tagetes oil. In the following we shall, therefore, deal exclusively with the oil derived from *Tagetes glandulifera*.

4 Guenther, *Soap* **15** (February 1939), 28.   6 *Parfums France* **14** (1936), 10.
5 *J. Chem. Soc.* **127** (1925), 2530.

0.187 to 0.263 per cent for plants growing in the Grasse region of Southern France, and harvested during the period of maturity (end of November).

Physicochemical Properties.—Oil of tagetes is a yellow-reddish liquid with a powerful, peculiar, somewhat disagreeable odor, reminiscent of rancid butter, spearmint, and—in great dilution—of apples. On standing, the oil readily polymerizes, until it turns into an almost solid gel. The oil, therefore, must be stored with great care, under complete exclusion of air and light.

An oil of tagetes procured by the author in Kenya Colony had these properties:

Specific Gravity at 15°.....	0.869
Optical Rotation..........	+2° 44'
Refractive Index at 20°....	1.4841
Acid Number.............	0
Saponification Number.....	13.5
Ester Number after Acetylation.................	71.1
Ketone Content, Calculated as Carvone............	44.6%
Solubility................	Insoluble in 10 vol. of 80% alcohol. Soluble in 0.5 to 1 vol. of 90% alcohol, cloudy with more

The oil distilled by Jones and Smith [7] from flowering plants growing near Brisbane (see above) had these properties:

Specific Gravity at 15.5°....................	0.8638
Specific Optical Rotation....................	+4° 0'
Refractive Index at 20°.....................	1.4820
Acid Number..............................	2
Saponification Number after Acetylation......	33

The oil distilled by Igolen [8] from plants growing in the Grasse region of Southern France (see above) exhibited the following values:

Specific Gravity at 15°............	0.922
Optical Rotation.................	+1° 30'
Refractive Index at 20°...........	1.5112
Acid Number.....................	1.96
Ester Number....................	16.81
Ester Number after Acetylation....	63.13
Solubility.......................	Soluble in 1 vol. of 90% alcohol

The same author [9] extracted fully matured tagetes plants with petroleum ether and obtained 0.278 per cent of a concrete which, on treatment with

[7] *J. Chem. Soc.* **127** (1925), 2530.   [9] *Ibid.*
[8] *Parfums France* **14** (1936), 10.

alcohol in the usual way, yielded 56.7 per cent of an alcohol-soluble, viscous absolute. Extraction of the plants with benzene gave a much larger yield of concrete, but of less floral odor than the concrete obtained by extraction with petroleum ether.

Chemical Composition.—The chemical composition of oil of *Tagetes glandulifera* was investigated by Jones and Smith,[10] and later by Jones,[11] who reported the presence of the following compounds:

d-Limonene. Identified by means of the tetrabromide m. 104°. The oil contains about 3 per cent of *d*-limonene.

Ocimene. Characterized by reduction to dimethyloctane, by conversion into *allo*-ocimene, and by reduction to dihydromyrcene (tetrabromide m. 88°). The oil contains about 30 per cent of ocimene.

Tagetone. Regarding this ketone, which occurs in the oil to the extent of 50 to 60 per cent, see Vol. II of this work, p. 384.

2,6-Dimethyl-7-octen-4-one. Aside from tagetone, the oil contains from 5 to 10 per cent of another ketone, $C_{10}H_{18}O$, b. 185°, optically slightly active. It forms a semicarbazone m. 92.5°, and a liquid dibromide. Reduction in the presence of catalysts yielded a saturated ketone, $C_{10}H_{20}O$, viz., 2,6-dimethyloctan-4-one. Jones and Smith succeeded in identifying the ketone $C_{10}H_{18}O$ as 2,6-dimethyl-7-octen-4-one:

$$H_3C \diagdown$$
$$CH \cdot CH_2 \cdot CO \cdot CH_2 \cdot CH \cdot CH{=}CH_2$$
$$H_3C \diagup \qquad\qquad | $$
$$\qquad\qquad\qquad CH_3$$

Production and Use.—Very little oil of tagetes is produced at present. It has been suggested as a modifier in hair lotions of the bay rum type.

OIL OF TANSY

Essence de Tanaise *Aceite Esencial Tanaceto* *Rainfarnöl*
Oleum Tanaceti

Tanacetum vulgare L. (fam. *Compositae*), our common tansy, is a perennial of strong aromatic odor and bitter taste. Introduced from Europe, it has been growing in this country for a long time. Having escaped cultiva-

---

[10] *J. Chem. Soc.* **127** (1925), 2530, 2538. [11] *Proc. Roy. Soc. Queensland* **45** (1933), 45.

tion, tansy occurs as a weed along waysides, fences, and in old fields from New England to Minnesota, and southward to Missouri and North Carolina. For the distillation of its oil, tansy-is now grown commercially in southern Michigan and northern Indiana, South Bend being the center of the industry.

The strong scented, herbaceous perennial, rising 2 or 3 ft., in some instances even 4 ft. high, has finely divided, fern-like leaves and yellow, button-like flowers which appear from July to September. The odor of the plant is strong and characteristic; the taste is warm, bitter, somewhat acrid, and aromatic. The leaves and flowering tops are in some demand for medicinal purposes and for aromatic bitters. The seeds are said to be effective as a vermifuge. The essential oil, which is contained chiefly in the flowers, has been known in Europe as an anthelmintic since medieval times.

Planting, Cultivation, and Harvest.—Tansy grows on almost any good soil, but most luxuriantly on rich muck soil well supplied with moisture. It may be propagated from seed, but more readily by division of the roots in early spring. The land is prepared as for peppermint, by plowing, fertilizing, and harrowing. Plants are set out with a cabbage planter, 18 in. apart, and in rows 3 ft. apart. If seed is used for planting, it should be sown in seedbeds very early in spring, or in the open; the seedlings are later transplanted to the fields. Rows must be cultivated regularly in order to keep the ground stirred. Tansy keeps out weeds and, therefore, requires little hand weeding. It is also a hardy plant, not easily subject to damage by insects, pests, or diseases.

Being a long-lived perennial, tansy does not have to be plowed under in fall; it may remain in the ground for twenty to twenty-five years.

The plants are cut during the period of flowering, about the middle of August. Harvesting is done with a wheat binder. After partial drying in the fields for 24 hr., the sheaves are loaded on trucks and hauled to the few distilleries which specialize in the distillation of this oil.

If sold as dried herb, the leaves and tops are separated from the stems and fully dried under sheds; by avoiding exposure to the sun, a bright green color is obtained. The plants lose about four-fifths of their weight by drying.

One acre produces from 9,000 to 12,000 lb. of semidried herb material, as employed for distilling. This corresponds to approximately 18 to 24 lb. of oil. On the other hand, one acre produces about 2,000 lb. of dry leaves and flowering tops, as sold to the crude drug trade.

Distillation.—The field distilleries are very similar to those used for distilling oils of peppermint, spearmint, or wormwood. Live steam, generated in a separate steam boiler, is blown through the herb material and the essential oil separated in a Florentine flask.

Approximately 3,000 lb. of the semidry herb are charged into a still; distillation of one batch lasts about 40 min. and yields approximately 6 lb. of oil. The yield of oil varies from 0.2 to 0.5 per cent.

Total Production.—Average total production of tansy oil in the Middle West amounts to only about 2,000 lb. per year. Because of limited consumption, there is always the danger of overproduction and resultant lowering of the oil prices to unprofitable levels. Therefore, the present few growers of tansy produce the oil only if prices are sufficiently attractive.

Physicochemical Properties.—Oil of tansy is a liquid of peculiar, aromatic odor and yellowish color which turns brown under the influence of air and light.

Gildemeister and Hoffmann [1] reported the following properties for commercial tansy oils from North America:

```
Specific Gravity at 15°.............. 0.925 to 0.935
Optical Rotation................... +24° 0' to +38° 0'
Refractive Index at 20°............. 1.457 to 1.462
Acid Number...................... Up to 1
Ester Number..................... Up to 16; in one case 42.9
Ester Number after Acetylation...... 28 to 64
Solubility........................ Soluble in 2 to 4 vol. of 70%
                                    alcohol; occasionally opal-
                                    escent to turbid on dilution
```

Oils distilled in Germany from fresh herb had specific gravities ranging from 0.925 to 0.940 at 15°; oils from dried herb had gravities up to 0.955. The oils were not clearly soluble in 70% alcohol (Gildemeister and Hoffmann).

North American tansy oils examined in the laboratories of Fritzsche Brothers, Inc., New York, exhibited properties ranging within these limits:

```
Specific Gravity at 15°/15°.............. 0.916 to 0.932
Optical Rotation....................... +27° 54' to +38° 50'
Refractive Index at 20°................. 1.4576 to 1.4635
Saponification Number.................. 4.7 to 26.1
Ester Number after Acetylation......... 29.2 to 69.1
Ketone Content, Calculated as Thujone
  (Hydroxylamine Hydrochloride Method) 50.0 to 67.3%
Solubility............................. Soluble in 2 to 3 vol. of 70% alcohol, clear
                                        to turbid with more. (A few oils were
                                        insoluble in 70% alcohol and required
                                        80% alcohol to yield clear solutions)
```

To improve their solubility a few lots of oil were redistilled. The following values were observed on these twice rectified oils:

[1] "Die Ätherischen Öle," 3d Ed., Vol. III, 994.

Specific Gravity at 15°/15°.............. 0.924 to 0.927
Optical Rotation...................... +33° 10' to +34° 10'
Refractive Index at 20°................. 1.4597 to 1.4610
Saponification Number.................. 4.7 to 19.6
Ester Number after Acetylation.......... 49.5 to 54.1
Ketone Content, Calculated as Thujone
  (Hydroxylamine Hydrochloride Method) About 65%
Solubility............................. Soluble in 2.5 to 3 vol. and more of 70%
  alcohol

The most important assay in the evaluation of tansy oil is the determination of the ketone (chiefly thujone) content which is best carried out by means of the hydroxylamine hydrochloride method. For details see Vol. I of the present work, p. 285.

Oil of tansy has also been produced, but only occasionally, in Hungary, where the plant grows wild in large patches, particularly on the plain near Hortobagy.[2] Three oils, produced and analyzed by de Bittera[3] had the following properties:

	I	II	III
Specific Gravity at 15°..........	0.9409	0.9208	0.9232
Optical Rotation...............	−6° 16'	−5° 42'	−6° 0'
Refractive Index at 20°.........	1.4588	1.4556	1.4547
Saponification Number..........	...	...	45.8
Solubility....................	Soluble in 70% alcohol		

In regard to odor and flavor, the Hungarian tansy oils were weaker than North American oils. Note the *laevo*rotation of these oils!

Six Hungarian oils described by Gildemeister and Hoffmann[4] also exhibited physicochemical properties, as well as an odor and flavor, differing from those of the North American oils:

Specific Gravity at 15°.............. 0.9213 to 0.9309
Optical Rotation................... −4° 50' to +15° 34'
Refractive Index at 20°............. 1.45622 to 1.47073
Acid Number...................... 1.2 to 4.7
Ester Number..................... 48.5 to 108.3
Ester Number after Acetylation...... 58.8 to 165.2
Solubility........................ Soluble in 2 to 2.4 vol. and more of
  70% alcohol. Separation of par-
  affins in the case of two oils

Klopfer[5] prepared terpeneless and sesquiterpeneless tansy oils, which were 1½ to 2 times stronger than the regular oils, and reported these values:

[2] Guenther, "Hungarian Essential Oils," *Am. Perfumer* **37** (July 1938), 42.
[3] *Ibid.*
[4] "Die Ätherischen Öle," 3d Ed., Vol. III, 994.
[5] *Ber. Schimmel & Co., Jubiläums Ausgabe* (1929), 175.

Specific Gravity at 15°......	0.930 to 0.935
Optical Rotation..........	$+38°\ 0'$ to $+40°\ 0'$
Acid Number..............	0
Ester Number.............	14 to 16
Solubility at 20°...........	Soluble in 1 to 1.2 vol. of 80% alcohol. Soluble in 5 to 7 vol. of 70% alcohol. Soluble in 20 to 25 vol. of 60% alcohol

These oils exhibited a very pronounced thujone odor.

Chemical Composition.—Investigating the chemical composition of tansy oil, Bruylants [6] isolated its chief constituent and declared it to be an aldehyde. Later, Semmler [7] proved that this compound was actually a ketone, identical with the thujone observed by Wallach [8] in oil of thuja.

The following compounds have been identified in oil of tansy:

β-Thujone. In 1904 Wallach [9] showed that thujone, the chief constituent of tansy oil, occurs in the oil in the dextrorotatory β- form. However, according to Short and Read,[10] β-thujone is actually d-isothujone, while α-thujone is l-thujone (cf. Vol. II of this work, p. 423). The natural products α- and β-thujone are a mixture of these two diastereoisomers in dynamic equilibrium.

l-Camphor and Borneol. About a century ago, Persoz,[11] and Vohl [12] observed the presence of camphor, after the oil had been oxidized with chromic acid. In order to decide whether the camphor was formed on oxidation of the oil, or whether it was actually present *per se*, Schimmel & Co.[13] fractionated an oil of tansy, after they had freed it from thujone, as much as possible, by means of the bisulfite compound. The fraction b. 205° separated a solid, crystalline mass, which was filtered on a suction filter and dissolved in 80 per cent alcohol. Odor and physicochemical properties pointed toward a mixture of camphor and borneol. On separation of this mixture by the method of Haller (see Vol. II of this work, p. 240), relatively large quantities of camphor and a little borneol were obtained. The camphor was identified by preparation of its oxime m. 116°. Determination of the optical rotation showed that the camphor was present not in the usual dextrorotatory, but in the very rare *laevorotatory*, form. Owing to the small quantity of borneol available, its rotation could not be determined.

Thujyl Alcohol(?). According to Bruylants, the occurrence of thujyl alcohol in oil of tansy is not improbable. However, the alcohol has not yet been identified as an actual constituent of the oil.

Terpenes(?). The same author also noted the presence of terpenes b. $\sim$160° in tansy oil. They may be pinene or camphene, but their identity still has to be established.

---

[6] *Ber.* **11** (1878), 449.
[7] *Ibid.* **25** (1892), 3343.
[8] *Liebigs Ann.* **272** (1893), 99.
[9] *Ibid.* **336** (1904), 267.
[10] *J. Chem. Soc.* (1938), 2016.
[11] *Compt. rend.* **13** (1841), 436. *Liebigs Ann.* **44** (1842), 313. *J. prakt. Chem.* **25** (1842), 55.
[12] *Arch. Pharm.* **124** (1853), 16. *Pharm. Zentr.* (1853), 318.
[13] *Ber. Schimmel & Co.*, October (1895), 35.

Use.—Oil of tansy is occasionally used as an effective anthelminticum. However, great care has to be exercised in the dosage, as the oil is quite toxic, an overdose causing vomiting, diarrhea, and collapse.

SUGGESTED ADDITIONAL LITERATURE

G. Schenck and W. H. Hein, "*Tanacetum vulgare,*" *Pharmazie* **4** (1949), 520; 5 (1950), 288. *Chem. Abstracts* **44** (1950), 10266.

OIL OF LEVANT WORMSEED *

Semen or *floris cinae,* an old-fashioned drug commonly known as "Levant wormseed" or simply "wormseed," actually does not consist of the dried seeds, but of the dried unexpanded flower heads of several *Artemisia* species, among them chiefly *Artemisia cina* Berg, and *A. maritima* L., small semi-shrubby perennials of the family *Compositae.* The former species grows wild and abundantly on the great plains of Iran, Turkestan, and Mongolia, the latter along the seacoasts and in salt marshes from England to Chinese Mongolia. The unexpanded flower heads contain santonin which in the past was used as a popular vermifuge particularly against *Ascaris lumbricoides,* but which lately has been replaced by less toxic and more efficient medicines (e.g., oil of chenopodium, the American wormseed oil). Formerly large quantities of Levant wormseed were collected by the natives of the Near East and shipped abroad for the extraction of santonin, a crystalline substance m. 170°, possessing the following structural formula:

(Santonin is the inner anhydride of santoninic acid.)

* Not to be confused with oil of American wormseed, derived from *Chenopodium ambrosioides* L. var. *anthelminticum* (L.) A. Gray (fam. *Chenopodiaceae*).

Aside from its principal constituent santonin, Levant wormseed also contains from 2 to 3 per cent of a volatile oil which in former years was occasionally obtained as a by-product in the commercial extraction of santonin.

Physicochemical Properties.—Levant wormseed oil is a yellowish liquid with a camphoraceous, rather unpleasant odor characteristic of the dried flowers, and reminiscent also of cineole. Gildemeister and Hoffmann[1] reported these properties for the oil:

Specific Gravity at 15°...... 0.915 to 0.940
Optical Rotation........... −1° 50′ to −7° 0′
Refractive Index at 20°..... 1.465 to 1.469
Solubility................. Soluble in 2 to 3 and more
           vol. of 70% alcohol

An oil of Levant wormseed obtained by Rutovski and Leonov[2] with a yield of 1.03 to 1.42 per cent exhibited the following values:

Specific Gravity at 25°/4°......... 0.9211
Optical Rotation................. −3° 11′
Refractive Index at 25°........... 1.4650
Acid Number..................... 2.8
Ester Number................... 12.1
Ester Number after Acetylation..... 38.0
Cineole Content:
   Old Resorcinol Method.......... 84.25%
   Modified Resorcinol Method...... 77.85%
Solubility...................... Soluble in 1.1 vol. of 70%
           alcohol; soluble in 0.5
           vol. of 80% alcohol

Chemical Composition.—The earlier investigations of the chemical composition of Levant wormseed oil carried out from 1841 to 1874 need not be described here because they resulted only in contradictions and no tangible data were obtained. In 1884 Wallach and Brass,[3] and Hell and Stürcke[4] almost simultaneously, but independently, succeeded in identifying cineole as the most important constituent of the oil. Since then the presence of the following compounds has been reported in oil of Levant wormseed:

Cineole. The chief constituent (see above).

dl-α-Pinene. Identified by Schindelmeiser[5] in the lowest boiling fractions of the oil. Nitrosochloride m. 103°–104°, hydrochloride m. 123°.

[1] "Die Ätherischen Öle," 3d Ed., Vol. III, 1005.
[2] *Arb. Wissensch. Chem. Pharm. Inst. Moscow* (1924), No. 10, 49.
[3] *Liebigs Ann.* **225** (1884), 291.　　　[5] *Apoth. Ztg.* **22** (1907), 876.
[4] *Ber.* **17** (1884), 1970.

Terpinene. In the middle fractions of the oil. Nitrosite m. 155°–157° (Schindel-meiser).

l-α-Terpineol. In the higher fractions. Nitrosochloride m. 103°, dipentene dihydro-chloride m. 48° (Schindelmeiser; later confirmed by Schimmel & Co.).[6]

α- or β-Terpinenol. The Schimmel chemists also noted in the alcohol fraction the presence of a terpinenol b. 208°–218°, which on treatment with a 3 per cent solution of sulfuric acid gave *cis*-terpin hydrate m. 116°–117°. It could not be ascertained whether the alcohol in question was α- or β-terpinenol.

A Sesquiterpene(?). In the higher boiling fractions b. 230°–260° Schimmel & Co. obtained a sesquiterpene b. 255°, d_{15} 0.9170, which did not yield a solid nitroso-chloride, and no hydrochloride and hydrobromide.

A Sesquiterpene Alcohol(?) or Paraffins(?). Crystals slowly separated from the distillation residue (b. about 280°) of the oil investigated by Schimmel & Co. They consisted either of a solid sesquiterpene alcohol or perhaps of the paraffin $C_{32}H_{66}$, m. 55°–58°, which Klobb, Garnier and Ehrwein [7] had isolated from the flowers of the Levant wormseed plant.

Use.—Oil of Levant wormseed is at present not produced on a commercial scale. The cineole, for the isolation of which the oil was formerly produced, can now be obtained much more economically from other sources —eucalyptus oils, e.g.

OIL OF WORMWOOD

Essence d'Absinthe   *Aceite Esencial Ajenjo*   *Wermutöl*
                          Oleum Absinthii

Botany and Occurrence.—*Artemisia absinthium* L. (fam. *Compositae*) the common or large species of wormwood, is a native of Europe, where it grows as a weed, and has also been extensively cultivated for many years. Prior to the prohibition of absinthe in France in 1915 the volatile oil distilled from wormwood was the principal and most active ingredient of that popular drink. (French manufacturers employed not only the oil from *Artemisia absinthium,* but also that of *Artemisia pontica* L., the small or Roman wormwood.) Since the prohibition of absinthe, wormwood cultivation in France has been greatly curtailed.

Wormwood is a shrubby, herbaceous, finely canescent, aromatic and much

---

[6] *Ber. Schimmel & Co.,* October (1908), 143.   [7] *Bull. soc. chim.* [4], **7** (1910), 940.

branched perennial growing 2 to 4 ft. tall. The flowers are yellowish, fertile, and occur in hemispheric panicled heads. The plant owes its intensely bitter taste to the presence of the glycoside absinthin, and another crystalline compound, anabsinthin. The strong odor, as well as the stimulant and toxic effects of the plant, are due to the volatile oil, a dark green, or yellowish-brown liquid, of peculiar, acrid taste. This oil is obtained by steam distillation of the overground parts of the plant during the flowering period.

Little is known regarding the date of the plant's introduction into North America. It has been naturalized, and now occurs as a weed on unused land, and along roadsides in many areas. It is also cultivated commercially in St. Joseph, Cass and Allegan counties in southern Michigan, and in St. Joseph County in Indiana, for distillation of the volatile oil. Total area of production is estimated at between 2,300 and 2,900 acres.[1] A small part of the harvest is marketed as dried herb for medicinal purposes, and for flavoring certain types of liqueurs and wines—vermouth, e.g. The flowering tops and leaves of the plant are simply stripped by hand from the stems, after the harvest, and prepared for the market by careful drying in the shade.

Planting and Cultivating.—Seeds may be sown broadcast in early fall, after a grain crop. However, it is preferable to start the plants from seeds sown in hotbeds early in the spring, or from cuttings of the young shoots taken in the spring and rooted in sand under glass or in the shade of a lath shed. Some growers plant seed in May, in a small plot of loose ground. The seeds are very small, and should be sown on the surface of the soil in cold frames, or seedbeds, and lightly covered with fine sand. The plants may be transplanted in moist weather at almost any time during the growing season. Mechanical setters can be used. The plants are set out in the field 18 in. apart, in rows 3 ft. apart. High ground, such as that used for corn or wheat, is preferred by many growers, because on such soil a wormwood planting requires much less hand weeding. Muck soil seems to produce an undesirable quality of oil, characterized by an abnormally low specific gravity. The soil must be kept absolutely free of weeds by cultivating at regular intervals during the first summer. A fair cutting of the herb may be expected the first year after planting, and full crops for two or three successive seasons.

Wormwood being a hardy perennial, a plantation lasts from 7 to 10 years. Production reaches its peak in the second or third year, but tapers off sharply during the fourth and fifth years.

---

[1] *Chemurgic Digest* **8**, Jan. (1949), 11.

Raising of wormwood is less hazardous than that of peppermint, because the plant is practically immune to diseases and not easily attacked by insects. Neither does it suffer from winter killing. The chief disadvantage of wormwood production lies in the fact that the growers obtain no crop during the first year of planting. Moreover, setting out of the young plants and weeding require considerable work, and are therefore relatively costly.

Harvesting.—In the latter part of July, one year after planting, the shrubs reach a height of about 3 ft. When in full bloom, they are cut with a scythe, or mowed down, and tied into sheaves with a wheat binder. The sheaves are then spread on the ground to dry for 24 hr., by which time a good deal of the moisture has evaporated. Finally, the sheaves are loaded on wagons or trucks and transported to the distilleries. The plant remains in bloom for about a week; for this reason, the harvesting of large areas is sometimes completed only after the flowering stage. Such delayed harvesting usually results in a lower yield of oil per acre, and also, it appears, in an inferior quality of oil.

Distillation.—In the producing areas, there are quite a number of field distilleries similar to those employed for the production of peppermint and spearmint oils. Live steam generated in a separate boiler is blown through the plant material, and the oil is collected in a Florentine flask. The average still holds about 3,000 lb. of herb material. Distillation of one charge requires about 2½ hr.; shorter periods of distillation produce oils of low specific gravity. Practically all of the oil distills in the first 2 hr. Nevertheless it is necessary to continue operation for another half-hour, in order to collect the last portions of oil, which contain important constituents (without which the quality of the oil would be inferior).

Under normal conditions, a charge of wormwood (3,000 lb.) yields from 8 to 12 lb. of oil, i.e., 0.27 to 0.40 per cent. Since one acre of wormwood produces about 6,000 lb. of semidry plant material, the yield of oil per acre ranges from 16 to 24 lb. Yields as high as 35 lb. per acre have been recorded.

Yield and quality of oil vary with condition of the herb—i.e., whether it is fresh and moist, clover dry, or completely dry—as well as with length of distillation. Distillation of wet herb, for example, results in a markedly poor yield; as a matter of fact, it is almost impossible to exhaust fresh, moist herb by steam distillation.

Yield and quality depend, in addition, upon the weather, and the stage of maturity the plant has reached at time of harvesting. Rabak's experiments [2] indicate that:

[2] *Ind. Eng. Chem.* **13** (1921), 536.

1. Yield of oil from the fresh herb during its flowering stage varies greatly from year to year, owing entirely to variant climatic conditions. Low precipitation, coupled with high temperature and much sunshine, affects the yield of oil favorably; opposite conditions result in a lower yield.

2. Drying of the plants before distillation results in loss of volatile oil by evaporation, and therefore in a reduction in the yield of oil, but apparently it promotes esterification in the oil. The content of esters in the oils from fresh herb over a period of years appears to be more constant than the content of terpene alcohols.

3. The highest yield of oil is obtained during the flowering period of the plant. Solubility of the oil in alcohol apparently is a criterion of the percentage of esters present. Likewise, specific gravity bears a close relationship to the ester content of the oils. The percentage of alcoholic constituents decreases as the plant approaches maturity.

It is to be regretted that Rabak gave no consideration to the thujone content of the oils he investigated, as the percentage of this main constituent of wormwood oil—which indeed determines its quality—is also considerably influenced by the stage of plant development, and probably by climatic and weather conditions as well. It appears that the thujone content is highest during the flowering period. As the plant matures, thujone is converted into thujyl alcohol and thujyl esters. Hence the plant should be distilled when in bloom and not permitted to go to seed. This fact is apparently not sufficiently appreciated by distillers, which may account for the wide variation in the properties of even the best oils.

So far as drying prior to distillation is concerned, it is well known that distillation of fresh plant material offers considerable difficulty, requiring long hours, and entailing high consumption of steam. On the other hand, complete drying in the field results in a loss of oil by evaporation. It appears, therefore, advisable to expose the cut plants in the field only until they are clover dry, at which stage they can be easily packed into the stills.

Physicochemical Properties.—Oil of wormwood is a dark green, occasionally bluish or brown, slightly viscous liquid, with the powerful odor characteristic of the plant. The flavor is bitter, harsh, and very persistent.

The various grades of wormwood oil vary in their physicochemical properties within quite wide limits, for the reasons explained above. The properties of the foreign oils differ from those of our domestic oils.

Pure North American oils examined over a period of years by Fritzsche Brothers, Inc., New York, had properties falling within this range:

Specific Gravity at 15°/15°.............. 0.917 to 0.951
Optical Rotation...................... Too dark to be determined, even in a
                                     short tube
Refractive Index at 20°................. 1.4600 to 1.4829
Saponification Number.................. 30.0 to 203.4
Ester Number after Acetylation.......... 56.4 to 237.0
Ketone Content, Calculated as Thujone
  (Hydroxylamine Hydrochloride Method) 6.9 to 69.2%
Solubility............................. Soluble in 1 to 1.5 vol. of 80% alcohol,
                                     clear to turbid with more. Some oils
                                     insoluble in 10 vol. of 80% alcohol

These limits, obviously quite wide, are those of pure oils distilled at various stages of plant development. The properties of the best North American oils, in the author's experience, lie within the following, more narrow limits:

Specific Gravity at 15°/15°.............. 0.921 to 0.937
Optical Rotation...................... Too dark to be determined, even in a
                                     short tube
Refractive Index at 20°................. 1.4600 to 1.4730
Saponification Number.................. 30.8 to 110.1
Ester Number after Acetylation.......... 77.0 to 141.4
Ketone Content, Calculated as Thujone
  (Hydroxylamine Hydrochloride Method) 40 to 69%
Solubility............................. Soluble in 1 to 1.5 vol. of 80% alcohol,
                                     clear to turbid with more. Some oils
                                     insoluble in 10 vol. of 80% alcohol

Because of the dark color of the oil, the optical rotation cannot be determined, even in a very short polariscopic tube. However, using 1 per cent alcoholic solutions of all sorts of commercial wormwood oils, Gildemeister and Hoffmann [3] found dextrorotations ranging from +40° to +70° (calculated for the undiluted oils). An oil distilled in Barrême (B.A.), France, from *wild* growing herb exhibited a laevorotation of −16° 40′ (determined in a 1 per cent alcoholic solution, and calculated for the undiluted oil).

The ketone content of the oil (calculated as thujone) is assayed by the hydroxylamine hydrochloride method. For details see Vol. I of the present work, p. 285.

As regards wormwood oils produced abroad, Gildemeister and Hoffmann [4] reported these properties for commercial oils from France, Algeria and Italy:

[3] "Die Ätherischen Öle," 3d Ed., Vol. III, 1012.
[4] *Ibid.*

	French Oils	Algerian Oils	Italian Oils
Specific Gravity at 15°......	0.901 to 0.954 (usually above 0.92)	0.905 to 0.939	0.918 to 0.943
Refractive Index at 20°.....	1.46684 (1 determination)	...	...
Acid Number..............	Up to 6.7	Up to 6.1	Up to 5.6
Ester Number.............	11 to 108 (in one case 135)	14 to 93	15 to 37
Ester Number after Acetylation....................	...	...	123.2
Solubility in 80% Alcohol...	Soluble in 1 to 2 vol., occasionally turbid with more alcohol. Some oils insoluble		
Solubility in 90% Alcohol...	Clearly soluble, either from the beginning or in 0.5 to 1 vol. In rare cases turbid on dilution		

The same authors also described the properties of wormwood oils distilled by Schimmel & Co. in Miltitz, Germany, from Hungarian herb material (I and II), native Miltitz material (III), and French plants grown in Miltitz (IV):

	I	II	III	IV
Specific Gravity at 15°............	0.8845	0.9125	0.932 to 0.954	0.9276 to 0.9331
Acid Number......	...	16.8	0.2 to 8.6	Up to 1.5
Ester Number......	35.0	75.4	76 to 185	38.8 to 65.3
Ester Number after Acetylation......	...	...	153 to 222	93.3 to 170.3
Solubility..........	Insoluble in 80% alcohol. Easily soluble in 90% alcohol	Insoluble in 80% alcohol. Soluble in 90% and in 95% alcohol; first clearly, but with turbidity on dilution	Usually soluble in 1 to 4 vol. of 80% alcohol. Soluble in all volumes of 90% alcohol; in a few cases 0.5 to 1 vol. required for clear solution	Soluble in 1 to 1.5 vol. of 80% alcohol, in one case turbid on dilution. Soluble in all volumes of 90% alcohol

The bulk of oil (III) distilled from 36° to 90° at 4 mm.; all fractions exhibited strong dextrorotation (+55° to +75°).

Gildemeister and Hoffmann [5] also reported on the properties of oils distilled in Barrême (B.A.), France, from wild growing plants, and from cultivated plants:

[5] *Ibid.*

	Oils from Wild Plants	*Oils from Cultivated Plants*
Specific Gravity at 15°.........	0.901 to 0.908	0.936 to 0.939
Acid Number.................	Up to 3.2	Up to 2.8
Ester Number.................	34.3 to 57.4	96.4 to 114
Ester Number after Acetylation. ...		164.5
Solubility....................	Even with 90% alcohol clearly soluble only in concentration	Not clearly soluble in 80% alcohol. Only one oil was clearly soluble in all volumes of 90% alcohol; in the case of the two other oils only the concentrated solutions were clear

The largest portion of the oils derived from wild growing plants distilled between 40° and 100° at 3 mm.; all fractions exhibited laevorotation ($-3°\,26'$ to $-21°\,10'$).

The bulk of the oils derived from cultivated plants distilled between 45° and 97° at 4 mm.; all fractions exhibited strong dextrorotation ($+51°\,28'$ to $+72°\,20'$). The presence of pinene could not be established with certainty.

Dorronsoro [6] investigated a Spanish wormwood oil distilled in Malaga from cultivated herb and found these properties:

Specific Gravity at 15°....................	0.9156
Ester Number...........................	27.64
Ester Content, Calculated as Thujyl Acetate	9.67%
Ester Number after Acetylation............	38.35
Total Alcohol Content, Calculated as Thujyl Alcohol..............................	10.47%
Free Alcohol Content, Calculated as Thujyl Alcohol..............................	3.10%
Ketone Content, Calculated as Thujone (Hydrogenation)..........................	57.50%
Solubility...............................	Soluble in 1 vol. of 90% alcohol, in 4.5 vol. of 80% alcohol, and in 15 vol. of 70% alcohol

In recent years attempts have been made to cultivate wormwood in Brazil. According to de Camargo Fonseca,[7] the essential oil content of wormwood leaves in Brazil varies with the season and the age of the leaves. On the basis of dry weight, young leaves contained 0.34 per cent and 0.38 per cent (2 samples) when collected in November, and 0.41 per cent (1 sample)

---

[6] *Mem. acad. cienc. exactas Madrid* **29** (1919).

[7] *Anais Faculdade Farm. e Odontol. Univ. São Paulo* **5** (1947), 65. *Chem. Abstracts* **42** (1948), 3138.

when collected in July. The corresponding values for the fully grown leaves were 0.15 per cent, 0.20 per cent, and 0.38 per cent, respectively.

As regards the oil from *wild* growing plants, mention has already been made that they exhibit laevorotation, the result probably of the presence of *l*-α-thujone (see "Chemical Composition," below). Two oils distilled by Roure-Bertrand Fils[8] from plants growing wild in the Alpes-Maritimes (Southern France) contained 9.0 and 5.5 per cent of esters, 7.0 and 4.3 per cent of combined alcohols, 71.9 and 76.3 per cent of free alcohols, and 8.4 and 3.0 per cent of thujone. The chief constituent of these oils, therefore, was thujyl alcohol; the thujone content was very low.

Chemical Composition.—The first investigation of the chemical composition of wormwood oil was undertaken more than a century ago by Leblanc,[9] who found that the chief constituent of the oil was a compound b. 205°, for which he proposed the empirical molecular formula $C_{10}H_{16}O$. This formula was confirmed by Cahours,[10] Schwanert,[11] and Gladstone.[12] Beilstein and Kupffer[13] assigned the name absynthole to the substance. Years later Semmler,[14] and Wallach[15] came to the conclusion that absynthole is identical with tanacetone or thujone.

The most thorough work on the chemical composition of oil of wormwood is that of Schimmel & Co.,[16] who identified several of the constituents listed below:

Phellandrene. A small quantity of phellandrene occurs in the fraction b. 158°–168° (Schimmel & Co.).

Pinene(?). Years ago Brühl[17] expressed the opinion that the terpene b. ∼160° observed earlier by Gladstone was *d*-pinene. Later Schimmel & Co. tried to identify pinene by means of the nitrosochloride, but did not obtain a sufficient quantity of crystals to determine their melting point.

α- and β-Thujone. In the fraction b. 200°–203° Schimmel & Co. identified thujone, by preparation of the tribromide m. 121°–122°. The ketone readily formed a bisulfite compound, but could not be isolated quantitatively from the complete oil by means of this compound. In 1904, Wallach[18] reported that the thujone present in wormwood oil consists chiefly of the dextrorotatory β- form. Ten years later Paolini and Lomonaco[19] observed the laevorotatory α- form in an Italian

[8] *Repts. Roure-Bertrand Fils*, April (1906), 36.
[9] *Compt. rend.* **21** (1845), 379. *Ann. chim. phys.* [3], **16** (1846), 333.
[10] *Compt. rend.* **25** (1847), 725.
[11] *Liebigs Ann.* **128** (1863), 110.
[12] *J. Chem. Soc.* **17** (1864), 1.
[13] *Liebigs Ann.* **170** (1873), 290.
[14] *Ber.* **25** (1892), 3350. Cf. *ibid.* **27** (1894), 895. *Ber. Schimmel & Co.*, October (1894), 51.
[15] *Liebigs Ann.* **286** (1895), 93.   [17] *Ber.* **21** (1888), 156.
[16] *Ber. Schimmel & Co.*, April (1897), 51.   [18] *Liebigs Ann.* **336** (1904), 268.
[19] *Atti accad. Lincei* Roma [5], **23** (1914), II, 123.

wormwood oil. The α- form probably is a constituent of *wild* growing wormwood also, because the oil derived from the latter source exhibits laevorotation. More recently Short and Read [20] showed the naturally occurring α- and β-thujones to be mixtures of the two diastereoisomers *l*-thujone and *d*-isothujone, in dynamic equilibrium. (For details, cf. Vol. II of the present work, p. 423).

Thujyl Alcohol, Free and as Acetate, Isovalerate, and Palmitate. In the fraction b. 210°–215° Schimmel & Co. characterized thujyl alcohol by careful oxidation, with chromic acid mixture, to thujone. Thujyl alcohol occurs in the oil free, as well as esterified with acetic, isovaleric, and palmitic acids. Quantitative saponification of an original and of an acetylized oil proved the oil to contain 17.6 per cent of thujyl acetate (corresponding to 13.9 per cent of thujyl alcohol), and 24.2 per cent of total alcohols (thujyl alcohol, free and esterified). Rabak [21] found as much as 47.5 per cent of esters (calculated as thujyl acetate) in an oil distilled from plants during the fruiting stage. (As regards the ratio of thujyl alcohol and its esters to thujone, see below).

In an Italian oil Paolini and Lomonaco [22] observed the presence of an isomeric thujyl alcohol, which they named δ-thujyl alcohol; it is identical with that found previously by Paolini and Divizia [23] in the reduction products of French wormwood oil. (Cf. Vol. II of this work, p. 248.)

A Hydrocarbon $C_{15}H_{20}$. Submitting the fraction $b_{0.5}$ 92°–100° of wormwood oil to chromatography on aluminum oxide, Sorm, Vonasek and Herout [24] obtained an orange colored bicyclic hydrocarbon with four double bonds, $C_{15}H_{20}$, b_8 127°, d_{20} 0.9525 n_D^{20} 1.5554.

Cadinene. In the fraction b. 260°–280°, which amounted to a substantial portion of the oil, Schimmel & Co. identified cadinene, by means of the hydrochloride m. 117°–118°.

Nerol. Elze [25] reported the presence of nerol in a wormwood oil of Spanish origin.

An Azulene(?). That the high boiling fractions (b. 270°–300°) of the oil contain a blue substance, probably identical with that occurring in the corresponding fraction of chamomile oil, was noted first by Gladstone,[26] and later confirmed by Beilstein and Kupffer.[2]

Attempting to isolate this azulene by chromatographic methods from the fraction $b_{0.5}$ 92°–100°, Sorm, Vonasek and Herout [28] recently obtained a colorless sesquiterpene or mixture of sesquiterpenes $C_{15}H_{24}$, and a blue azulene. Between the two layers Sorm et al. noted an orange-colored fraction which, on repeated chromatographic treatment, yielded a hydrocarbon $C_{15}H_{20}$, viz., "Chamazulenogene" b_8 127°. This bicyclic substance contains four double bonds and is perhaps a dihydro derivative of chamazulene, or a hydrocarbon with the same structure and containing exocyclic double bonds.

[20] *J. Chem. Soc.* (1938), 2016.
[21] *Ind. Eng. Chem.* **13** (1921), 536.
[22] *Atti accad. Lincei* Roma [5], **23** (1914), II, 123.
[23] *Ibid.* [5], **21** (1920), I, 570.
[24] *Collection Czech. Chem. Communs.* **14** (1949), 91. *Chem. Abstracts* **44** (1950), 5847.
[25] *Chem. Ztg.* **34** (1910), 857.
[26] *J. Chem. Soc.* **17** (1864), 1. *Jahresber. Chem.* (1863), 549.
[27] *Liebigs Ann.* **170** (1873), 290.
[28] *Collection Czech. Chem. Communs.* **14** (1949), 91. *Chem. Abstracts* **44** (1950), 5847.

Roark [29] observed formic and salicylic acids in the saponification lye of a wormwood oil from Wisconsin, but Schimmel & Co.[30] questioned whether these two acids are normal constituents of wormwood oil.

The ratio of thujyl alcohol and its esters to thujone changes with the stage of plant development. According to Charabot,[31] the thujone originally present in the plant disappears as the plant matures, being partly converted into thujyl alcohol. Charabot found the following percentages of thujyl alcohol (combined, free, and total) and thujone in two oils distilled at different stages of plant development:

| | Specific Gravity at 24° | Ester Content (%) | Thujyl Alcohol Content | | | Thujone Content (%) |
			Combined (%)	Free (%)	Total (%)	
No. 1.....	0.9307	9.7	7.6	9.0	16.6	43.1
No. 2.....	0.9253	13.1	10.3	9.2	19.5	35.0

Use.—Oil of wormwood is employed, together with tinctures of the dried herb, as an important flavoring ingredient in alcoholic liqueurs of the absinthe type, and in vermouth. Many well-known brands of liqueurs contain small quantities. The effect is tonic, stomachic, and stimulant. The primary use of the oil, however, is in certain well-known medicinal preparations, externally applied for the relief of rheumatic pains. If taken internally in excessive dosage, it becomes an active narcotic poison. A half ounce may produce insensibility, convulsions, or even death. Emetics, stimulants, and demulcants are recommended as antidotes.

[29] *Midland Drugg. Pharm. Rev.* **45** (1911), 237.
[30] *Ber. Schimmel & Co.*, October (1911), 96.
[31] *Compt. rend.* **130** (1900), 923. *Bull. soc. chim.* [3], **23** (1900), 474.

INDEX

Numbers in *italics* indicate main entries; in the case of a plant species, they indicate references to monographs (or sections of monographs) dealing with the *oil* of the species.

Wherever possible, synonymous names for chemical compounds have been brought together under one main heading, such heading being in each case that employed in Vol. II of this series. Cross-references are made from the several synonyms. Thus: in the text, *Methyleugenol* may also be called *Eugenol methyl ether*. In this index, the latter name is cross-referred to *Methyleugenol*. The researcher interested in tracing the occurrence of any specific chemical compound in various oils should find this system helpful.